Collected Works of A.M. Turing

PURE MATHEMATICS

Collected Works of A.M. Turing

Pure Mathematics
Edited by J.L. BRITTON

Mathematical Logic
Edited by R.O. GANDY and C.E.M. YATES

Mechanical Intelligence
Edited by D.C. INCE

Morphogenesis
Edited by P.T. SAUNDERS

NORTH-HOLLAND
AMSTERDAM · LONDON · NEW YORK · TOKYO

Collected Works of A.M. Turing

PURE MATHEMATICS

Edited by

J.L. BRITTON
King's College, London, United Kingdom

with a section on Turing's statistical work by I.J. GOOD
Virginia Polytechnic Institute and State University, Blacksburg, VA, USA

1992
NORTH-HOLLAND
AMSTERDAM · LONDON · NEW YORK · TOKYO

ELSEVIER SCIENCE PUBLISHERS B.V.
Sara Burgerhartstraat 25
P.O. Box 211, 1000 AE Amsterdam, Netherlands

Distributors for the United States and Canada:

ELSEVIER SCIENCE PUBLISHING COMPANY INC.
655 Avenue of the Americas
New York, NY 10010, USA

ISBN: 0 444 88059 3

Library of Congress Cataloging-in-Publication Data

Turing, Alan Mathison, 1912-1954.
Pure mathematics / edited by J.L. Britton.
p. cm. -- (Collected works of A.M. Turing)
Includes bibliographical references and index.
ISBN 0-444-88059-3
1. Mathematics. I. Britton, J.L., 1927- . II. Title.
III. Series: Turing, Alan Mathison, 1912-1954. Works. 1990.
QA3.T87 1992
510--dc20 90-36182
CIP

Transferred to digital printing 2005

Acknowledgement is gratefully made to the following for permission to reprint previously published articles by A.M. Turing (and others):

Biometrika Trustees for "Studies in the History of Probability and Statistics. XXXVII. A.M. Turing's Statistical Work in World War II", by I.J. Good, Biometrika 66 (2) (1979) 393–396.

London Mathematical Society for "Equivalence of Left and Right Almost Periodicity", J. London Math. Soc. 10 (1935) 284–285;

and for "A Method for the Calculation of the Zeta-function", Proc. London Math. Soc. (2) 48 (1943) 180–197;

and for "Some Calculations of the Riemann Zeta-function", Proc. London Math. Soc. (3) 3 (1953) 99–117;

and for "On the Difference $\pi(x) - \mathrm{li}\, x$", by A.M. Cohen and M.J.E. Mayhew, Proc. London Math. Soc. (3) 18 (1968) 691–713.

Oxford University Press for "Rounding-off Errors in Matrix Processes", Quart. J. Mech. Appl. Math. 1 (1948) 287–308.

Penguin Books Ltd. for "Solvable and Unsolvable Problems", Science News 31 (1954) 7–23.

Princeton University Press for "Finite Approximations to Lie Groups", Ann. of Math. 39(1) (1938) 105–111;

and for "The Word Problem in Semi-groups with Cancellation", Ann. of Math. 52(2) (1950) 491–505;

and for "An Analysis of Turing's 'The Word Problem in Semi-groups with Cancellation'", by W.W. Boone, Ann. of Math. 67(1) (1958) 195–202.

Wolters-Noordhoff Publishing for "The Extensions of a Group", Compositio Math. 5 (1938) 357–367.

PREFACE

It is not in dispute that A.M. Turing was one of the leading figures in twentieth-century science. The fact would have been known to the general public sooner but for the Official Secrets Act, which prevented discussion of his wartime work. At all events it is now widely known that he was, to the extent that any single person can claim to have been so, the inventor of the "computer". Indeed, with the aid of Andrew Hodges's excellent biography, *A.M. Turing: the Enigma*, even non-mathematicians like myself have some idea of how his idea of a "universal machine" arose – as a sort of byproduct of a paper answering Hilbert's *Entscheidungsproblem*. However, his work in pure mathematics and mathematical logic extended considerably further; and the work of his last years, on morphogenesis in plants, is, so one understands, also of the greatest originality and of permanent importance.

I was a friend of his and found him an extraordinarily attractive companion, and I was bitterly distressed, as all his friends were, by his tragic death – also angry at the judicial system which helped to lead to it. However, this is not the place for me to write about him personally.

I am, though, also his legal executor, and in fulfilment of my duty I have organised the present edition of his works, which is intended to include all his mature scientific writing, including a substantial quantity of unpublished material. The edition will comprise four volumes, i.e.: *Pure Mathematics*, edited by Professor J.L. Britton; *Mathematical Logic*, edited by Professor R.O. Gandy and Professor C.E.M. Yates; *Mechanical Intelligence*, edited by Professor D.C. Ince; and *Morphogenesis*, edited by Professor P.T. Saunders.

My warmest thanks are due to the editors of the volumes, to the modern archivist at King's College, Cambridge, to Dr. Arjen Sevenster and Mr. Jan Kastelein at Elsevier (North-Holland), and to Dr. Einar H. Fredriksson, who did a great deal to make this edition possible.

P.N. FURBANK

ALAN MATHISON TURING – CHRONOLOGY

1912 Born 23 June in London, son of Julius Mathison Turing of the Indian Civil Service and Ethel Sara née Stoney
1926 Enters Sherborne School
1931 Enters King's College, Cambridge as mathematical scholar
1934 Graduates with distinction
1935 Is elected Fellow of King's College for dissertation on the Central Limit Theorem of Probability
1936 Goes to Princeton University where he works with Alonzo Church
1937 (January) His article "On Computable Numbers, with an Application to the Entscheidungsproblem" is published in *Proceedings of the London Mathematical Society*
Wins Procter Fellowship at Princeton
1938 Back in U.K. Attends course at the Government Code and Cypher School (G.C. & C.S.)
1939 Delivers undergraduate lecture-course in Cambridge and attends Wittgenstein's class on Foundations of Mathematics
4 September reports to G.C. & C.S. at Bletchley Park, in Buckinghamshire, where he heads work on German naval "Enigma" encoding machine
1942 Moves out of naval Enigma to become chief research consultant to G.C. & C.S.
In November sails to USA to establish liaison with American codebreakers
1943 January–March at Bell Laboratories in New York, working on speech-encypherment
1944 Seconded to the Special Communications Unit at Hanslope Park in north Buckinghamshire, where he works on his own speech-encypherment project *Delilah*
1945 With end of war is determined to design a prototype "universal machine" or "computer". In June is offered post with National Physical Laboratory at Teddington and begins work on ACE computer
1947 Severs relations with ACE project and returns to Cambridge
1948 Moves to Manchester University to work on prototype computer
1950 Publishes "Computing Machinery and Intelligence" in *Mind*
1951 Is elected FRS. Has become interested in problem of morphogenesis
1952 His article "The Chemical Basis of Morphogenesis" is published in *Philosophical Transactions of the Royal Society*
1954 Dies by his own hand in Wimslow (Cheshire) (7 June)

INTRODUCTION

This is one of four volumes covering Turing's mathematical works. The other volumes are concerned with mathematical logic, computer science and mathematical biology, respectively.

This division of Turing's work into four parts is to some extent arbitrary. For example, in the present volume, the papers on matrices (1948) and on the zeta-function (1953) have strong connexions with computer science (the latter paper even including a technical description of hardware), while the papers on the word problem (1950 and II) and the popular article (1954) might have been included in the volume on mathematical logic. And of course, Turing's masterpiece, *On Computable Numbers, with an Application to the Entscheidungsproblem* (1937) could equally well be classified as mathematical logic or computer science.

The excellent biography of Turing by Andrew Hodges (HODGES 1983) may profitably be read in conjunction with the four volumes.

One could say that Turing's first published work, though with limited circulation, was his fellowship dissertation of 1935. It is interesting that this was on statistics. Later, he always looked out for any statistical aspects of the problem under consideration. Elementary statistics occurs, for instance, in the papers (1948, III and V).

Moreover, he made further contributions to statistics. These are described in this volume by Professor I.J. Good, who for a period during the war was Turing's main statistical assistant. This statistical section is divided into two parts: his paper of 1979 on Turing's statistical work and an introduction to that paper, specially written for this volume. This throws some interesting new light on the work during the war on cryptanalysis.

The paper on permutation groups (III) is also related to cryptanalysis; its terminology indicates that it was motivated by a study of the Enigma machine (cf. the biography). This is perhaps the most interesting of the unpublished papers from today's standpoint.

The originals of the unpublished papers (I)–(IV) and the dissertation are in the archives at King's College, Cambridge. I have been unable to locate the original of (V) although Note (3.40) of the biography suggests that this is also at King's College, Cambridge, possibly due to a confusion between (IV) and (V), both of which are on the distribution of prime numbers. However, I had access to a photocopy of (V); unfortunately, in some places not everything on the page has been copied.

The originals are in many places hard to decipher, except for the dissertation in which the handwriting is quite neat. I made hand copies of (I)–(IV) and then a typed version of these for the printers.

We now briefly review the papers by Turing included in this volume and the two related papers by BOONE (1958) and by COHEN and MAYHEW (1968). Technical summaries of Turing's longer papers appear later in the volume.

Equivalence of Left and Right Almost Periodicity (1935)

Turing called the result of this paper a "small-scale discovery" but it was surely a promising beginning to have noticed something that Von Neumann, already enormously successful, had missed.

The topic of almost periodicity has various aspects but the one being considered here is concerned with complex-valued functions on an abstract group.

This is a short and easily readable paper.

Finite Approximations to Lie Groups (1938 A)

Let G be an abstract group and a metric space such that the group operations (product and inverse) are continuous. It is known that without loss of generality the metric can be chosen such that

$$D(ax, ay) = D(x, y),$$

for all x, y, a in G. Call G *approximable* if to each $\varepsilon > 0$ we have a finite group H_ε which is a subset, not in general a subgroup, of G and (i) each x in G is within distance ε of some element of H_ε; (ii) if a, $b \in H_\varepsilon$, then $D(a \circ b, ab) < \varepsilon$, where $a \circ b$ is the product in H_ε and ab is the product in G.

S. Ulam had proposed the problem: Which groups are approximable?

In this paper Turing shows that if a connected Lie group is approximable, then it is compact and Abelian.

The proof is interesting; it involves an interplay between representations of finite groups and representations of compact groups.

The Extensions of a Group (1938 B)

Recall that a group G is an *extension* of a group N by a group B if N is a normal subgroup of G and G/N is isomorphic to B. Let A be the group of automorphisms of N and I the group of inner automorphisms. Then there is a homomorphism $\theta : G/N \to A/I$ given by $gN \to \hat{g}I$ $(g \in G)$, where $\hat{g}$ is the automorphism of N defined by $\hat{g}(n) = g^{-1}ng$ $(n \in N)$.

Now assume we are given a group N, a group B and a homomorphism $X: B \to A/I$. When does an extension G of N by B exist such that, if the isomorphic groups G/N and B are identified, then θ is just X? This is the problem considered in this paper.

This was an important contribution to the theory of extensions of a not necessarily Abelian group as it stood at that time.

A Method for the Calculation of the Zeta-function (1943)

Some Calculations of the Riemann Zeta-function (1953)

It is probably not widely known that Turing did some important theoretical work on the Riemann zeta-function besides his computational work on this function.

Recall that the Riemann zeta-function $s \to \zeta(s)$, $s = \sigma + it$, is analytic in $\mathbb{C}$ except for a pole at $s = 1$. When $\sigma > 1$ we have

$$\zeta(s) = \sum_{n=1}^{\infty} n^{-s} \quad \text{and} \quad \zeta(s) = \prod (1 - p^{-s})^{-1},$$

where the product is over all prime numbers p. The function has infinitely many zeros in the strip $0 < \sigma < 1$ while outside the strip its only zeros are $s = -2, -4, -6, \ldots$. The Riemann hypothesis (conjecture) is that all the zeros in the strip lie on the line $\sigma = \frac{1}{2}$.

The objective of the first paper (written in 1939 but not published until 1943) is to obtain a practical method for calculating $\zeta(\frac{1}{2} + it)$. (There is an elaborate piece of contour integration, done with bravura.)

The other paper (1953) describes how in 1950 the Manchester University computer was used to investigate the zeros of the zeta-function for $2\pi(63)^2 \leqslant t \leqslant 2\pi(64)^2$ and $0 < t < 1540$ in the hope of finding zeros (in the strip) off the critical line. The first part of this paper gives the theoretical basis of the method while the second part gives a brief description of the hardware and states some of the practical computing strategy. It gives some insight into the excitement and frustrations of the early days of computing.

Thanks are due to Dr. D.R. Heath-Brown for an assessment of Turing's work on the zeta-function; this appears later, after the technical summary and notes on the 1953 paper. (See also EDWARDS 1974.)

Rounding-off Errors in Matrix Processes (1948)

This is an enjoyable paper on numerical analysis with Turing "thinking aloud" about solving linear equations and inverting matrices, starting from scratch.

The interest is on inverting $n \times n$ matrices when n is large and whether, in carrying out any of the standard methods for inversion, errors build up exponentially with n. Consideration is given to statistical bounds for errors but the emphasis is on absolute bounds.

My colleague Frank D. Burgoyne says: "Turing's paper was one of the earliest attempts to examine the error analysis of the various methods of solving linear equations and inverting matrices. His analysis was basically sound. The main importance of the paper was that it was published at the dawn of the modern computing era, and it gave indications of which methods were "safe" when solving such problems on a computer."

The Word Problem in Semi-groups with Cancellation (1950)

A *semi-group* is a set within which is defined an associative product. A *semi-group with cancellation* (SWC) is a semi-group S such that for all a, b, c in S $ab=ac$ implies $b=c$ and $ba=ca$ implies $b=c$. Let K denote "semi-group, SWC or group". A *K-presentation* is a pair (S,D), where S is a finite set of symbols $s_1,\dots,s_n$ and D is a finite set of formal equations $U_i=V_i$ $(i=1,\dots,r)$ where U_i and V_i are words in the symbols; a *word* means a finite string of symbols in the semi-group or SWC case, but means an expression of the form $s_{i_1}^{e_1}\dots s_{i_k}^{e_k}$ where $e_j=\pm 1$ $(j=1,\dots,k)$ in the group case. We say that $W_1=W_2$, where W_1, W_2 are words, is a *relation* for (S,D) if, whenever X is a K containing elements $s_1,\dots,s_n$ such that $U_i=V_i$ in X for all i, then $W_1=W_2$ in X. In particular each $U_i=V_i$ is called a *defining relation* (or fundamental relation (F.R.)).

We say that the *word problem is solvable for* (S,D) if there is an algorithm which will determine of any pair of words W_1, W_2 whether or not $W_1=W_2$ is a relation for (S,D).

For any (S,D) one can construct $[S,D]$, which is a K and which is unique up to isomorphism; it contains elements $s_1,\dots,s_n$, each element of it is a word in $s_1,\dots,s_n$ and $W_1=W_2$ is a relation for (S,D) if and only if we have $W_1=W_2$ in $[S,D]$.

In 1947, Post and Markov independently showed that there is a semi-group presentation with unsolvable word problem. The problem of extending this result to groups received a lot of attention but proved difficult. In the present paper, Turing considers the half-way house of semi-groups with cancellation; undoubtedly it influenced both Novikov who finally obtained the result for groups in 1955 and Boone, who proved the result independently at about the same time.

Solvable and Unsolvable Problems (1954)

This is an article from the Penguin *Science News* which would attract the same kind of reader as, say, the *Scientific American* but which was in paperback form and was published quarterly. The article is an enjoyable and successful piece of popular exposition.

A Note on Normal Numbers (I)

Normal numbers are defined as follows. Let α be a real number and let $t \geqslant 2$ be an integer. Consider the expansion of α in scale t and let the part after the decimal point be $.\,\alpha_1\alpha_2\alpha_3\ldots$. Let γ be an ordered set of numbers from $\{0, 1, \ldots, t-1\}$ and let $l(\gamma)$ be the number of elements in γ. Denote by $S(\alpha, t, \gamma, R)$ the number of occurrences of the block γ in $\alpha_1\alpha_2\ldots\alpha_R$. Then α is *normal* if

$$R^{-1}S(\alpha, t, \gamma, R) \to t^{-r} \quad \text{as } R \to \infty,$$

for all γ, t, where $r = l(\gamma)$.

It is known that almost all numbers are normal but it is not easy to give an example of one. In this unpublished paper Turing discusses the construction of normal numbers.

The Word Problem in Compact Groups (II)

As mentioned earlier, if (S, D) is a group presentation, the corresponding group $[S, D]$ may have unsolvable word problem. If, however, we restrict attention to presentations such that $[S, D]$ is a compact group, is the word problem always solvable? Turing answers this in the affirmative by making use of an important theorem of Tarski (1948) in mathematical logic.

On Permutation Groups (III)

Although this paper was evidently motivated by Turing's study of the Enigma machine (see Hodges 1983), it is essentially an important piece of pure mathematics. Turing was led to consider what turns out to be a formidable problem on permutation groups, which is as follows.

Consider permutations of the objects $a_1, \ldots, a_T$. Let R be the T-cycle $(a_1\,a_2\ldots a_T)$. For any permutation U, let $H(U)$ denote the group of all permutations of the form

$$R^{t_0}UR^{t_1}\ldots UR^{t_p}, \quad \sum t_i = 0.$$

$H(U)$ is called *exceptional* if it is not the symmetric or alternating group. The problem is to find all exceptional groups or at least to find all U such that $H(U)$ is exceptional.

Besides employing his usual ingenuity, Turing has to perform some really extensive calculations in order to solve the problem for the cases $T=1,2,\ldots,8$.

Clearly this problem is a challenge to present day workers in permutation groups. Computer scientists also may find it interesting to see if they can check Turing's results and extend his calculations beyond $T=8$.

The Difference $\psi(x)-x$ (IV)

Recall that Chebyshev's ψ-function is defined by $\psi(x)=\sum \log p$, where the sum is over all prime numbers p and positive integers m such that $p^m \leqslant x$. We have $\psi(x) \sim x$ as $x \to \infty$; this is equivalent to the prime number theorem that $\pi(x) \sim x/\log x$ as $x \to \infty$, where $\pi(x)$ is the number of primes less than or equal to x. In this paper it is shown that for some x ($\exp 99 < x < \exp 10^{428}$) $(\psi(x)-x)/x^{1/2} > 1.0001$. It is known that $\psi(x)-x$ changes sign infinitely often.

A.M. Turing and S. Skewes, *On a Theorem of Littlewood* (V)

In spite of the joint authorship, it seems probable that this unpublished paper was written by Turing alone.

Let $\pi(x)$ denote as usual the number of primes $\leqslant x$ and let

$$\operatorname{li} x = \lim_{h \to 0+} \left(\int_0^{1-h} + \int_{1+h}^{x} \right) \left(\frac{1}{\log u} \right) du.$$

We have $\pi(x) \sim \operatorname{li} x$ as $x \to \infty$. The Riemann hypothesis is equivalent to $\pi(x) - \operatorname{li} x = O(x^{1/2} \log x)$ as $x \to \infty$, but here the interest is on the sign of $\pi(x) - \operatorname{li} x$. Littlewood proved that

$$\frac{\pi(x) - \operatorname{li} x}{(x^{1/2}/\log x) \log\log\log x}$$

has positive limit superior and negative limit inferior as $x \to \infty$; thus $\pi(x) - \operatorname{li} x$ changes sign infinitely often. It is negative for $2 < x < 10^7$. The main part of this paper is devoted to a proof that $\pi(x) - \operatorname{li} x$ is positive for some x ($2 < x < \exp\exp 661$).

The second half of the paper, which was undoubtedly written by Turing alone, contains some informal and speculative passages as well as a substantial theorem. Attention here centres on the consequences for the theory of $\pi(x) - \operatorname{li} x$ if an actual zero of the zeta-function off the critical line were found by means of a computer.

W.W. Boone, *An Analysis of Turing's "The Word Problem in Semi-groups with Cancellation"*, (1958)

This is a careful study of Turing's paper (1950) and the author confirms that Turing's proof is essentially correct.

A.M. Cohen and M.J.E. Mayhew, *On the Difference* $\pi(x) - \operatorname{li} x$, (1968)

This is a considerably corrected and amplified version of the first part of the TURING and SKEWES paper (V). They show that the proof in that paper is essentially correct but the conclusion is that $\pi(x) - \operatorname{li} x$ is positive for some x, $2 < x < \exp\exp 1236$ (rather than $\exp\exp 661$).

Acknowledgements

Thanks are due first to the authors and publishers who gave their permission to reprint the papers in this volume and to King's College, Cambridge, for allowing access to the unpublished papers.

For their assistance and advice I should like to thank Sir Michael Atiyah, the late J. Frank Adams, Frank D. Burgoyne, Alan M. Cohen, Martin Edjvet, Michael Halls, Roger Heath-Brown, Wilfrid Hodges, David L. Johnson, Deane Montgomery, S. Sankaran and G.K. Sankaran.

Finally my thanks are due to Jack Good for agreeing to write a special article for this volume and, on his behalf, I thank Donald Michie and Shaun Wylie.

J.L. BRITTON
19th May, 1989

POSTSCRIPT

I was a research student at Manchester University from September 1951 to August 1953. Turing's position was Deputy Director of the Computing Machine Laboratory; thus he was not formally a member of the mathematics staff but in practice he was a star member of it. The department was a very strong one: M.H.A. Newman, M.J. Lighthill, M.S. Bartlett, Kurt Mahler, G.I. Camm, B.H. Neumann, C.R. Illingworth, F.G. Friedlander, Walter Ledermann, Graham Higman, H.G. Hopkins, G.E.H. Reuter, A.H. Stone, Eric Wild, Samuel Levine, D.S. Jones, P.J. Hilton, J.E. Moyal, F.D. Kahn, G.E. Wall, J.A. Green, M.B. Glauert, Yael N. Dowker, R.K. Livesley and A.M. Walker.

I was introduced to Turing by my supervisor, Bernhard Neumann, in 1951. I can recall little of Turing. It was only towards the end of my stay that I first became interested in his work.

Turing did not give any lecture courses. I was friendly with his research student, Ivor Jones, whose field was logic.

The only lecture by Turing that I attended was one he gave to the student mathematical society. It was entitled "On large numbers". Figuring in the lecture were the following large numbers M and N. A hypothetical bird flies, once each year, to the top of Mount Everest and removes one grain of "sand"; M is the number of seconds needed to level the mountain. The other number N was such that $1/N$ is the probability that "this piece of chalk will jump from my hand and write a line of Shakespeare on the board before falling to the ground". Turing had, of course, numerical estimates for M and N.

Bernhard Neumann told me the following story. At one time (probably 1949) he believed he had discovered a proof of the solvability of the word problem for groups; upon mentioning this to Turing, he was disconcerted to learn that Turing had just completed a proof of its unsolvability. Both of them urgently re-examined their proofs and both proofs were found to be wrong.

J.L.B.

REMARKS ON TURING'S DISSERTATION

Turing's Fellowship dissertation was circulated to the electors of fellowships at King's College, Cambridge, namely Messrs. Ingham, Keynes, Braithwaite, Matthews, McCombie, Beves, F.L. Lucas and the Provost. The date of the election was March 16th, 1935. According to the biography, Philip Hall and A.C. Pigou also acted as electors.

The dissertation was entitled "On the Gaussian error function". It contains a proof of the Central Limit Theorem, which Turing discovered without knowing that a proof already existed; it was originally proved by Lindeberg in 1922.

Although the dissertation is in immediately publishable form, it seems inappropriate to disturb its present status by reprinting it in this volume. For biographical interest, however, the preface is reprinted below.

Preface

The object of this paper is to give a rigorous demonstration of the "limit theorem of the theory of probability". I had completed the essential part of it by the end of February 1934 but when considering publishing it I was informed that an almost identical proof had been given by Lindeberg[1]. The only important differences between the two papers is that I have introduced and laid stress on a type of condition which I call quasi-necessary (§8). We have both used "distribution functions" (§2) to describe errors instead of frequency functions (Appendix B) as was usual formerly. Lindeberg also uses (D) of §12 and Theorem 6 or their equivalents.

Since reading Lindeberg's paper I have for obvious reasons made no alterations to that part of the paper which is similar to his (viz. §9 to §13), but I have added elsewhere remarks on points of interest and the appendices.

So far as I know the results of §8 have not been given before. Many proofs of the completeness of the Hermite functions are already available (footnote p.33) but I believe that that given in Appendix A is original. The remarks in Appendix B are probably not new. Appendix C is nothing more than a rigorous deduction of well-known facts. It is only given for the sake of logical completeness and it is of little consequence whether it is original or not.

---

[1] *Math. Z.* **15** (1922).

My paper originated as an attempt to make rigorous the "popular" proof mentioned in Appendix B. I first met this proof in a course of lectures by Prof. Eddington. Variations of it are given by Czüber, Morgan, Crofton and others. Beyond this I have not used the work of others or other sources of information in the main body of the paper, except for elementary matter forming part of one's general mathematical education, but in the appendices I may mention Liapounoff's papers which I discuss there.

I consider §9 to §13 is by far the most important part of this paper, the remainder being comment and elaboration. At a first reading therefore §8 and the appendices may be omitted.

CONTENTS

EQUIVALENCE OF LEFT AND RIGHT ALMOST PERIODICITY

A. M. TURING*.

In his paper "Almost periodic functions in a group", J. v. Neumann† has used independently the ideas of left and right periodicity. I shall show that these are equivalent.

$f(x)$ is a complex-valued function of a variable x which runs through an arbitrary group $\mathfrak{G}$. $f(x)$ is said to be right almost periodic (r.a.p.) if for each $\epsilon > 0$ we can find a finite set $b_1, \ldots, b_m$ of elements of $\mathfrak{G}$ such that to each t of $\mathfrak{G}$ there corresponds a $\mu = \mu(t)$ satisfying

$$|f(xt) - f(xb_\mu)| < \epsilon \quad \text{for all } x \in \mathfrak{G}. \tag{D}$$ ⟦1⟧

The definition of left almost periodicity is obtained from this by replacing the inequality (D) by

$$|f(tx) - f(b_\mu x)| < \epsilon.$$

Suppose now that $f(x)$ is r.a.p., then to prove $f(x)$ l.a.p. it is sufficient to find, for each $\epsilon > 0$, a finite number of elements $c_1, \ldots, c_n$ of $\mathfrak{G}$ such that to each s of $\mathfrak{G}$ there corresponds a $\nu = \nu(s)$ satisfying

$$|f(sb_\pi) - f(c_\nu b_\pi)| < \epsilon \quad \text{for each } \pi; \tag{K}$$

---

* Received 23 April, 1935; read 25 April, 1935.

† J. v. Neumann, *Trans. American Math. Soc.*, 36 (1934), 445–492.

for then, by the r.a.p. property of $f(x)$,

$$|f(sb_\mu)-f(st)| < \epsilon,$$

$$|f(c_\nu b_\mu)-f(c_\nu t)| < \epsilon,$$

where $\mu = \mu(t)$.

Putting $\pi = \mu(t)$ in the inequality (K), we have

$$|f(st)-f(c_\nu t)| < 3\epsilon \quad \text{for each } t,$$

i.e. $f(x)$ is l.a.p.

To prove the existence of the elements $c_1, \ldots, c_n$ let us introduce a space R of m complex dimensions. Consider the set S of points P_y of R whose coordinates are $[f(yb_1), \ldots, f(yb_m)]$ (y runs through $\mathfrak{G}$). $f(x)$ being r.a.p. is bounded*; S is therefore bounded and can be covered with a finite number of spheres of diameter ϵ each containing some point of S. Let the finite set of elements of S obtained in this way be $P_{c_1}, \ldots, P_{c_n}$; then for each s of $\mathfrak{G}$ there is a $\nu = \nu(s)$ with P_s distant less than ϵ from P_{c_ν}; hence, for each μ,

$$|f(sb_\mu)-f(c_\nu b_\mu)| < \epsilon,$$

i.e. $c_1, \ldots, c_n$ have the required property.

Thus $f(x)$ is r.a.p. implies that $f(x)$ is l.a.p. and the converse follows similarly or by the use of the inverse group. v. Neumann's theory can
⟦2⟧ now be used to show that each l.a.p. function has a unique left mean. Previously it was necessary to suppose $f(x)$ to be both l.a.p. and r.a.p. The theory of a.p. functions in a group can now be taken over to sets of objects which admit transitive transformations by the group. Let $\mathfrak{A}$ be a set of objects admitting (left-) transformations by the group $\mathfrak{G}$. Represent the elements of $\mathfrak{A}$ by small Gothic letters. Then to a function $f(\mathfrak{y})$ in $\mathfrak{A}$ corresponds a function $f(y)$ in $\mathfrak{G}$ defined by $f(y) = f(\mathfrak{y})$, whenever $y\mathfrak{t} = \mathfrak{y}$, $\mathfrak{t}$ being some fixed element of $\mathfrak{A}$. $f(\mathfrak{y})$ may be said to be a.p. if $f(y)$ is l.a.p., and will then have a unique left mean.

King's College,
Cambridge.

---

* Putting $x = e$ in (D), we have

$$|f(t)-f(b_\mu)| < \epsilon.$$

Then $$|f(t)| < \epsilon + \max\{|f(b_1)|, \ldots, |f(b_m)|\}.$$

⟦2⟧

ANNALS OF MATHEMATICS
Vol. 39, No. 1, January, 1938

FINITE APPROXIMATIONS TO LIE GROUPS

BY A. M. TURING

(Received April 28, 1937; revised September 29, 1937)

A certain sense in which a finite group may be said to approximate the structure of a metrical group will be discussed. On account of Jordan's theorem on ⟦1⟧
finite groups of linear transformations[1] it is clear that we cannot hope to approximate a general Lie group with finite subgroups. I shall show that we cannot approximate even with groups which are 'approximately subgroups': in fact the only approximable Lie groups are the compact Abelian groups. The key to the situation is again afforded by Jordan's theorem, but it is not immediately applicable. It is necessary to find representations of the approximating groups whose degree depends only on the group approximated.

Approximability of metrical groups. Suppose G is a group with a metric D invariant under left transformations i.e. $D(ax, ay) = D(x, y)$ for all x, y, a
of G. Let H_ϵ be a finite subset of G in which is defined a second product with ⟦2⟧
respect to which it forms a group (if a and b are in H_ϵ their product as elements of G will be written ab; the product of them as elements of H_ϵ will be written $a_\circ b$, the inverse of a as an element of H_ϵ is written $[a]^{-1}$ and the identities of G and H_ϵ are written e, e_ϵ), and suppose each element x of G is within distance ϵ
of an element $r(x)$ of H_ϵ, and for each a, b of H_ϵ $D(a_\circ b, ab) > \epsilon$. Then H_ϵ ⟦3⟧
will be said to be an ϵ-approximation to G.

A group is said to be approximable if it has an ϵ-approximation for each $\epsilon > 0$.

Immediately from the definition we see that an approximable group is totally ⟦4⟧
bounded i.e. conditionally compact. It is therefore possible to find a metric which is both left and right invariant and equivalent to the given metric, in the sense that the class of open sets is the same for either metric. In future therefore we shall suppose that our metric is both ways invariant, and we shall denote the distance between x and y by $D(x, y)$.

It has been shown by J. v. Neumann[2] that with a conditionally compact

---

[1] This theorem states that a finite group of linear transformations has an Abelian self conjugate subgroup whose index does not exceed a certain bound depending only on the degree.

[2] J. v. Neumann, *Zum Haarschen Mass in topologischen Gruppen*, Compositio Mathematica, vol. 1 (1934) pp. 106–114; or alternatively, J. v. Neumann, *Almost periodic functions in a group*, Transactions of the American Mathematical Society, vol. 36 (1934) pp. 445–492, (remember that every continuous function in a conditionally compact group is a.p.). If the reader prefers to restrict the group in some way and use some other mean he has only to verify the inequality (1).

group we can define a mean for each continuous (complex-valued) function in the group, in such a way that (denoting the mean of $f(x)$ by $\int_G f(x)\,dx$)

$$\int_G (f(x) + g(x))\,dx = \int_G f(x)\,dx + \int_G g(x)\,dx$$

$$\int_G f(ax)\,dx = \int_G f(xa)\,dx = \int_G f(x)\,dx$$

and so that if $\epsilon > 0$ (and $f(x)$ is continuous), then there is a finite set of elements
⟦5⟧ $a_1, a_2, \cdots, a_N$ of G such that

$$\left| \frac{1}{N} \sum_{i=1}^{N} f(xa_i) - \int_G f(x)\,dx \right| < \epsilon. \tag{1}$$

Before proceeding to the proofs of our main theorems we shall establish some elementary inequalities following immediately from our definition. Suppose the function $r(x)$ belongs to an ϵ-approximation H_ϵ to G, then

$$D(r(x)_0 r(y), xy) \leqq D(r(x)_0 r(y), r(x)r(y)) + D(r(x)r(y), xr(y)) + D(xr(y), xy) < 3\epsilon \tag{2}$$

and for any a, c of H_ϵ

$$D(c_0 a_0 [c]^{-1}, cac^{-1}) < 4\epsilon \tag{3}$$

for

$$\begin{aligned} D(c_0 a_0 [c]^{-1}, cac^{-1}) &< D(ca[c]^{-1}, cac^{-1}) + 2\epsilon \\ &= D([c]^{-1}c, e) + 2\epsilon \\ &\leqq D([c]^{-1}c, [c]_0^{-1}c) + D(e_\epsilon, e) + 2\epsilon \\ &\leqq D(e_\epsilon, e) + 3\epsilon \\ &= D(e_\epsilon^2, e_\epsilon) + 3\epsilon \\ &\leqq D(e_{\epsilon 0} e_\epsilon, e_\epsilon) + 4\epsilon = 4\epsilon. \end{aligned}$$
⟦6⟧

⟦7⟧ THEOREM 1. *Let G be an approximable group with a true continuous representation by matrices of degree n. Then it may be approximated by finite groups with true representations of the same degree n.*

LEMMA. *If H_η is an η-approximation (of order h_η) to the group G and if $f(x)$ is a continuous function in G such that*

$$|f(x) - f(x')| < \Delta \quad \text{when} \quad D(x, x') < \eta$$

then

$$\left| \frac{1}{h_\eta} \sum_{a \in H_\eta} f(a) - \int_G f(x)\,dx \right| \leqq 2\Delta. \tag{4}$$

⟦4⟧

We put

$$\int_G f(x)\,dx = A \qquad \frac{1}{h_\eta}\sum_{a \in H_\eta} f(a) = B$$

then given $\epsilon > 0$ there are $a_1, a_2, \cdots, a_N$ such that

$$\left|\frac{1}{N}\sum_{i=1}^{N} f(xa_i) - A\right| < \epsilon \tag{5}$$

for each element x of G. If in (5) we successively put x equal to each member of H_η and combine the resulting inequalities we obtain

$$\left|\frac{1}{Nh_\eta}\sum_{i=1}^{N}\sum_{c \in H_\eta} f(ca_i) - A\right| < \epsilon,$$

but $D(ca_i, c_0 r(a_i)) < 2\eta$, so that $|f(ca_i) - f(c_0 r(a_i))| < 2\Delta$ and therefore ⟦8⟧

$$\left|\frac{1}{Nh_\eta}\sum_{i=1}^{N}\sum_{c \in H_\eta} f(c_0 r(a_i)) - A\right| < \epsilon + 2\Delta. \tag{6}$$

However

$$\frac{1}{h_\eta}\sum_{c \in H_\eta} f(c_0 r(a_i)) = B$$

so that (6) yields (4) since ϵ was arbitrary.

PROOF OF THE THEOREM. Without loss of generality we may suppose that ⟦9⟧ the given representation of G does not contain any irreducible component more than once. Let $\chi(x)$ be the character of the representation. This function will satisfy

$$\chi(x) = \int_G \chi(xy)\overline{\chi(y)}\,dy \tag{7}$$

$$\chi(x) = \chi(cxc^{-1}) \tag{8}$$

$$|\chi(x)| \leqq n \tag{9}$$

and since it is the character of a true representation

$$\chi(x) \neq \chi(e) = n \quad \text{if} \quad x \neq e.$$

Let $\epsilon > 0$. Then for some α, $1 > \alpha > 0$, $|\chi(x) - n| > \alpha$ when $D(x, e) \geqq \frac{1}{4}\epsilon$. ⟦10⟧
Now let η be so chosen that $\epsilon/16 > \eta > 0$ and

$$|\chi(x) - \chi(x')| < \alpha/(50n^2) \quad \text{when} \quad D(x, x') < 4\eta \tag{10}$$

$$|\chi(ay)\overline{\chi(y)} - \chi(ay')\overline{\chi(y')}| < \alpha/(50n) \text{ all } a, \text{ when } D(y, y') < 2\eta \tag{11}$$ ⟦11⟧

and take a corresponding η-approximation H_η. If we put

$$\varphi(a) = \frac{1}{h_\eta}\sum_{c \in H_\eta} \chi(c_0 a_0 [c]^{-1}) \tag{12}$$

then

$$(13) \qquad |\varphi(a) - \chi(a)| \leqq \frac{1}{h_\eta} \sum_{c \in H_\eta} |\chi(c_0 a_0 [c]^{-1}) - \chi(cac^{-1})| < \frac{\alpha}{50n^2}$$

for $D(c_0a_0[c]^{-1}, cac^{-1}) < 4\eta$ by (3) and therefore each summand is less than $\alpha/(50n^2)$. We have

$$(14) \qquad \left| \frac{1}{h_\eta} \sum_{b \in H_\eta} \varphi(a_0 b)\overline{\varphi(b)} - \chi(a) \right| \leqq \frac{1}{h_\eta} \left| \sum_{b \in H_\eta} (\varphi(a_0 b)\overline{\varphi(b)} - \chi(a_0 b)\overline{\chi(b)}) \right| + \frac{1}{h_\eta} \left| \sum_{b \in H_\eta} (\chi(a_0 b) - \chi(ab))\overline{\chi(b)} \right| + \left| \frac{1}{h_\eta} \sum_{b \in H_\eta} \chi(ab)\overline{\chi(b)} - \int_G \chi(ay)\overline{\chi(y)}\, dy \right|.$$

⟦12⟧ Applying the lemma to $\chi(ay)\overline{\chi(y)}$ and making use of (11) we have

$$(15) \qquad \left| \frac{1}{h_\eta} \sum_{b \in H_\eta} \chi(ab)\overline{\chi(b)} - \int_G \chi(ay)\overline{\chi(y)}\, dy \right| < \frac{2\alpha}{50n}$$

and from (9), (10) we obtain

⟦13⟧ $$(16) \qquad \frac{1}{h_\eta} \left| \sum_{b \in H_\eta} (\chi(a_0 b) - \chi(ab))\overline{\chi(b)} \right| < \frac{\alpha}{50n}.$$

Finally

⟦14⟧ $$(17) \qquad \frac{1}{h_\eta} \left| \sum_{b \in H_\eta} (\varphi(a_0 b)\overline{\varphi(b)} - \chi(a_0 b)\overline{\chi(b)}) \right| \leqq \frac{1}{h_\eta} \sum_{b \in H_\eta} |(\varphi(a_0 b) - \chi(a_0 b))\overline{\varphi(b)}| + \frac{1}{h_\eta} \sum_{b \in H_\eta} |(\overline{\varphi(b)} - \overline{\chi(b)})\chi(a_0 b)| < \frac{2\alpha}{50n}$$

by (9) and (13). Combining (14), (15), (16), (17),

$$(18) \qquad \left| \frac{1}{h_\eta} \sum \varphi(a_0 b)\overline{\varphi(b)} - \chi(a) \right| < \frac{\alpha}{10n}$$

$$(19) \qquad \left| \frac{1}{h_\eta} \sum \varphi(a_0 b)\overline{\varphi(b)} - \varphi(a) \right| < \frac{\alpha}{8n}.$$

Now $\varphi(a) = \varphi(c_0 a_0 [c]^{-1})$ for each a, c of H_η. This function is therefore expressible as a sum of characters

$$\varphi(a) = \sum_{\lambda=1}^{M} \alpha_\lambda \chi^{(\lambda)}(a),$$

$\chi^{(1)}(a), \cdots, \chi^{(M)}(a)$ being the characters of the different irreducible representations of H_η. From the general theory of representations

⟦15⟧ $$\frac{1}{h_\eta} \sum_{b \in H_\eta} \chi^{(\lambda)}(a_0 b)\overline{\chi^{(\mu)}(b)} = \delta_{\lambda\mu}\chi^{(\lambda)}(a)$$

⟦6⟧

(19) therefore becomes ⟦16⟧

$$\left| \sum_{\lambda=1}^{M} \alpha_\lambda(\bar{\alpha}_\lambda - 1)\chi^{(\lambda)}(a) \right| < \frac{\alpha}{8n}.$$

Squaring each side of this inequality and summing over H_η,

$$\frac{1}{h_\eta}\sum_{\lambda=1}^{M}\sum_{a \in H_\eta} |\alpha_\lambda|^2 \, |1 - \alpha_\lambda|^2 \, |\chi^{(\lambda)}(a)|^2 = \sum_{\lambda=1}^{M} |\alpha_\lambda|^2 \, |1 - \alpha_\lambda|^2 < \frac{\alpha^2}{64n^2}.$$

If we define $\xi(a)$ by

$$\xi(a) = \sum_{|1-\alpha_\lambda| > |\alpha_\lambda|} \chi^{(\lambda)}(a)$$

it will satisfy

$$\frac{1}{h_\eta}\sum_{a \in H_\eta} \xi(a_0 b)\overline{\xi(b)} = \xi(a) \tag{20}$$

and

$$\begin{aligned}\frac{1}{h_\eta}\sum_{a \in H_\eta} |\xi(a) - \varphi(a)|^2 &= \sum_{\lambda=1}^{M} \operatorname{Min}\,(|\alpha_\lambda|^2, |1 - \alpha_\lambda|^2) \\ &\leqq 4\sum_{\lambda=1}^{M} |\alpha_\lambda|^2 \, |1 - \alpha_\lambda|^2 < \frac{\alpha^2}{16n^2}.\end{aligned} \tag{21}$$

We now wish to infer from the inequality (21) that the functions $\varphi(a)$ and $\xi(a)$ differ only slightly at each point of H_η. This is possible on account of the relations (19), (20).

$$\begin{aligned}&\left|\frac{1}{h_\eta}\sum_{b \in H_\eta} (\xi(a_0 b)\overline{\xi(b)} - \varphi(a_0 b)\overline{\varphi(b)})\right| \\ &\leqq \sum_{b \in H_\eta} \frac{1}{h_\eta}\left|(\xi(a_0 b) - \varphi(a_0 b))\overline{\xi(b)}\right| + \sum_{b \in H_\eta} \frac{1}{h_\eta}\left|(\overline{\xi(b)} - \overline{\varphi(b)})\varphi(a_0 b)\right| \\ &\leqq \left\{\frac{1}{h_\eta}\sum_{b \in H_\eta} |\xi(b) - \varphi(b)|^2\right\}^{\frac{1}{2}}\left(\left\{\frac{1}{h_\eta}\sum_{b \in H_\eta} |\xi(b)|^2\right\}^{\frac{1}{2}} + \left\{\frac{1}{h_\eta}\sum_{b \in H_\eta} |\varphi(b)|^2\right\}^{\frac{1}{2}}\right) \\ &< \frac{\alpha}{4n}(n + n) = \frac{1}{2}\alpha,\end{aligned} \tag{22}$$

since $|\xi(b)| \leqq n$ and $|\varphi(b)| \leqq n$ for each b of H_η. Now combine (18), (20), ⟦17⟧
(22), and we have

$$|\xi(a) - \chi(a)| < \frac{1}{2}\alpha + \frac{\alpha}{10n} < \alpha.$$

This implies that $\xi(e_\eta) = \chi(e) = n$ and that if $D(a, e) \geqq \frac{1}{4}\epsilon$ then $\xi(a) \neq \chi(e) =$ ⟦18⟧
$\xi(e_\eta)$. $\xi(a) = \xi(e_\eta)$ only for elements of a certain self-conjugate subgroup N ⟦19⟧
entirely contained within distance $\frac{1}{4}\epsilon$ of the identity of G. The factor group has a true representation of degree n, and I shall show that it can be taken as a

[[20]] ϵ-approximation to G. We choose an element in each coset of N as a representative of that coset and define the function $v(a)$ (a in H_η) to be the representative of the coset in which a lies. The totality of elements $v(a)$ we call K.
[[21]] Putting $v(a) \otimes v(b) = v(a_0 b)$, K forms a group with respect to the product $\otimes$. For each a of H_η there is an element m of N for which $v(a) = a_0 m$ and therefore

$$D(a, v(a)) \leqq D(a, am) + D(am, a_0 m) < \tfrac{1}{4}\epsilon + \eta.$$

[[22]] Consequently if we put $R(x) = v(r(x))$ we have

$$D(R(x), x) \leqq D(v(r(x)), r(x)) + D(r(x), x) < (\tfrac{1}{4}\epsilon + \eta) + \eta < \epsilon$$

and

[[23]] $$D(v(a) \otimes v(b), v(a)v(b)) \leqq D(v(a_0 b), a_0 b) + D(a_0 b, ab) + D(ab, v(a)v(b))$$
$$< 3(\tfrac{1}{4}\epsilon + \eta) + \eta < \epsilon,$$

which shows that K is an ϵ-approximation to G.

[[24]] THEOREM 2. *An approximable Lie group is compact and Abelian.*

LEMMA. *A closed subgroup of a connected group cannot have a finite index greater than* 1.

Suppose H is a closed subgroup of G and has index i, $1 < i < \infty$. Then $G - H$ is not void and is closed, being the sum of a finite number of closed sets, the cosets of H. G is the sum of two closed disjoint sets neither of which is void, and therefore is not connected.

If G is a compact Lie group it cannot have a closed subgroup of positive measure different from the whole group.

PROOF OF THE THEOREM. An approximable Lie group is complete and conditionally compact, i.e. it is compact, and is therefore a group of linear transformations,[3] of degree n say. By theorem 1 we can approximate it by finite groups H_ϵ of linear transformations of degree n. But by Jordan's theorem[4] each finite group of linear transformations has an Abelian subgroup whose index does not exceed a certain bound $Z(n)$ depending only on the degree. Let A_ϵ be this Abelian subgroup in H_ϵ. Then there is a finite number $c_1, c_2, \cdots c_N$ ($N \leqq Z(n)$) of elements of H_ϵ such that every element of H_ϵ is of the form $c_{i0} a$ where a is in A_ϵ. For any x of G we have

$$D(x, r(x)) < \epsilon$$

$$r(x) = c_{i0} a, \; a \in A_\epsilon, \; i \leqq N$$

$$D(c_{i0} a, c_i a) < \epsilon.$$

[3] J. v. Neumann, *Die Einführung analytischer Parameter in topologischen Gruppen*, Annals of Mathematics, vol. 34 (1933), pp. 170-190.

[4] A. Speiser, *Theorie der Gruppen von endlicher Ordnung*, (Berlin 1927) 2nd ed., p. 215.

Hence every element of G is of the form $c_i ad$ where d is within distance 2ϵ of the identity of G and $i \leqq N$. The points ad must therefore form a set E_ϵ of measure $1/Z(n)$ at least. Now put $x = ad\ y = a'd'$:

$$
\begin{aligned}
D(xy, yx) &= D(ada'd', a'd'ad) \\
&\leqq 2D(d, d') + D(a_0 a', a_0' a) + D(aa', a_0 a') + D(a_0' a, a'a) \qquad (23) \\
&< 6\epsilon.
\end{aligned}
$$
⟦25⟧

In the product group $G \times G$ we have therefore a set $E_\epsilon \times E_\epsilon$ of pairs (x, y) of measure $1/(Z(n))^2$ at least, in which $D(xy, yx) < 6\epsilon$. Now take a sequence ϵ_i tending to 0, and put $F_i = \sum_{j \geqq i} E_{\epsilon_j} \times E_{\epsilon_j}$, $E = \prod F_i$. For each $i \leqq N$,

$$mF_i \geqq m(E_{\epsilon_i} \times E_{\epsilon_i}) \geqq \frac{1}{(Z(n))^2}.$$

Then $mE \geqq 1/(Z(n))^2$ since the F_i are a decreasing sequence. If $(x, y) \in E$ then for each i, $(x, y) \in F_i$, i.e. $x \in E_{\epsilon_j}$, $y \in E_{\epsilon_j}$ for some $j \geqq i$. Then by (23), $D(xy, yx) < 6\epsilon_j \leqq 6\epsilon_i$: but i was arbitrary so that $D(xy, yx) = 0$, $xy = yx$.

Now let N_x be the set of those y for which $xy = yx$, i.e. the normaliser of x. Then

$$\int_G mN_x \, dx \geqq mE \geqq \frac{1}{(Z(n))^2}.$$
⟦26⟧

Consequently $mN_x > 0$ in an x-set of positive measure. But if $mN_x > 0$ we have $N_x = G$ by the lemma, for N_x is certainly closed. This shows that the centre of G is of positive measure, and again applying the lemma we see that G is Abelian.

PRINCETON UNIVERSITY.

The extensions of a group

by

A. M. Turing
Cambridge, England

A group $\mathfrak{G}$ is said to be an extension of $\mathfrak{N}$ by $\mathfrak{G}'$ if $\mathfrak{N}$ is a self conjugate subgroup of $\mathfrak{G}$ and $\mathfrak{G}/\mathfrak{N} \cong \mathfrak{G}'$. The problem of finding the extensions of $\mathfrak{N}$ by $\mathfrak{G}'$ has been investigated by Schreier [1]) and by Baer [2]).

Let $\mathfrak{A}$ be the automorphism group of $\mathfrak{N}$ and $\mathfrak{J}$ the subgroup of inner automorphisms. Then to each coset γ of $\mathfrak{N}$ in $\mathfrak{G}$ there corresponds a coset $X(\gamma)$ of $\mathfrak{J}$ in $\mathfrak{A}$, such that if $c \in \gamma$ then the automorphism induced by c in $\mathfrak{N}$ belongs to $X(\gamma)$. $X(\gamma)$ is a [[1]]
homomorphism of $\mathfrak{G}'$ in $\mathfrak{A}/\mathfrak{J}$. Baer's investigations are concerned [[2]]
with finding all possible groups $\mathfrak{G}$ when $\mathfrak{N}$, $\mathfrak{G}'$ and the homomorphism $X(\gamma)$ are given. As a first step towards the solution of this problem the possible structures of $\mathfrak{G}/\mathfrak{Z}(\mathfrak{N})$ are found, where $\mathfrak{Z}(\mathfrak{N})$ denotes the centre of $\mathfrak{N}$; it then only remains to solve the original problem in the case where $\mathfrak{N}$ is Abelian. This case is treated entirely differently. In the present paper it is proposed to show how Baer's method for the case when $\mathfrak{N}$ is Abelian can be used for any group $\mathfrak{N}$.

In a practical determination of all extensions with given characteristics it is necessary to find the structure of the relation group of the factor group $\mathfrak{G}'$. This is so even when $\mathfrak{N}$ is Abelian. The problem is considered in the second half of the paper.

As an illustration the theory is applied to the case of extensions of an arbitrary group by a cyclic group.

§ 1. *Extensions with given automorphisms.*

The problem of finding extensions of a group by a given group inducing given classes of automorphisms is best treated

1) O. SCHREIER, Über die Erweiterung von Gruppen [Monats. f. Math. u. Phys. **34** (1926), 165—180].

2) R. BAER, Erweiterung von Gruppen und ihren Isomorphismen [Math. Zeitschr. **38** (1934), 375—416].

by reducing it to another problem which is both described and solved in the following

THEOREM 1.

$\mathfrak{N}$, $\mathfrak{F}$ are given groups [3]) *and $\mathfrak{R}$ is a self conjugate subgroup*
⟦3⟧ *of $\mathfrak{F}$; χ_a is a homomorphism of $\mathfrak{F}$ into the group of automorphisms of $\mathfrak{N}$ (i.e. $\chi_a(\mathfrak{b})$ as a function of $\mathfrak{b}$ is an automorphism of $\mathfrak{N}$ and satisfies*

$$\chi_{ab}(\mathfrak{b}) = \chi_b(\chi_a(\mathfrak{b})) \qquad (1,\ b,\ a,\ \mathfrak{b})$$

for all b, a of $\mathfrak{F}$ and $\mathfrak{b}$ of $\mathfrak{N}$) and $\mathfrak{a}(r)$ is a homomorphism of $\mathfrak{R}$ into $\mathfrak{N}$. Then there is a group [3]) *$\mathfrak{G}$ in which $\mathfrak{N}$ is a self conjugate subgroup, and a homomorphism $\mathfrak{w}(r)$ of $\mathfrak{F}$ into $\mathfrak{G}$ which satisfies:*

a)
$$\left.\begin{array}{ll} \chi_a(\mathfrak{b}) = (\mathfrak{w}(a))^{-1}\mathfrak{b}\mathfrak{w}(a) & (\mathrm{I},\ a,\ \mathfrak{b}) \\ \mathfrak{w}(r) = \mathfrak{a}(r) & (\mathrm{II},\ r) \end{array}\right\} \qquad (2)$$

(for all a in $\mathfrak{F}$, r in $\mathfrak{R}$ and $\mathfrak{b}$ in $\mathfrak{N}$),

b) every coset of $\mathfrak{N}$ in $\mathfrak{G}$ contains an element of $\mathfrak{w}(\mathfrak{F})$,

c) $\mathfrak{w}(\mathfrak{R}) = \mathfrak{N} \cap \mathfrak{w}(\mathfrak{F})$

if and only if

$$\left.\begin{array}{ll} \chi_a(\mathfrak{a}(r)) = \mathfrak{a}(a^{-1}ra) & (\mathrm{I},\ a,\ r) \\ \chi_r(\mathfrak{b}) = (\mathfrak{a}(r))^{-1}\mathfrak{b}\mathfrak{a}(r) & (\mathrm{II},\ r,\ \mathfrak{b}) \end{array}\right\} \qquad (3)$$

(for all a in $\mathfrak{F}$, r in $\mathfrak{R}$ and $\mathfrak{b}$ in $\mathfrak{N}$).

The relevance of this theorem to the original extension problem can be seen from the

COROLLARY:

$\mathfrak{F}$ is a free [4]) *group with a self conjugate subgroup $\mathfrak{R}$ ($\mathfrak{F}/\mathfrak{R} \cong \mathfrak{G}'$,*
⟦4⟧ *say) and $X(\gamma)$ is a homomorphism of $\mathfrak{F}/\mathfrak{R}$ into the classes of automorphisms of a given group $\mathfrak{N}$. Let χ_a be any homomorphism of $\mathfrak{F}$ into the automorphisms of $\mathfrak{N}$ for which χ_a belongs to the class $X(\alpha)$ whenever a belongs to the coset α of $\mathfrak{R}$ in $\mathfrak{F}$. Then an extension*

---

[3]) Elements of $\mathfrak{F}$ are denoted by italic letters, elements of $\mathfrak{N}$ and $\mathfrak{G}$ by German letters, and elements of $\mathfrak{F}/\mathfrak{R}$ by Greek letters. e, $\mathfrak{e}$, ε are the identities of these groups and E is the identity of Φ in § 2.

[4]) It is essential that $\mathfrak{F}$ should be free if the conditions are to be necessary. A trivial example shows that at least $\mathfrak{F}$ cannot be arbitrary if we require $\mathfrak{a}(r)$ to be related to a function $\mathfrak{w}(a)$ as in the theorem. Let $\mathfrak{G}$ be the cyclic group $\{\mathfrak{b}\}$ of order 4, $\mathfrak{F}$ the cyclic group $\{g\}$ of order 2, and let $\mathfrak{N}$ be $\{\mathfrak{b}^2\}$ and $\chi_g(\mathfrak{b}^2) = \mathfrak{b}^2$. Then we should have to have $\mathfrak{a}(e) = \mathfrak{e}$. But $\mathfrak{a}(e) = \mathfrak{a}(g^2) = \mathfrak{b}^2$.

⟦12⟧

$\mathfrak{G}$ of $\mathfrak{N}$ by $\mathfrak{G}'$ in which the coset α of $\mathfrak{N}$ induces the class $X(\alpha)$ can be found if and only if there is a homomorphism $\mathfrak{a}(r)$ of $\mathfrak{R}$ in $\mathfrak{N}$ satisfying (3). ⟦5⟧

Proof of the theorem. The necessity of the conditions is trivial. $(3, \mathrm{I}, a, r)$ follows from $(2, \mathrm{I}, a, \mathfrak{a}(r))$, $(2, \mathrm{II}, r)$, $(2, \mathrm{II}, a^{-1}ra)$ and the fact that $\mathfrak{w}$ is a homomorphism: $(3, \mathrm{II}, r, \mathfrak{b})$ follows immediately from $(2, \mathrm{I}, r, \mathfrak{b})$ and $(2, \mathrm{II}, r)$. ⟦6⟧

For the sufficiency we have to construct the group $\mathfrak{G}$. The elements of this group are to be all classes of equivalent pairs $(a, \mathfrak{a})$ (a in $\mathfrak{F}$, $\mathfrak{a}$ in $\mathfrak{N}$), $(a', \mathfrak{a}')$ being equivalent to $(a, \mathfrak{a})$ if and only if $a^{-1}a'$ belongs to $\mathfrak{R}$ and $\mathfrak{a}(a^{-1}a') = \mathfrak{a}\,\mathfrak{a}'^{-1}$.
This is an equivalence relation, for

1) If $a^{-1}a' \in \mathfrak{R}$ and $\mathfrak{a}(a^{-1}a') = \mathfrak{a}\,\mathfrak{a}'^{-1}$
then
$$a'^{-1}a \in \mathfrak{R} \text{ and } \mathfrak{a}(a'^{-1}a) = \mathfrak{a}'\mathfrak{a}^{-1},$$
i.e. the relation is symmetric.

2) $a^{-1}a = e \in \mathfrak{R}, \quad \mathfrak{a}(a^{-1}a) = \mathfrak{e} = \mathfrak{a}\,\mathfrak{a}^{-1},$

i.e. the relation is reflexive.

3) If $a^{-1}a' \in \mathfrak{R}$, $a'^{-1}a'' \in \mathfrak{R}$,
$$\mathfrak{a}(a^{-1}a') = \mathfrak{a}\,\mathfrak{a}'^{-1} \text{ and } \mathfrak{a}(a'^{-1}a'') = \mathfrak{a}'\,\mathfrak{a}''^{-1},$$
then
$$a^{-1}a'' = (a^{-1}a')(a'^{-1}a'') \in \mathfrak{R}$$
$$\mathfrak{a}(a^{-1}a'') = \mathfrak{a}(a^{-1}a')\,\mathfrak{a}(a'^{-1}a'') = \mathfrak{a}\,\mathfrak{a}'^{-1}\mathfrak{a}'\,\mathfrak{a}''^{-1} = \mathfrak{a}\,\mathfrak{a}''^{-1},$$
i.e. the relation is transitive.

The product of two pairs is defined by
$$(a, \mathfrak{a})(b, \mathfrak{b}) = (ab, \chi_b(\mathfrak{a})\mathfrak{b}).$$
This will be a valid definition of a product of classes of pairs if we can shew that if $(a', \mathfrak{a}')$ is equivalent to $(a, \mathfrak{a})$ and $(b', \mathfrak{b}')$ is equivalent to $(b, \mathfrak{b})$ then $(a', \mathfrak{a}')(b', \mathfrak{b}')$ is equivalent to $(a', \mathfrak{a})(b, \mathfrak{b})$ i.e. that if
$$\left.\begin{aligned} & a^{-1}a' \in \mathfrak{R},\ b^{-1}b' \in \mathfrak{R} \\ & \mathfrak{a}(a^{-1}a') = \mathfrak{a}\,\mathfrak{a}'^{-1} \text{ and } \mathfrak{a}(b^{-1}b') = \mathfrak{b}\mathfrak{b}'^{-1} \end{aligned}\right\} \qquad (4)$$
then
$$(ab)^{-1}(a'b') \in \mathfrak{R} \text{ and } \mathfrak{a}((ab)^{-1}(a'b')) = (\chi_{b'}(\mathfrak{a})\mathfrak{b})(\chi_b(\mathfrak{a}'), \mathfrak{b}')^{-1}.$$ ⟦7⟧

⟦13⟧

Now

$$(ab)^{-1}(a'b') = b^{-1}(a^{-1}a')b \cdot b^{-1}b' \in \mathfrak{N}$$

and

$$\begin{aligned}\mathfrak{a}\big((ab)^{-1}(a'b')\big) &= \chi_b(\mathfrak{a}\,\mathfrak{a}'^{-1})\mathfrak{b}\mathfrak{b}'^{-1}\\ &= \chi_b(\mathfrak{a})\mathfrak{b}\cdot\mathfrak{b}'^{-1}\cdot(\mathfrak{b}\mathfrak{b}'^{-1})^{-1}\chi_b(\mathfrak{a}'^{-1})\mathfrak{b}\mathfrak{b}'^{-1}\\ &= \big(\chi_b(\mathfrak{a})\mathfrak{b}\big)\cdot\big(\chi_{b'}(\mathfrak{a}')\mathfrak{b}'\big)^{-1}\end{aligned}$$

by (4), $(3, \text{II}, b^{-1}b', \chi_b(\mathfrak{a}'))$ and (1).
The product is also associative, for

$$\begin{aligned}\big((a, \mathfrak{a})(b, \mathfrak{b})\big)(c, \mathfrak{c}) &= \big(ab, \chi_b(\mathfrak{a})\mathfrak{b}\big)(c, \mathfrak{c})\\ &= \big(abc, \chi_c(\chi_b(\mathfrak{a})\mathfrak{b})\mathfrak{c}\big)\\ (a, \mathfrak{a})\big((b, \mathfrak{b})(c, \mathfrak{c})\big) &= (a, \mathfrak{a})(bc, \chi_c(\mathfrak{b})\mathfrak{c})\\ &= \big(abc, \chi_{bc}(\mathfrak{a})\chi_c(\mathfrak{b})\mathfrak{c}\big)\end{aligned}$$

and these two expressions are equal on account of (1) and the fact that χ_c is an automorphism.

$(e, \mathfrak{e})$ is an identity and $(a^{-1}, \chi_{a^{-1}}(\mathfrak{a}^{-1}))$ is an inverse to $(a, \mathfrak{a})$. Consequently with this product our classes of pairs form a group.

$\mathfrak{w}(a)$ is defined by

$$\mathfrak{w}(a) = (a, \mathfrak{e})$$

and is clearly a homomorphism. To show that (2, II) is satisfied
⟦8⟧ it is only necessary to verify that $(r, \mathfrak{e})$ is equivalent to $(e, \mathfrak{a}(r))$.
As regards (2, I) we have

$$\begin{aligned}(\mathfrak{w}(a))^{-1}\mathfrak{b}\mathfrak{w}(a) &= (a^{-1}, \mathfrak{e})\mathfrak{b}(a, \mathfrak{e})\\ &= (a^{-1}, \mathfrak{e})(a, \chi_a(\mathfrak{b}))\\ &= (e, \chi_a(\mathfrak{b})) = \chi_a(\mathfrak{b}).\end{aligned}$$

Since

$$\begin{aligned}(g, \mathfrak{b}) &= (g, \mathfrak{e})(e, \mathfrak{b})\\ &= \mathfrak{w}(g)\mathfrak{b}\end{aligned}$$

condition b is satisfied.

The condition that an element $\mathfrak{b}$ of $\mathfrak{N}$ should also lie in $\mathfrak{w}(\mathfrak{F})$ is that $(e, \mathfrak{b})$ be equivalent to some pair $(a, \mathfrak{e})$. This means that a is in $\mathfrak{R}$ and $\mathfrak{a}(a) = \mathfrak{b}$. Hence $\mathfrak{N} \cap \mathfrak{w}(\mathfrak{F}) \subseteqq \mathfrak{w}(\mathfrak{R})$. This is satis-
⟦9⟧ fied since $\mathfrak{w}(r) = \mathfrak{a}(r)$ so that $\mathfrak{w}(\mathfrak{R}) \subseteqq \mathfrak{N}$.

Proof of the corollary. Suppose $\mathfrak{G}$ is a group with the required properties, $e_1, e_2, \ldots, e_n$ a set of free generators of $\mathfrak{F}$ and ω the function determining the homomorphism of $\mathfrak{F}$ on $\mathfrak{G}'$. Let $\mathfrak{e}_1$,

⟦14⟧

$\mathfrak{e}_2, \ldots, \mathfrak{e}_n$ be elements of $\mathfrak{G}$ with the property that $\mathfrak{e}_i$ is in the coset $\omega(e_i)$ of $\mathfrak{N}$ and that

$$\chi_{e_i}(\mathfrak{b}) = \mathfrak{e}_i^{-1}\mathfrak{b}\mathfrak{e}_i \text{ for all } \mathfrak{b} \text{ in } \mathfrak{N}.$$ ⟦10⟧

Then if $\mathfrak{w}$ be a homomorphism of $\mathfrak{F}$ into $\mathfrak{G}$ satisfying $\mathfrak{w}(e_i) = \mathfrak{e}_i$ ⟦11⟧
for all i, we shall have

$$\chi_a(\mathfrak{b}) = (\mathfrak{w}(a))^{-1}\mathfrak{b}\mathfrak{w}(a) \text{ for all } \mathfrak{b} \text{ in } \mathfrak{N}.$$

If we put $\mathfrak{w}(r) = \mathfrak{a}(r)$ for elements r of $\mathfrak{R}$ then by the first half of theorem 1 (whose proof makes no use of b, c) the conditions (3) must hold.

If on the other hand we have a function $\mathfrak{a}(r)$ satisfying (3) we can form the group $\mathfrak{G}$ of theorem 1 which is easily seen to have the required properties. ⟦12⟧

For specific applications the corollary to theorem 1 is more useful in the form of

THEOREM 2.

$\mathfrak{N}$ is a given group. $\mathfrak{F}$ is a free group with the generators $e_1, e_2, \ldots, e_n$. $\mathfrak{R}$ is the least self conjugate subgroup of $\mathfrak{F}$ containing $r_1, r_2, \ldots, r_l$. ω maps $\mathfrak{F}$ homomorphically on $\mathfrak{G}'$, the elements mapped on the identity being those of $\mathfrak{R}$. χ_a is a homomorphism of $\mathfrak{F}$ into the automorphisms of $\mathfrak{N}$, and $X(\omega(a))$ is the class of automorphisms containing χ_a. $\mathfrak{r}_1^*, \mathfrak{r}_2^*, \ldots, \mathfrak{r}_l^*$ are elements of $\mathfrak{N}$ such that ⟦13⟧

$$\chi_{r_i}(\mathfrak{b}) = \mathfrak{r}_i^{*-1}\mathfrak{b}\mathfrak{r}_i^* \text{ for all } \mathfrak{b} \text{ of } \mathfrak{N}. \tag{5}$$

Then there is an extension $\mathfrak{G}$ of $\mathfrak{N}$ by $\mathfrak{G}'$ realising the classes $X(\alpha)$ of automorphisms if and only if[5]) *there are elements $\mathfrak{z}_1, \mathfrak{z}_2, \ldots, \mathfrak{z}_l$ of the centre of $\mathfrak{N}$ for which the equation*

$$\prod_{i=1}^{N} \chi_{a_i}\left(\mathfrak{z}_{t_i}^{\tau_i}\right) = \prod_{i=1}^{N} \chi_{a_i}\left(\mathfrak{r}_{t_i}^{*\tau_i}\right) \tag{6}$$

holds whenever the corresponding equation

$$\prod_{i=1}^{N} a_i^{-1} r_{t_i}^{\tau_i} a_i = e \tag{7}$$

holds in $\mathfrak{F}$.

If the extension $\mathfrak{G}$ exists, then by the corollary to theorem 1 there is a homomorphism $\mathfrak{a}(r)$ of $\mathfrak{R}$ into $\mathfrak{N}$, satisfying (3). This homomorphism is completely determined by its values for

[5]) That the elements $\mathfrak{z}_i$ do not always exist can be seen from an example given by Baer (loc. cit., 415). Another example will be given in § 3.

⟦15⟧

$r_1, r_2, \ldots, r_l$. Let us put $\mathfrak{a}(r_i) = \mathfrak{r}_i$ and $\mathfrak{z}_i = \mathfrak{r}_i^{-1}\mathfrak{r}_i^*$. $\mathfrak{z}_i$ is certainly
[[14]] in the centre, for $\mathfrak{r}_i$ and $\mathfrak{r}_i^*$ induce the same automorphism. Denoting the expression on the left hand side of (7) by p we must have $\mathfrak{a}(p) = \mathfrak{a}(e) = \mathfrak{e}$. But if we make use of (3, I) and the fact that $\mathfrak{a}(r)$ is a homomorphism we obtain

$$\prod_{i=1}^{N} \chi_{a_i}\left(\mathfrak{r}_{t_i}^{\tau_i}\right) = \mathfrak{e} \tag{8}$$

which is equivalent to (6).

Now suppose that we are given the elements $\mathfrak{z}_i$ i.e. that we are given $\mathfrak{r}_i$ inducing the automorphisms χ_{r_i} and satisfying (8). Then if we put

[[15]]
$$\mathfrak{a}\left(\prod_{i=1}^{N} b_i^{-1} r_{s_i}^{\tau_i} b_i\right) = \prod_{i=1}^{N} \chi_{b_i}\left(r_{s_i}^{\sigma_i}\right),$$

we have a definition of $\mathfrak{a}(r)$ which can be seen to be unique on account of (8), and to be a homomorphism of $\mathfrak{R}$ in $\mathfrak{N}$.

If

$$r = \prod_{i=1}^{N} b_i^{-1} r_{s_i}^{\sigma_i} b_i$$

then

$$\begin{aligned}\mathfrak{a}(a^{-1}ra) &= \mathfrak{a}\left(\prod_{i=1}^{N} (b_i a)^{-1} r_{s_i}^{\sigma_i} b_i a\right)\\ &= \prod_{i=1}^{N} \chi_{b_i a}\left(\mathfrak{r}_{s_i}^{\sigma_i}\right) = \chi_a(\mathfrak{a}(r)),\end{aligned}$$

i.e. (3, I) is satisfied. Also

$$\begin{aligned}\chi_{a^{-1}r_i a}(\mathfrak{b}) &= \chi_a\left(\chi_{r_i}\left(\chi_{a^{-1}}(\mathfrak{b})\right)\right)\\ &= \chi_a(\mathfrak{r}_i^{-1})\,\mathfrak{b}\,\chi_a(\mathfrak{r}_i)\end{aligned}$$

so that (3, II, $a^{-1}r_i a$, $\mathfrak{b}$) is satisfied. But if (3, II, r, $\mathfrak{b}$) and (3, II, s, $\mathfrak{b}$) are satisfied for all $\mathfrak{b}$ then (3, II, rs, $\mathfrak{b}$) is satisfied for all $\mathfrak{b}$. Consequently (3, II) is satisfied and the corollary to theorem 1 applies.

[[16]] In the cases when the centre of $\mathfrak{N}$ consists either of the identity alone or of the whole group there is always a solution of the equations (6). The expressions on the right hand sides of these equations always represent centre elements, so that in the case where the centre consists of the identity alone, there is a solution by putting $\mathfrak{z}_i = \mathfrak{e}$ for each i. If $\mathfrak{N}$ is Abelian we put $\mathfrak{z}_i = \mathfrak{r}_i^*$. For the general case we have to be able to find all the relations (7).

[[16]]

§ 2. *The relations between the relations of a group.*

Suppose $\mathfrak{F}$ is a free group with the generators $e_1, \ldots, e_n$ and $\mathfrak{R}$ is the least self conjugate subgroup containing certain elements $r_1, r_2, \ldots, r_l$. The factor group will be called $\mathfrak{G}'$. As has been shewn it is important in the extension problem to be able to express the structure of $\mathfrak{G}'$ in terms of relations between the conjugates of the relations $r_1, r_2, \ldots, r_l$. This problem has been solved by Reidemeister [6]). It is necessary to repeat his conclusions to obtain another theorem on extensions (theorem 4).

Precisely the problem may be stated as follows. $\mathfrak{R}$ is generated by all elements of $\mathfrak{F}$ of form $a^{-1}r_ia$; it may therefore be regarded as the factor group $\Phi/\mathbf{P}$ of the the free group Φ with the generators $E_{i,a}$ with respect to some self conjugate subgroup $\mathbf{P}$. The problem ⟦17⟧
is to find a set of elements of Φ whose conjugates generate $\mathbf{P}$. $\mathbf{P}$ contains for instance all elements of form

$$E_{i,\,ab^{-1}r_jb}\, E_{j,b}^{-1}\, E_{i,a}^{-1}\, E_{j,b}. \tag{9}$$

If our method for finding the relations $\mathbf{P}$ is to be constructive it is necessary that the structure of the original group $\mathfrak{G}'$ should be known, or what amounts to the same, that we have a constructive method for determining whether a given member of $\mathfrak{F}$ is a member of $\mathfrak{R}$. If this is the case we can find a constructive function v_a defined for all a in $\mathfrak{F}$, constant in each coset of $\mathfrak{R}$, taking its value in that coset and satisfying $v_e = e$. These elements are a set of representatives of the cosets of $\mathfrak{R}$. If we put $v_a^{-1}a = r_a$ then r_a is a relation (member of $\mathfrak{R}$) for each a.

We define r_{a,e_i} by the condition

$$v_a e_i = v_{ae_i}\, r_{a,e_i}.$$

Then $\mathfrak{R}$ is generated by the relations $b^{-1}r_{a,e_i}b$. For if $\overline{\mathfrak{R}}$ is the group generated by these and contains r_c, then

$$\left.\begin{aligned} r_{ce_i} &= r_{c,e_i}\, e_i^{-1}\, r_c\, e_i \in \overline{\mathfrak{R}} \\ \text{and} \qquad & \\ r_{ce_i^{-1}} &= e_i\, r_{ce_i^{-1},e_i}\, r_c\, e_i^{-1} \in \overline{\mathfrak{R}}. \end{aligned}\right\} \tag{10}$$

⟦18⟧

But $\overline{\mathfrak{R}}$ contains $r_e = e$; it therefore contains r_c for each c.

Now suppose that for each r_{v_c,e_i} we have chosen an element R_{v_c,e_i} of Φ corresponding to it in the homomorphism τ of Φ on $\mathfrak{R}$ and let us define automorphisms χ_a by

---

[6]) K. Reidemeister, Knoten und Gruppen [Hamb. Abhandl. 5 (1926), 8—23].

⟦17⟧

$$\chi_a(E_{i,b}) = E_{i,ba}. \tag{11}$$

Then we may define R_c recursively by the equations

$$\left.\begin{aligned} R_e &= E \\ R_{ce_i} &= R_{v_c, e_i}\,\chi_{e_i}(R_c) \\ R_{ce_i^{-1}} &= \chi_{e_i^{-1}}(R_{v_c, e_i}\,R_c) \end{aligned}\right\} \tag{12}$$

⟦19⟧

so that if either $k = e_i$ or $k = e_i^{-1}$ we shall have

⟦20⟧ $$R_{ck} = R_{v_c k}\,\chi_k(R_c). \tag{13}$$

Our definition will be valid if and only if we always have

$$R_{(ce_i)e_i^{-1}} = R_{(ce_i^{-1})e_i} = R_c.$$

It may easily be verified that this is so.

Since the equations (13) and

$$r_{ck} = r_{v_c k}\,k^{-1}\,r_c\,k \tag{14}$$

hold whenever k is a generator or its inverse, R_c must correspond
⟦21⟧ to r_c in τ. Now for all b, i we have

$$r_{v_b r_i} = r_i$$

and therefore $R_{v_b r_i}$ and $E_i (= E_{i,e})$ must belong to the same coset of **P**. I.e.

$$R_{v_b r_i}\,E_i^{-1} \tag{15}$$

must belong to **P**. By operating with the automorphisms χ_a we see that all elements of form

$$\chi_a(R_{v_b r_i}\,E_i^{-1}) \tag{16}$$

⟦22⟧ belong to **P**. The structure of $\mathfrak{R}$ may now be decribed by

Theorem 3.

The group of relations **P** *of* $\mathfrak{R}$ *is the least self conjugate subgroup of* Φ *containing all elements of form* (9) *and* (16).

Only a sketch is given for the proof of this theorem. The first step is to shew that

$$R_{ax} = R_{v_a x}\,\chi_x(R_a) \tag{17}$$

for all x, a. For this purpose we consider the set Ξ of all x such that (17) holds for all a. Then we can shew that xy belongs to Ξ if x and y belong to it. The generators e_i, e_i^{-1} belong to Ξ
⟦23⟧ by (13).

⟦18⟧

Now let $\overline{\mathbf{P}}$ be the self conjugate subgroup of Φ generated by (9), (16). We shew that $R_{v_a r}$ is independent of a modulo $\overline{\mathbf{P}}$ and that if we allow $K(r)$ to stand for the coset $\overline{\mathbf{P}}$ of containing R_r, then $K(r)$ gives a homomorphism of $\mathfrak{R}$ on $\Phi/\overline{\mathbf{P}}$. But there is certainly a homomorphism τ' of $\Phi/\overline{\mathbf{P}}$ on $\mathfrak{R}$ determined by τ and satisfying $\tau'(K(r)) = r$. This is only possible if τ' and K are isomorphisms and $\mathbf{P} = \overline{\mathbf{P}}$. ⟦24⟧

To prove these properties of K we call $\mathbf{H}$ the totality of relations r for wich $R_{v_a r}$ is independent of a modulo $\overline{\mathbf{P}}$, and we shew successively

I) if r, s belong to $\mathbf{H}$ then $R_{v_a rs} \in K(r)K(s)$,

II) if r belongs to $\mathbf{H}$ then $R_{v_a r^{-1}} \in K(r)^{-1}$,

III) if r belongs to $\mathbf{H}$ and k is a generator or its inverse then $R_{v_a k^{-1} rk} \in \chi_k(K(r))$ (with an obvious adaptation of the meaning of χ_k).

For the proof of I), II), III), (17) is essential and so is the invariance of $\overline{\mathbf{P}}$ under the automorphisms χ_a.

Now let us return to the extension problem. For this purpose the only significant relations are those given by (15). In fact we have

THEOREM 4.

In theorem 2 we may replace the condition (6) by the condition that if the homomorphism ϑ of Φ on $\mathfrak{R}$ be determined by ⟦25⟧

$$\vartheta(E_{i,a}) = \chi_a(\mathfrak{r}_i^* \mathfrak{z}_i^{-1}) \tag{18}$$

then we must have $\vartheta(X) = \mathfrak{e}$ for all X of form (15).

The homomorphism ϑ transforms the automorphism χ_a of § 2 into the automorphism χ_a of § 1. I.e.

$$\vartheta(\chi_a(X)) = \chi_a(\vartheta(X)).$$ ⟦26⟧

In particular if $\vartheta(X) = \mathfrak{e}$ we shall have

$$\vartheta(\chi_a(X)) = \chi_a(\mathfrak{e}) = \mathfrak{e}$$

so that the elements of form (16) are mapped on the identity by ϑ. It remains to shew that $\vartheta(Z) = \mathfrak{e}$ for all Z of form (9). ⟦27⟧

$$\begin{aligned}&\vartheta(E_{i,ab^{-1}r_j b} E_{j,b}^{-1} E_{i,a}^{-1} E_{j,b})\\ &= \chi_{ab^{-1}r_j b}(\mathfrak{r}_i)\,\chi_b(\mathfrak{r}_j^{-1})\,\chi_a(\mathfrak{r}_i^{-1})\,\chi_b(\mathfrak{r}_j)\\ &= \chi_b\left(\chi_{ab^{-1}r_j}(\mathfrak{r}_i)\,\mathfrak{r}_j^{-1}\,\chi_{ab^{-1}}(\mathfrak{r}_i^{-1})\,\mathfrak{r}_j\right)\\ &= \chi_b(\mathfrak{e}) \text{ by (5). We have put } \mathfrak{r}_i^* \mathfrak{z}_i^{-1} = \mathfrak{r}_i.\end{aligned}$$ ⟦28⟧

⟦19⟧

Thus ϑ maps the whole of $\mathbf{P}$ on the identity and the conditions
[29] of theorem 2 are satisfied.

§ 3. *Cyclic extensions.*

When $\mathfrak{G}'$ is a cyclic group of order n we take $\mathfrak{F}$ to be the free group with the single generator a and $\mathfrak{R}$ to be the subgroup generated by $q \equiv a^n$. The representative elements v_b may be
[30] taken to be

$$e,\ a,\ a^2, \ldots,\ a^{n-1}.$$

We easily find that

$$r_{a^p,a} = e \text{ if } p \not\equiv -1 \pmod{n}$$
$$r_{a^{n-1},a} = q.$$

If Q_{a^p} be the element of Φ corresponding to $a^{-p}qa^p$ we have

$$R_{a^p,a} = E,$$
$$R_{a^{n-1},a} = Q$$

and from the equations (12) we obtain

$$\left.\begin{aligned} R_{a^p} &= E \\ R_{a^{n+p}} &= Q_{a^p}. \end{aligned}\right\} \quad (0 \leqq p \leqq n-1)$$

The expressions (15) are therefore

$$Q_{a^p}Q^{-1} \quad (p = 0,\ 1, \ldots,\ n-1). \tag{19}$$

If ϑ is a homomorphism of Φ and $\vartheta(Q_a^{-1}Q) = \mathfrak{e}$, then

$$\vartheta(Q_{a^p}) = \vartheta(Q_{a^{p-1}}) = \ldots = \vartheta(Q). \tag{20}$$

Now making use of (19), (20), theorem 4 for a cyclic extension becomes

THEOREM 5.

[31] *A is a class of automorphisms of a group $\mathfrak{N}$. 4^{w} is the first power of A which is the class of inner automorphisms and ξ is an arbitrary automorphism out of A. $\mathfrak{x}^*$ is an element of $\mathfrak{N}$ which induces the inner automorphism ξ^n. Then there is an extension of*
[32] *$\mathfrak{N}$ by the cyclic group of order n realising the classes A, $A^2, \ldots$ of automorphisms if and only if there is an element $\mathfrak{z}$ in the centre of $\mathfrak{N}$ satisfying*

$$\xi(\mathfrak{z})\mathfrak{z}^{-1} = \xi(\mathfrak{x}^*)\mathfrak{x}^{*-1}. \tag{21}$$

We have put

$$\vartheta(Q) = \mathfrak{r}^*\mathfrak{z}^{-1}$$

$$\vartheta(Q_a) = \xi(\mathfrak{r}^*\mathfrak{z}^{-1}).$$ ⟦33⟧

If we put $\xi(\mathfrak{z})\mathfrak{z}^{-1} = \varphi(\mathfrak{z})$ then φ is a (possibly improper) automorphism of the centre $\mathfrak{z}$ of $\mathfrak{N}$. For some groups all of the automorphisms φ are improper. This is the case for all cyclic groups whose order is a power of 2. In these cases we shall have unrealisable classes of automorphisms if $\xi(\mathfrak{r}^*)\mathfrak{r}^{*^{-1}}$ is not in $\varphi(\mathfrak{Z})$. Suppose for instance that $\mathfrak{N}$ is the dihedral group D_{10} of order 20, generated by a, b with the relations

$$a^{10} = b^2 = (ab)^2 = e.$$

The centre of this group consists of e and a^5. We define the automorphism ξ by

$$\xi(a) = a^7$$

$$\xi(b) = ba^5,$$

then

$$\xi^2(a) = a^{-1} = b^{-1}ab,$$

$$\xi^2(b) = b = b^{-1}bb.$$

$\mathfrak{r}^*$ can therefore be taken to be b. The equation (21) becomes

$$\xi(\mathfrak{z})\mathfrak{z}^{-1} = \xi(b)b^{-1} = a^5,$$

but $\xi(\mathfrak{z})\mathfrak{z}^{-1} = e$ for both centre elements.

(Received March 22nd, 1937.)

⟦21⟧

A METHOD FOR THE CALCULATION OF THE ZETA-FUNCTION

By A. M. TURING.

[Received 7 March, 1939.—Read 16 March, 1939.]

An asymptotic series for the zeta-function was found by Riemann and has been published by Siegel*, and applied by Titchmarsh† to the calculation of the approximate positions of some of the zeros of the function. It is difficult to obtain satisfactory estimates for the remainders with this asymptotic series, as may be seen from the first of these two papers of Titchmarsh, unless t is very large. In the present paper a method of calculation will be described, which, like the asymptotic formula, is based on the approximate functional equation; it is applicable for all values of s. It is likely to be most valuable for a range of t where t is neither so small that the Euler-Maclaurin summation method can be used (*e.g.* $t > 30$) nor large enough for the Riemann-Siegel asymptotic formula (*e.g.* $t < 1000$).

Roughly speaking, the method is to use the approximate functional equation for the zeta-function, with the remainder expressed as an integral, which for the moment we write as $\int_{-\infty}^{\infty} h(x)\,dx$. We approximate to the integral by the obvious sum $\sum_{k=-K}^{K} \frac{1}{\kappa} h\left(\frac{k}{\kappa}\right)$ and we find that, if certain modifications are made in the "main series", this gives a remarkably accurate result; when the number of terms taken is $T = 2K+1$ the error is of the order of magnitude of $e^{-\frac{1}{4}\pi T}$. The theta-functions give another case of this phenomenon. We have the identity

$$\sum_{n=-\infty}^{\infty} e^{-\pi n^2 x} = \frac{1}{\sqrt{x}} \sum_{n=-\infty}^{\infty} e^{-\pi n^2/x}$$

* C. L. Siegel, "Über Riemanns Nachlass zur analytischen Zahlentheorie", *Quell. Gesch. Math.*, B, 2 (1931), 45–80.

† E. C. Titchmarsh, "The zeros of the Riemann zeta-function", *Proc. Royal Soc.* (A), 151 (1935), 234–255; also 157 (1936), 261–263.

for any positive x. Hence

$$\left|\frac{1}{\kappa}\sum_{n=-\infty}^{\infty} e^{-\pi n^2/\kappa^2} - \int_{-\infty}^{\infty} e^{-\pi v^2} dv\right| = 2\sum_{n=1}^{\infty} e^{-\pi n^2 \kappa^2} < \frac{2}{1-e^{-2\pi}} e^{-\pi\kappa^2} \quad (\text{if } \kappa \geqslant 1),$$

$$\left|\frac{1}{\kappa}\sum_{n=-K}^{K} e^{-\pi n^2/\kappa^2} - \int_{-\infty}^{\infty} e^{-\pi v^2} dv\right| < \frac{2}{1-e^{-2\pi}} e^{-\pi\kappa^2} + \frac{\kappa e^{-\pi K^2/\kappa^2}}{\pi K}$$

$$< e^{-\pi K}\left(\frac{2}{1-e^{-2\pi}} + \frac{1}{\pi}\right) \quad (\text{if } K = \kappa^2).$$

As we have proved it above, this inequality depends entirely on the special form of the function $e^{-\pi u^2}$. But we can also prove it in this way. We integrate the function $e^{-\pi u^2}/(1-e^{-2\pi i u\kappa})$ round the rectangle with the vertices $\pm R \pm i\kappa$. In the limit $R \to \infty$ we obtain, by the theorem of residues,

$$\left(\int_{-\infty-i\kappa}^{\infty-i\kappa} + \int_{\infty+i\kappa}^{-\infty+i\kappa}\right) \frac{e^{-\pi u^2}}{1-e^{-2\pi i u\kappa}} du = \sum_{k=-\infty}^{\infty} \frac{1}{\kappa} e^{-\pi k^2/\kappa^2},$$

i.e.

$$\sum_{k=-\infty}^{\infty} \frac{1}{\kappa} e^{-\pi k^2/\kappa^2} - \int_{-\infty-i\kappa}^{\infty-i\kappa} e^{-\pi u^2} du = \int_{-\infty-i\kappa}^{\infty-i\kappa} \frac{e^{-\pi u^2} e^{-2\pi i u\kappa}}{1-e^{-2\pi i u\kappa}} du + \int_{\infty+i\kappa}^{-\infty+i\kappa} \frac{e^{-\pi u^2} e^{2\pi i u\kappa}}{1-e^{2\pi i u\kappa}} du.$$

The path of integration on the left-hand side of the equation can be replaced by the real axis, while the right-hand side is less in modulus than

$$\frac{2}{1-e^{-2\pi\kappa^2}} \int_{-\infty}^{\infty} \left|\exp\left[-\pi(u-i\kappa)^2 - 2\pi i(u-i\kappa)\kappa\right]\right| du$$

$$\leqslant \frac{2}{1-e^{-2\pi\kappa^2}} \int_{-\infty}^{\infty} \exp(-\pi u^2 - \pi\kappa^2)\, du \leqslant \frac{2e^{-\pi\kappa^2}}{1-e^{-2\pi}} \quad (\text{if } \kappa \geqslant 1).$$

This argument can be used in more general cases, but in moving the path of integration we may encounter singularities of the integrand, which will modify the result.

We base our calculation on an integral representation of the zeta-function due to Riemann*.

1. *Evaluation of a definite integral.* Let

$$G(u) = \int_{0 \nearrow 1} \frac{\exp(i\pi z^2 + 2\pi i u z)\, dz}{e^{i\pi z} - e^{-i\pi z}},$$

where $0 \nearrow 1$ signifies that the integration is along a line from $-\epsilon\infty$ to $\epsilon\infty$ cutting the real axis between 0 and 1, and $\epsilon = e^{\frac{1}{4}\pi i}$. We denote

* Siegel, *loc. cit.*, 24.

a similar integral over $-1 \nearrow 0$ by $G_1(u)$. Then, by a change of variables, ⟦1⟧
we obtain

$$G_1(u) = G(u-1)\, e^{-2\pi i u},$$

and, by the theorem of residues,

$$G(u) = G_1(u)+1.$$

Multiplying the identity

$$\frac{1}{e^{i\pi z}-e^{-i\pi z}} = e^{-i\pi z} + \frac{e^{-2\pi i z}}{e^{i\pi z}-e^{-i\pi z}}$$

by $\exp(i\pi z^2 + 2\pi i u z)$, and integrating over $0 \nearrow 1$, we obtain ⟦2⟧

$$G(u) = \epsilon\, e^{-i\pi(u-\frac{1}{2})^2} + G(u-1).$$

Combining these results we have

$$G(u) = \frac{1}{1-e^{-2\pi i u}} - \frac{e^{-i\pi u^2}}{e^{i\pi u}-e^{-i\pi u}}. \tag{1.1}$$

2. *An integral representation of the zeta-function.* We integrate $z^{-s}/(1-e^{-2\pi i z})$ round a curve L which may be taken to consist of a straight line from $i\infty$ to $\frac{1}{2}$, a semicircle from $\frac{1}{2}$ to $-\frac{1}{2}$ lying in the lower half-plane, and a straight line from $-\frac{1}{2}$ to $i\infty$. We define z^{-s} so that it has its usual value on the positive real axis, and is continuous except on the positive imaginary axis. The integral round any part of a circle with 0 as centre tends to 0 as the radius tends to infinity through appropriate values, ⟦3⟧
provided that $\mathbf{R}s > 1$, and therefore in this case we have

$$\int_L \frac{z^{-s}\,dz}{1-e^{-2\pi i z}} = 2\pi i \text{ (sum of residues of } \frac{z^{-s}}{1-e^{-2\pi i z}} \text{ at integers other than 0)}$$

$$= \sum_{m=1}^{\infty} (m^{-s} + e^{i\pi s} m^{-s}) = \zeta(s)(1+e^{i\pi s}).$$

This gives the analytic continuation of $\zeta(s)$ over the whole plane except possibly for even integers. Now, by (1.1), ⟦4⟧

$$\int_L u^{-s}\left(\frac{1}{1-e^{-2\pi i u}} - \frac{e^{-i\pi u^2}}{e^{i\pi u}-e^{-i\pi u}}\right) du = \int_L u^{-s}\left[\int_{0\nearrow 1} \frac{\exp(i\pi z^2 + 2\pi i u z)\,dz}{e^{i\pi z}-e^{-i\pi z}}\right] du$$

$$= \int_{0\nearrow 1} \frac{e^{i\pi z^2}}{e^{i\pi z}-e^{-i\pi z}} \left(\int_L u^{-s} e^{2\pi i u z}\, du\right) dz$$

$$= -2(2\pi)^{s-1} \sin s\pi\, \Gamma(1-s)\, e^{\frac{1}{2}is\pi} \int_{0\nearrow 1} \frac{e^{i\pi z^2} z^{s-1}}{e^{i\pi z}-e^{-i\pi z}}\, dz,$$ ⟦5⟧

i.e.

$$\zeta(s)(1+e^{i\pi s}) = \int_L \frac{e^{-i\pi u^2} u^{-s}\, du}{e^{i\pi u}-e^{-i\pi u}} - 2(2\pi)^{s-1} \sin s\pi\, \Gamma(1-s)\, e^{\frac{1}{2}is\pi} \int_{0\nearrow 1} \frac{e^{i\pi z^2} z^{s-1}\, dz}{e^{i\pi z}-e^{-i\pi z}}.$$

⟦6⟧ In the first integral we may replace L by the two lines $0\downarrow 1$ and $-1\uparrow 0$, and we get

$$\int_L \frac{e^{-i\pi u^2} u^{-s}\, du}{e^{i\pi u}-e^{-i\pi u}} = \int_{0\downarrow 1} \frac{e^{-i\pi u^2} u^{-s}\, du}{e^{i\pi u}-e^{-i\pi u}} + \int_{-1\uparrow 0} \frac{e^{-i\pi u^2} u^{-s}\, du}{e^{i\pi u}-e^{-i\pi u}}$$

⟦7⟧

$$= \int_{0\downarrow 1} \left(\frac{e^{-i\pi u^2} u^{-s}}{e^{i\pi u}-e^{-i\pi u}} - \frac{e^{i\pi u^2} u^{-s} e^{i\pi s}}{e^{-i\pi u}-e^{i\pi u}}\right) du.$$

⟦8⟧ The curve may now be replaced by $0\nwarrow 1$ if the sign is changed, and we thus obtain

$$-\int_{0\nwarrow 1} (1+e^{i\pi s}) \frac{e^{-i\pi u^2} u^{-s}\, du}{e^{i\pi u}-e^{-i\pi u}}.$$

The zeta-function is then expressed in the form

$$\zeta(s) = -\int_{0\nwarrow 1} \frac{e^{-i\pi u^2} u^{-s}\, du}{e^{i\pi u}-e^{-i\pi u}} - 2(2\pi)^{s-1} \sin \tfrac{1}{2}s\pi\, \Gamma(1-s) \int_{0\nearrow 1} \frac{e^{i\pi z^2} z^{s-1}\, dz}{e^{i\pi z}-e^{-i\pi z}} \qquad (2.1)$$

and the calculation of $\zeta(s)$ is reduced to that of the integral

$$I(s) = \int_{0\nearrow 1} \frac{e^{i\pi z^2} z^{-s}\, dz}{e^{i\pi z}-e^{-i\pi z}} = \int_{0\nearrow 1} h(z)\, dz; \qquad (2.2)$$

⟦9⟧ for

$$\overline{I(s)} = -\int_{0\nwarrow 1} \frac{e^{-i\pi z^2} z^{-\bar{s}}\, dz}{e^{-i\pi z}-e^{i\pi z}}.$$

If we multiply both sides of (2.1) by $\Gamma(\frac{1}{2}s)\pi^{-\frac{1}{2}s}$, and make use of the relation

⟦10⟧

$$\Gamma(1-s) = \frac{2^{-s}\pi^{\frac{1}{2}}\Gamma(\frac{1}{2}-\frac{1}{2}s)}{\sin \frac{1}{2}s\pi\, \Gamma(\frac{1}{2}s)},$$

we obtain

⟦11⟧

$$\zeta(s)\Gamma(\tfrac{1}{2}s)\pi^{-\frac{1}{2}s} = \Gamma(\tfrac{1}{2}s)\pi^{-\frac{1}{2}s}\overline{I(\bar{s})} - \Gamma(\tfrac{1}{2}-\tfrac{1}{2}s)\pi^{\frac{1}{2}s-\frac{1}{2}} I(1-s), \qquad (2.3)$$

and on the critical line this is equal to

⟦12⟧

$$-2\Re\,\Gamma(\tfrac{1}{2}s)\pi^{-\frac{1}{2}s} I(\bar{s}). \qquad (2.4)$$

For points on this line there is, therefore, only one real integral to calculate. For points not on the line there are four real integrals.

3. *The method of calculation.* Let μ be a positive real number lying between the integers m and $m+1$. Then

$$I(s) = \int_{0\nearrow 1} h(z)\, dz = \int_{m\nearrow m+1} h(z)\, dz - \sum_{r=1}^{m} r^{-s}.$$

⟦26⟧

Now let κ be a positive real number and put

$$g(z) = \frac{h(z)}{1-\exp[-2\pi\kappa\epsilon(z-\mu)]}.$$

The function g has simple poles at the integer points other than 0, and at the points $p_k = \mu + \epsilon k/\kappa$, where k is an integer; otherwise it is regular except on the closed positive imaginary axis. The residue at the non-zero integer r is

$$\frac{r^{-s}}{2\pi i\{1-\exp[-2\pi\kappa\epsilon(r-\mu)]\}},$$

and at p_k it is

$$\frac{h(p_k)}{2\pi\kappa\epsilon}.$$

The line on which the poles p_k lie, taken as running from left to right, will be called P.

Let J and J' be two curves going from the third quadrant to the first and from the first to the third respectively, J being entirely on the right of P and J' entirely on its left and on the right of the origin. Suppose also that there is a positive real number a such that, at sufficiently great distance from the origin, the curves lie in the region where either $a < \arg z < \frac{1}{2}\pi - a$ or $\frac{1}{2}\pi - a > \arg z > -\pi + a$, and that the length of curve ⟦13⟧
with $|z| < R$ is $O(R)$. Then it is easily seen that

$$\int_{J+J'} g(z)\,dz = 2\pi i \text{ (sum of residues of } g \text{ between } J \text{ and } J')$$

$$= \Sigma \frac{r^{-s}}{1-\exp[-2\pi\kappa\epsilon(r-\mu)]} + \sum_k \frac{\epsilon}{\kappa} h(p_k),$$

where the first sum is taken over the integers r lying between J and J'. Now

$$\int_J g(z)\,dz = \int_J \big(g(z)-h(z)\big)\,dz + \int_J h(z)\,dz$$

$$= \int_J h(z)\frac{\exp[-2\pi\kappa\epsilon(z-\mu)]}{1-\exp[-2\pi\kappa\epsilon(z-\mu)]}\,dz + \int_P h(z)\,dz, \qquad ⟦14⟧$$

and

$$\int_{J'} g(z)\,dz = -\int_{J'} h(z)\frac{\exp[2\pi\kappa\epsilon(z-\mu)]}{1-\exp[2\pi\kappa\epsilon(z-\mu)]}\,dz.$$

If the curves J and J' are always distant more than $\frac{1}{4}\kappa^{-1}$ from the line P, then we have on them

$$\left|\frac{1}{1-\exp[2\pi\kappa\epsilon\,\mathrm{sg}(z)(z-\mu)]}\right| < 1{\cdot}27, \qquad ⟦15⟧$$

where sg(z) has the value 1 or -1 according as z is to the left or to the right of P. We can now collect our results in the form

⟦16⟧
$$I(s) = \sum_k \frac{\epsilon}{\kappa}\, h(p_k) - \sum_{r=1}^{\infty} r^{-s}\theta_r + R_0,$$

where θ_r has the value 1 if r is on the left of J', the value

$$\{1-\exp[2\pi\kappa\epsilon(r-\mu)]\}^{-1}$$

if r is between J and J', and the value 0 otherwise. The remainder R_0 satisfies

⟦17⟧
$$|R_0| < \int_{J+J'} 1{\cdot}27 \exp[-\sqrt{2\pi\kappa\mu}\, \mathrm{sg}(z)] \frac{|z|^{-\sigma}}{|e^{i\pi z}-e^{-i\pi z}|} e^{\Re\phi(z)} |dz|, \quad (3.2)$$

where
$$\phi(z) = i\pi z^2 + 2\pi\kappa\epsilon z\, \mathrm{sg}(z) - it \log z. \quad (3.3)$$

We may also write the formula for $I(s)$ in the form

$$I(s) = \sum_{k=-\infty}^{\infty} \frac{\epsilon}{\kappa} \frac{e^{i\pi p_k^2} p_k^{-s}}{e^{i\pi p_k} - e^{-i\pi p_k}} - \sum_{r=1}^{\infty} \frac{r^{-s}}{1-\exp[2\pi\kappa\epsilon(r-\mu)]} + R, \quad (3.4)$$

where $R = R_0 + R_1$, $p_k = \mu + \epsilon k/\kappa$ and

⟦18⟧
$$|R_1| < 1{\cdot}27 \sum r^{-s} \exp[-\sqrt{2}\,\pi\kappa|r-\mu|],$$

the summation being over positive integers not between J and J'. By expressing $I(s)$ in this form we can eliminate from the numerical calculation any reference to the position of the curves J, J', and the remainder is not appreciably increased. In § 5 we choose the curves so that $R_1 = 0$. Of course, in the calculation the factor $\{1-\exp[2\pi\kappa\epsilon(r-\mu)]\}^{-1}$ will be put equal to 1 or to 0 except for a comparatively small number of terms.

In estimating the remainder we suppose that $\sigma \geqslant 0$, but this is not necessary.

4. *General remarks on the estimation of the remainder.* Suppose that

$$U = \int_{C_0} e^{w(z)} k(z)\, dz.$$

Then, for any curve C deformable into C_0 in the domain of regularity of $e^{w(z)} k(z)$,

$$|U| \leqslant \int_C e^{\Re w(z)} |k(z)|\, dz. \quad (4.1)$$

Now suppose that $\Re w(z)$ has large and fast variations, while $|k(z)|$ is comparatively steady. Then the value of the integral (4.1) is

⟦28⟧

principally affected by the maximum value of $\mathfrak{R}w(z)$ on C, and a good inequality for $|U|$ is obtained by minimising this maximum. It is easily seen that if there is a curve for which the maximum is minimised, and if z_0 is a point at which the maximum on this curve is attained, then $w'(z_0) = 0$, *i.e.* z_0 is a "saddle point" of w. Suppose that in the neighbourhood of the saddle point the curve is $z = z_0 + le^{i\alpha}$, l being an arc-length parameter. Then the contribution to the integral from the neighbourhood of the saddle point is approximately

$$e^{\mathfrak{R}w(z_0)} |k(z_0)| \int_{-\infty}^{\infty} |\exp [\tfrac{1}{2} w''(z_0) e^{2i\alpha} l^2]| \, dl = \frac{(2\pi)^{\frac{1}{2}} |k(z_0)| e^{\mathfrak{R}w(z_0)}}{\sqrt{\{-\mathfrak{R}[w''(z_0) e^{2i\alpha}]\}}}.$$

We naturally choose α to be $\frac{1}{2}\pi - \frac{1}{2} \arg w''(z_0)$, and then the expression becomes

$$\sqrt{\left\{\frac{2\pi}{w''(z_0)}\right\}} \, |k(z_0) e^{w(z_0)}|.$$

In the estimation of R_0 we could take $w(z)$ to be either $\phi(z)$ or $\phi(z) \pm i\pi z$, in the latter case different signs being taken on the two curves J, J'. With the first of these forms the analysis is simpler, but the second gives a better result; we deal only with the simpler form.

We actually use the idea of saddle-point integration in the following form. Suppose that we have a curve C with arc-length parameter l beginning at $l = 0$. Then

(*a*) if $\mathfrak{R}\psi'(z) \dfrac{dz}{dl} \leqslant -al$ on the curve, where $a > 0$, then

$$e^{\mathfrak{R}\psi[z(l)]} \leqslant e^{-\frac{1}{2}al^2} e^{\mathfrak{R}\psi[z(0)]} \qquad [[19]]$$

on the curve, and consequently

$$\int_C e^{\mathfrak{R}\psi[(z(l)]} |dz| \leqslant \sqrt{\left(\frac{\pi}{2a}\right)} \, e^{\mathfrak{R}\psi[z(0)]}. \qquad [[20]]$$

This enables us to estimate the contribution to the integral from the neighbourhood of the saddle-point; we may estimate the contribution from the remainder of the curve by one or more applications of

(*b*) if $\mathfrak{R}\psi'(z) \dfrac{dz}{dl} \leqslant -a < 0$ on the curve, then $e^{\mathfrak{R}\psi[(z(l)]} \leqslant e^{-al} e^{\mathfrak{R}\psi[z(0)]}$ and

$$\int_C e^{\mathfrak{R}\psi[z(l)]} |dz| \leqslant \frac{1}{a} e^{\mathfrak{R}\psi[z(0)]}.$$

5. *Detailed estimation of the remainder.* For the two curves J, J' we have two different saddle-points z_0, z_0', these being zeros of $\phi'(z)$ on the

two sides of the line P. We may put

$$\phi'(z) = 2\pi i z + 2\pi\kappa\epsilon - it/z = \frac{2\pi i}{z}(z-z_0)(z-z_1)$$

on the left of P, and

⟦21⟧ $$\phi'(z) = 2\pi i z - 2\pi\kappa\epsilon - it/z = \frac{2\pi i}{z}(z-z_0')(z-z_1')$$

on the right of P; the points z_0, z_0' are in the right half-plane and z_1, z_1'
⟦22⟧ in the left. If we put further $\tau = t/2\pi$, $\rho = \kappa\tau^{-\frac{1}{2}}$, $\zeta = z\tau^{-\frac{1}{2}}$, we may write these equations as

$$(\zeta-\zeta_0)(\zeta-\zeta_1) = \zeta^2+\bar{\epsilon}\rho\zeta-1, \quad (\zeta-\zeta_0')(\zeta-\zeta_1') = \zeta^2-\bar{\epsilon}\rho\zeta-1.$$

The roots satisfy

⟦23⟧ $$\zeta_0+\zeta_1' = 0, \quad \zeta_1+\zeta_0' = 0, \quad \arg\zeta_0+\arg\zeta_0' = 0, \quad \tfrac{1}{4}\pi > \arg\zeta_0' > 0.$$

There is a cubic curve independent of ρ on which all four roots lie. We shall need a number of other properties of the roots, and we mention them as we require them; they are mostly inequalities which can be proved by straightforward but laborious methods.

The behaviour of a number of functions of ρ is shown in Fig. 1.

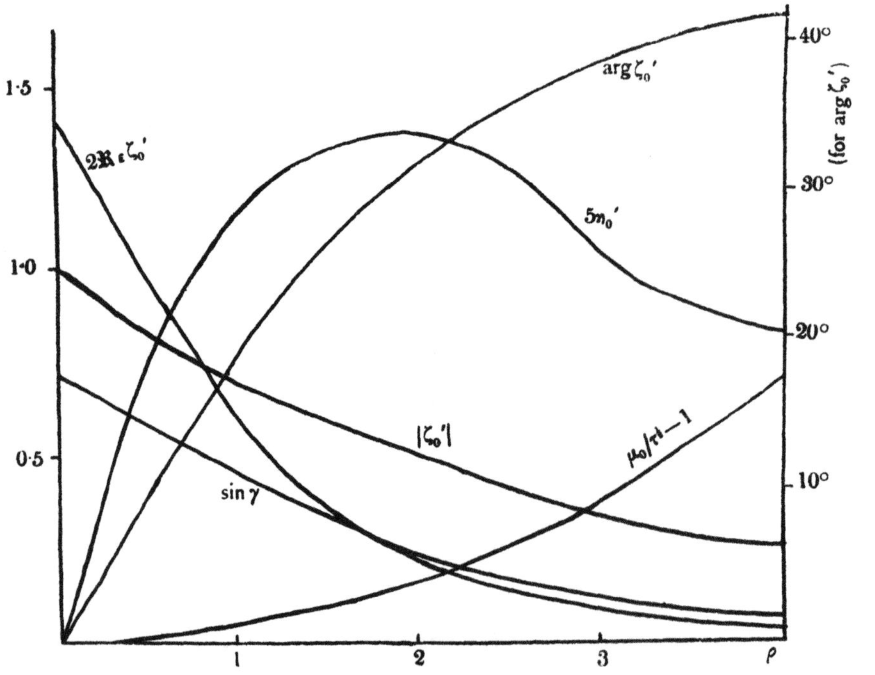

Fig. 1.

⟦30⟧

We choose the curves of integration J, J' as follows. J consists of three straight parts J_1, J_2, J_3, of which J_1 is a straight line from $-\epsilon\infty$ through z_0 to b, where $b = z_0 - \frac{1}{2}y_0(1+i)$; J_2 joins b to $b+\beta$, where β is ⟦24⟧
real and positive; and J_3 joins $b+\beta$ to $\epsilon\infty$, passing through a half odd integer. As β tends to infinity the contribution to the remainder from J_3 tends to zero and the contribution from J_2 tends to a limit. We may ⟦25⟧
therefore omit J_3 and suppose J to consist of J_1 together with J_2, taken as extending to infinity; there are then no poles on the right of J to contribute to R_1.

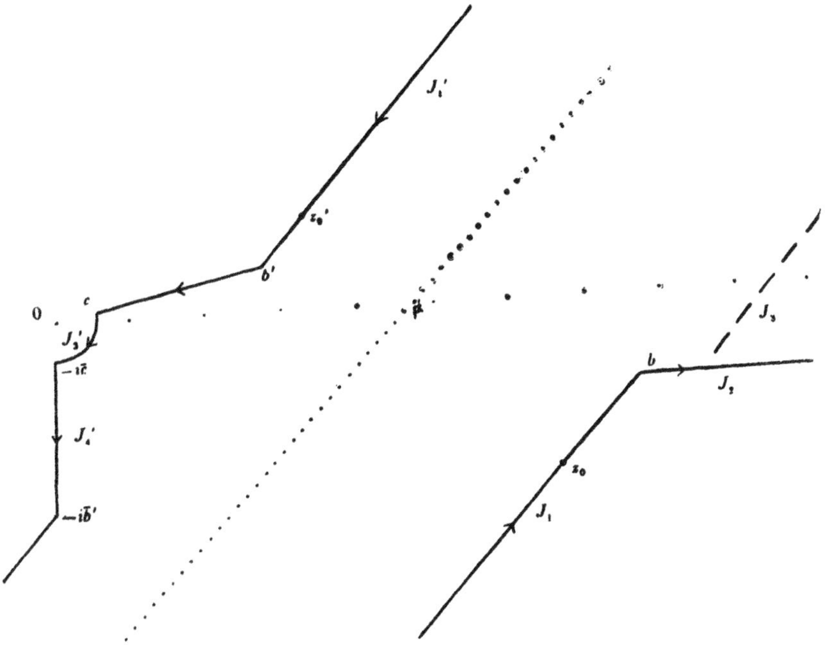

Fig. 2.

The curve J' consists of the four parts J_1', J_2', J_3', J_4', of which J_2' is further divided into J_5' and J_6'. Let $b' = z_0' - \frac{1}{2}y_0'(1+i)$. Then J_1' ⟦26⟧
is a straight line from $\epsilon\infty$ to b' through z_0'; J_2' is a straight line from b' to c, where c/b' is real, $0 < c/b' < 1$, and $|c| = \min(\frac{1}{2}, \frac{1}{3}|z_0'|)$. The curve J_3' is part of a circle lying in the half-plane $\mathfrak{R}\mathfrak{e}z > 0$ and joining c to $-i\bar{c}$. We obtain J_4' by reflecting J_1' and J_2' in the line $\mathfrak{I}\mathfrak{e}z = 0$ and reversing its direction. We divide J_2' into J_5' and J_6', of which J_5' is the part for which $y > \frac{1}{3}y_0'$. Both of these parts are of positive length, J_5' being of length $\frac{1}{6}|z_0'|$ at least. ⟦27⟧

[[28]] On J_1 and J_2 we have $|z| > \tau^{\frac{1}{2}} \mathfrak{R} \epsilon\zeta$ and

[[29]]
$$\left| \frac{1}{e^{i\pi z} - e^{-i\pi z}} \right| \leqslant \tfrac{1}{2} \operatorname{cosech} |\tfrac{1}{2}\pi y_0|.$$

The contribution to R_0 from J is therefore at most

$$\tfrac{1}{2}(1{\cdot}27)\, \tau^{-\frac{1}{4}\sigma} (\mathfrak{R}\epsilon\zeta_0)^{-\sigma} \exp\left[\sqrt{2\pi\kappa\mu}\right] \operatorname{cosech} |\tfrac{1}{2}\pi y_0| \int_J e^{\mathfrak{R}\phi(z)} |dz|. \tag{5.1}$$

[[30]] On J_1 we have $0 > \arg(z - z_1)/z > -\cos^{-1}\frac{1}{3}$, a simple consequence of the facts that the distance of J_1 from 0 is greater than $2^{-\frac{1}{2}}|z_1|$ and that $\arg z_1 < -\frac{3}{4}\pi$. Also

$$|2\pi(z - z_1)(z - z_0)/z| > \pi |z - z_0|,$$

so that

$$\mathfrak{R}\, \frac{2\pi i (z - z_1)(z - z_0)}{z} \frac{dz}{dl} < -\tfrac{1}{3}\pi l = -\tfrac{1}{3}\pi |z - z_0|,$$

if l always decreases as z_0 is approached. Then, applying (a) of § 4, we have

$$\int_{J_1} e^{\mathfrak{R}\phi(z)} |dz| < \sqrt{6}\, e^{\mathfrak{R}\phi(z_0)}. \tag{5.2}$$

Also
$$\mathfrak{R}\phi(b) < \mathfrak{R}\phi(z_0) - \tfrac{1}{6}\pi y_0^2. \tag{5.3}$$

On J_2, $dz/dl = 1$ and

$$\mathfrak{R}\phi'(z) = \mathfrak{R}\left(2\pi i z - 2\pi\kappa\epsilon - \frac{2\pi i \tau}{z}\right) = -\pi y_0 - \sqrt{2}\,\pi\kappa - \frac{\pi\tau y_0}{x^2 + \frac{1}{4}y_0^2}.$$

[[31]] But $\tau < x_0^2$, since $\xi_0 > 1$ for all ρ, and therefore

[[32]]
$$\mathfrak{R}\phi'(z) \frac{dz}{dl} < \pi(-\sqrt{2}\,\kappa - 2y_0) = -2\pi y_0'. \tag{5.4}$$

Consequently, by (b) of § 4, (5.3) and (5.4),

[[33]]
$$\int_{J_2} e^{\mathfrak{R}\phi(z)} |dz| < \frac{e^{\mathfrak{R}\phi(z)}}{2\pi y_0'} < \frac{\exp\left[\mathfrak{R}\phi(z_0) - \frac{1}{6}\pi y_0^2\right]}{2\pi y_0'}. \tag{5.5}$$

Now let us turn to J'. If z is a point of $J_1' + J_2'$, then $-i\bar{z}$ is the corresponding point of J_4', and

$$\mathfrak{R}\phi(-i\bar{z}) = \mathfrak{R}[i\pi(-i\bar{z})^2 + 2\kappa\epsilon(-i\bar{z}) - it \log z - it \log(-i\bar{z}/z)]$$
$$\leqslant \mathfrak{R}[-i\pi(iz)^2 + 2\kappa\bar{\epsilon} i z - it \log z - \tfrac{1}{2}t\pi],$$

i.e.
$$\mathfrak{R}\phi(-i\bar{z}) \leqslant \mathfrak{R}\phi(z) - \tfrac{1}{2}t\pi.$$

Also $|-i\bar{z}| = |z|$ and

[[34]]
$$|\operatorname{cosec}(-i\pi z)| < |\operatorname{cosec} \pi z|,$$

[[32]]

as may be proved by the use of the product representation of the sine function, remembering that $|\arg z| < \frac{1}{4}\pi$. Consequently

$$\int_{J'} \frac{|z|^{-\sigma}}{|e^{i\pi z}-e^{-i\pi z}|}\, e^{\Re\phi(z)}\,|dz| < (1+e^{-\frac{1}{2}t\pi}) \int_{J_1'+J_2'+J_3'} \frac{|z|^{-\sigma}}{|e^{i\pi z}-e^{-i\pi z}|}\, e^{\Re\phi(z)}\,|dz|. \qquad (5.6)$$

On J_1' we have $0 > \arg(z-z_1')/z > \gamma-\frac{1}{2}\pi$, where $\gamma = \frac{1}{4}\pi - \arg\zeta_0'$, and ⟦35⟧

$$|2\pi(z-z_1')(z-z_0')/z| > 2\pi|z-z_0'|.$$

Therefore
$$\Re\phi'(z)\,\frac{dz}{dl} < -2\pi\sin\gamma$$

on J_1', if l decreases as z_0' is approached from either side. Applying (*a*) of § 4, we have

$$\int_{J_1'} e^{\Re\phi(z)}\,|dz| < \frac{e^{\Re\phi(z_0')}}{(\sin\gamma)^{\frac{1}{2}}} \qquad (5.7)$$

and
$$\Re\phi(b) < \Re\phi(z_0') - \tfrac{1}{2}\pi y_0'^2 \sin\gamma. \qquad (5.8)$$

On J_2'

$$\Re\phi'(z)\,\frac{dz}{dl} = \Re\,\frac{-2\pi i z}{|z|}\left(z+\bar{\epsilon}\kappa-\frac{\tau}{z}\right)$$ ⟦36⟧

$$\leqslant \Re\,\frac{-2\pi i b'}{|b'|}\left(b'+\bar{\epsilon}\kappa-\frac{\tau}{b'}\right)$$ ⟦37⟧

$$= \Re\,\frac{-2\pi i}{|b'|}(b'-z_0')(b'-z_1'). \qquad (5.9)$$

But $\arg(z_0'-b') = \frac{1}{4}\pi$ and ⟦38⟧

$$0 > \arg(b'-z_1') > \arg(-z_1') = \gamma-\tfrac{1}{4}\pi,$$

so that

$$\tfrac{1}{2}\pi+\gamma < \arg\frac{-2\pi i}{|b'|}(b'-z_0')(b'-z_1') < \tfrac{3}{4}\pi.$$

Also
$$\left|-2\pi i(b'-z_0')(b'-z_1')/|b'|\right| > \sqrt{2}\,\pi y_0',$$

so that

$$\Re\,\frac{-2\pi i}{|b'|}(b'-z_0')(b'-z_1') < -\sqrt{2}\,\pi y_0'\sin\gamma, \qquad (5.10)$$

and applying (*b*) of § 4, and using (5.9), (5.10) and (5.8), we obtain

$$\int_{J_2'} e^{\Re\phi(z)}\,|dz| < (\sqrt{2}\,\pi y_0'\sin\gamma)^{-1}\exp\left[\Re\phi(z_0') - \tfrac{1}{2}\pi y_0'^2\sin\gamma\right]. \qquad (5.11)$$ ⟦39⟧

⟦33⟧

Also

[[40]] $$\mathfrak{R}\phi(\tfrac{2}{3}b') < \mathfrak{R}\phi(z_0') - \tfrac{1}{2}\pi y_0'^2 \sin\gamma - \tfrac{1}{6}\sqrt{2}\,\pi y_0' |z_0'| \sin\gamma. \qquad (5.12)$$

On J_3'

$$\left| \arg\left\{ \frac{-2\pi i}{z} (z-z_0')(z-z_1') \frac{dz}{dl} \right\} \right| < 2\sin^{-1}\tfrac{1}{3} < \tfrac{1}{4}\pi < \tfrac{1}{2}\pi - \gamma;$$

so that on both J_6' and J_3'

[[41]] $$\mathfrak{R}\, \frac{2\pi i}{z} (z-z_0')(z-z_1') \frac{dz}{dl} < -\sqrt{2}\,\pi y_0' \sin\gamma.$$

This gives, using also (5.12),

$$\int_{J_6'+J_3'} e^{\mathfrak{R}\phi(z)} |dz|$$

$$< (\sqrt{2}\,\pi y_0' \sin\gamma)^{-1} \exp[\mathfrak{R}\phi(z_0') - \tfrac{1}{2}\pi y_0'^2 \sin\gamma - \tfrac{1}{6}\sqrt{2}\,\pi y_0' |z_0'| \sin\gamma]. \qquad (5.13)$$

On J_1' we have

[[42]] $$\left| \frac{z^{-\sigma}}{e^{i\pi z} - e^{-i\pi z}} \right| < \tfrac{1}{2}\tau^{-\frac{1}{2}\sigma} |\tfrac{1}{2}\zeta_0'|^{-\sigma} \operatorname{cosech} |\tfrac{1}{2}\pi y_0'|; \qquad (5.15)$$

on J_5'

[[43]] $$\left| \frac{z^{-\sigma}}{e^{i\pi z} - e^{-i\pi z}} \right| < \tfrac{1}{2}\tau^{-\frac{1}{2}\sigma} |\tfrac{1}{3}\zeta_0'|^{-\sigma} \operatorname{cosech} |\tfrac{1}{3}\pi y_0'|; \qquad (5.16)$$

and on J_6', J_3'

[[44]] $$\left| \frac{z^{-\sigma}}{e^{i\pi z} - e^{-i\pi z}} \right| < \frac{1}{2\pi} [\min(\tfrac{1}{2}, \tfrac{1}{3}\tau^{\frac{1}{2}} |\zeta_0'|)]^{-\sigma-1} \operatorname{cosec} \arg \zeta_0'. \qquad (5.17)$$

We may now collect our results to give an inequality for $|R|$. We use (5.1), (5.2), (5.5), (5.6), (5.7), (5.11), (5.13), (5.15), (5.16) and (5.17), and we make the exponents $\mathfrak{R}\phi(z_0) + \sqrt{2}\,\pi\kappa\mu$ and $\mathfrak{R}\phi(z_0') - \sqrt{2}\,\pi\kappa\mu$ more explicit by the use of the relation $z_0^2 = \bar{\epsilon}\kappa z_0 + \tau$:

$$|R| = |R_0| < 0.635\, (\mathfrak{R}\epsilon\zeta_0)^{-\sigma} \tau^{-\frac{1}{2}\sigma} \operatorname{cosech} |\tfrac{1}{2}\pi y_0| \left\{ 2{\cdot}45 + \frac{0{\cdot}160}{y_0'} e^{-\frac{1}{2}\pi y_0^2} \right\}$$

[[45]] $$\times \exp[-\pi\kappa\mathfrak{R}\epsilon(z_0 - 2\mu) + 2\pi\tau \arg\zeta_0] + 0{\cdot}635(1 + e^{-\frac{1}{2}\pi t}) \tau^{-\frac{1}{2}\sigma} |\zeta_0'|^{-\sigma}$$

$$\times \Big\{ (\sin\gamma)^{-\frac{1}{2}} 2^{\sigma} \operatorname{cosech} |\tfrac{1}{2}\pi y_0'| + (\sqrt{2}\,\pi y_0' \sin\gamma)^{-1} \exp[-\tfrac{1}{2}\pi y_0'^2 \sin\gamma]$$

$$\times \Big[3^{\sigma} \operatorname{cosech} |\tfrac{1}{3}\pi y_0'| + \frac{1}{\pi} |\zeta_0'|^{\sigma} \tau^{\frac{1}{2}\sigma} [\min(\tfrac{1}{2}, \tfrac{1}{3}\tau^{\frac{1}{2}} |\zeta_0'|)]^{-\sigma-1} \operatorname{cosec} \arg\zeta_0'$$

[[46]] $$\times \exp[-\tfrac{1}{6}\sqrt{2}\,\pi y_0' |z_0'| \sin\gamma] \Big] \Big\} \exp[-\pi\kappa\mathfrak{R}\epsilon(2\mu - z_0') + 2\pi\tau \arg\zeta_0']. \qquad (5.18)$$

[[34]]

At this point it may be as well to repeat the definitions of the various quantities appearing on the right-hand side of this inequality, in terms of s, κ:

$$\sigma = \mathfrak{R}s, \quad t = \mathfrak{I}s, \quad \tau = t/2\pi, \quad \rho = \kappa\tau^{-\frac{1}{2}}, \quad \epsilon = e^{\frac{1}{4}i\pi}.$$

The complex numbers ζ_0', ζ_1', ζ_1, ζ_0 are the roots of the equation

$$(\zeta^2-1)^2 = -i\rho^2\zeta^2$$

lying respectively in the first, second, third and fourth quadrants, and

$$\zeta = \xi+i\eta, \quad z = x+iy = \tau^{\frac{1}{2}}\zeta, \quad \gamma = \tfrac{1}{4}\pi - \arg\zeta_0',$$

where ξ, η, x, y are real.

The estimate (5.18) for R is somewhat complicated. It may be simplified considerably when ρ is not large. I give an estimate for the case $\rho \leqslant \frac{1}{2}$; we then have

⟦47⟧

$$\left.\begin{array}{ll} \sqrt{2}\,\mathfrak{R}\epsilon\zeta_0' > 1, & \mathfrak{R}\epsilon(1-\zeta_0') > 0{\cdot}45\rho, \\ -\eta_0 > \rho/2\sqrt{2}, & \sin\gamma > 0{\cdot}55, \\ \eta_0' > 0{\cdot}29\rho, & |\zeta_0'| > 0{\cdot}81. \end{array}\right\} \qquad (5.19)$$

The result is that, for $\rho \leqslant \frac{1}{2}$, $t \geqslant 25$,

⟦48⟧

$$\begin{aligned} |R| = |R_0| < \tau^{-\frac{1}{2}\sigma}\Big[&0{\cdot}76\,.\,2^{\frac{1}{2}-\frac{1}{2}\sigma}\operatorname{cosech}0{\cdot}78(\kappa/\sqrt{2}) \\ &\times\left\{2{\cdot}45+0{\cdot}40(\kappa/\sqrt{2})^{-1}\exp\left(-0{\cdot}13(\kappa/\sqrt{2})^2\right)\right\}e^{-A}+0{\cdot}71(0.81)^{\frac{1}{2}-\sigma} \\ &\times\Big[1{\cdot}91\,.\,2^{\sigma-\frac{1}{2}}\operatorname{cosech}0{\cdot}64(\kappa/\sqrt{2})+1{\cdot}00(\kappa/\sqrt{2})^{-1}\exp\left(-0{\cdot}14(\kappa/\sqrt{2})^2\right) \\ &\times\Big\{1{\cdot}74\,.\,3^{\sigma-\frac{1}{2}}\operatorname{cosech}0{\cdot}42(\kappa/\sqrt{2})+1{\cdot}7(1{\cdot}62)^{\sigma-\frac{1}{2}}\tau^{\frac{1}{2}\sigma+\frac{1}{2}} \\ &\qquad\times(\kappa/\sqrt{2})\exp\left(-0{\cdot}13\tau^{\frac{1}{2}}(\kappa/\sqrt{2})\right)\Big\}\Big]e^{-B}\Big], \end{aligned} \qquad (5.20)$$

where

$$A = \pi\kappa\mathfrak{R}\epsilon(z_0-2\mu)+2\pi\tau\arg\zeta_0', \quad B = \pi\kappa\mathfrak{R}\epsilon(2\mu-z_0')-2\pi\tau\arg\zeta_0'.$$

In the case $\rho \geqslant \frac{1}{2}$ the estimation of the remainder from (5.18) may be made easier by the use of the following inequalities, which are valid for all positive ρ:

$$\mathfrak{R}\epsilon\zeta_0' > \frac{1}{\rho^3+\rho^2+\rho+\sqrt{2}}, \quad \eta_0' > \frac{0{\cdot}65\rho}{\rho^2+2}, \quad \sin\gamma > \frac{1}{\rho^2+\frac{1}{2}\rho+\sqrt{2}}. \qquad (5.21)$$

6. *Choice of parameters. The remainder with a finite series.* The remainder R as given by (5.18) is the sum of two terms in which the main factors are e^{-A} and e^{-B}. The most favourable choice of the parameter μ is presumably approximately that which makes these two factors equal. Calling this value μ_0, we have

$$\mu_0 = \frac{1}{2\sqrt{2}}\left[\mathbf{R}\epsilon(z_0'+z_0)+\frac{4\tau^{\frac{1}{2}}}{\rho}\arg\zeta_0'\right]$$

[[49]] $$= \frac{\tau^{\frac{1}{2}}}{\sqrt{2}}\left[\mathbf{R}\sqrt{\left(i+\tfrac{1}{4}\rho^2\right)}+\frac{2}{\rho}\arg\zeta_0'\right].$$

[[50]] As ρ tends to infinity, $\mu_0 \sim \kappa/2\sqrt{2}$; and, as ρ tends to 0, $\mu_0 \sim \tau^{\frac{1}{2}}$. Also

$$\mathbf{R}\epsilon(\mu_0-z_0') = \tfrac{1}{4}\left[\mathbf{R}\epsilon(z_0-3z_0')+\frac{4\tau^{\frac{1}{2}}}{\rho}\arg\zeta_0'\right]$$

$$= \tfrac{1}{2}\kappa\left[1-\mathbf{R}\sqrt{\left(\tfrac{1}{4}+\frac{i}{\rho^2}\right)}+\frac{2}{\rho^2}\arg\zeta_0'\right].$$

As ρ tends to 0, the factor

$$1-\mathbf{R}\sqrt{\left(\tfrac{1}{4}+\frac{i}{\rho^2}\right)}+\frac{2}{\rho^2}\arg\zeta_0'$$

[[51]] tends to 1; and, as ρ tends to infinity, this factor tends to $\frac{1}{2}$. For all positive
[[52]] values of ρ it is greater than $\frac{1}{2}$, and therefore

$$\mathbf{R}\epsilon(\mu_0-z_0') > \tfrac{1}{4}\kappa.$$

Also $$\mathbf{R}\epsilon(z_0-\mu_0) = \tfrac{1}{4}\left[\mathbf{R}\epsilon(3z_0-z_0')-\frac{4\tau^{\frac{1}{2}}}{\rho}\arg\zeta_0'\right],$$

$$= \tfrac{1}{2}\kappa\left[1+\mathbf{R}\sqrt{\left(\tfrac{1}{4}+\frac{i}{\rho^2}\right)}-\frac{2}{\rho^2}\arg\zeta_0'\right]$$

[[53]] and we have $$\mathbf{R}\epsilon(z_0-\mu_0) > \tfrac{1}{2}\kappa.$$

If for μ we choose $\mu_0 \pm \delta$, where $\delta > 0$, then the greater of the exponential factors e^{-A}, e^{-B} is

[[54]] $$\exp\left[-\tfrac{1}{2}\pi\kappa^2+\sqrt{2}\,\pi\kappa\delta\right].$$

The values of μ which we may choose are restricted only by the condition that the curves J, J' must be distant at least $1/(4\kappa)$ from P. If $\kappa \geqslant \sqrt{2}$, we may then choose μ in the interval $(\mu_0, \mu_0+\frac{1}{2})$; and, if $\kappa \geqslant 2$, we may
[[55]] choose it in the interval $(\mu_0-\frac{1}{2}, \mu_0+\frac{1}{2})$. However, for such small values of κ it will probably be best to choose μ rather close to μ_0. We need not consider the case of smaller values of κ than $\sqrt{2}$, since, as will appear,

it is not advantageous to take such small values, even when there is only one term taken from the series $\Sigma h(p_k)$.

When ρ is small, μ_0 is close to $\tau^{\frac{1}{2}}$, and it is therefore probably simplest to choose a value of μ which is close to $\tau^{\frac{1}{2}}$ without actually calculating μ_0. The inequality

$$0 < \mu_0/\tau^{\frac{1}{2}} - 1 < \rho^2/3 \quad (0 < \rho < 1)$$

should then be of value.

For large values of κ (*e.g.* $\kappa > 3$) we shall do well to choose μ to be either an integer or a half odd integer. In the case that μ is an integer the function g has a double pole at μ; in place of the terms

$$\frac{\epsilon}{\kappa} h(\mu) - \frac{\tau^{-s}}{1-\exp[2\pi\kappa\epsilon(\mu-\mu)]}$$

we have therefore to put the residue at 0 of

$$\frac{2\pi i(-)^{\mu}(z+\mu)^{-s}\exp[i\pi(z+\mu)^2 - i\pi z]}{(1-e^{-2\pi i z})(1-e^{-2\pi\kappa\epsilon z})}$$

and this is equal to

$$\mu^{-s}\left\{-\tfrac{1}{2} + \frac{\sigma}{2\pi\kappa\epsilon\mu} - \frac{\epsilon}{\kappa}\left(\mu - \frac{t}{2\pi\mu}\right)\right\}.$$

In the practical applications of course we take only a finite number of terms from the series $\Sigma(\epsilon/\kappa)h(p_k)$. We therefore want an estimate of the error arising from this. If we decide what is the greatest total error η we can admit in our result we can proceed in this way. We choose κ so that $|R| < \frac{1}{2}\eta$ and then take sufficiently many terms of the series for the error from this second source not to exceed $\frac{1}{2}\eta$. Now let us estimate this second remainder. For this purpose we prove the following

LEMMA. *The function* $|e^{i\pi z^2} z^{-it}|$ *has only one maximum on the line* P.

Put $z = \mu\big(1+\theta(1+i)\big)$, $a = t/(2\pi\mu^2)$; then θ is real and

$$\log|e^{i\pi z^2} z^{-it}| = \Re\left[i\pi\big(1+\theta(1+i)\big)^2 - 2\pi i a \log\left\{\mu\big(1+\theta(1+i)\big)\right\}\right]\mu^2.$$

Let us abbreviate the right-hand side to $H(\theta)$. Then

$$H'(\theta) = 2\pi\mu^2\left(-1-2\theta+\frac{a}{(1+\theta)^2+\theta^2}\right) = -2\pi\mu^2\,\frac{\big((1+\theta)^2+\theta^2\big)(1+2\theta)-a}{(1+\theta)^2+\theta^2}.$$

Now

$$\frac{d}{d\theta}\Big[\big((1+\theta)^2+\theta^2\big)(1+2\theta)-a\Big] = 3(2\theta+1)^2+1 > 0,$$

and therefore $\big((1+\theta)^2+\theta^2\big)(1+2\theta)-a$ cannot vanish for more than one value of θ; it clearly vanishes for at least one value. Then $H(\theta)$ has just one stationary value, which is easily seen to be a maximum. This completes the proof of the lemma. If $a < 1$ the value θ giving the maximum satisfies $0 > \theta > a^{\frac{1}{3}}-1$.

Put $u_k = \kappa^{-1}|h(p_k)|$; then, if $\sigma \geqslant 0$,

⟦56⟧ $$\sum_{k=K+1}^{\infty} u_{-k} < \frac{(\mu/\sqrt{2})^{-\sigma}}{1-\exp[\sqrt{2}\pi(K+1)/\kappa]} \sum_{k=K+1}^{\infty} \exp\left(-\frac{\pi k}{\sqrt{2}\kappa}\right) |e^{i\pi p_{-k}^2} p_{-k}^{-it}|.$$

If we suppose that $\Im(p_{-K-1})$ is less than the value of y for which the maximum of $|e^{i\pi z^2} z^{-it}|$ occurs, and if $k \geqslant K+1$, then

$$|e^{i\pi p_{-k}^2} p_{-k}^{-it}| \leqslant |e^{i\pi p_{-K-1}^2} p_{-K-1}^{-it}|$$

and therefore

$$\sum_{K+1}^{\infty} u_{-k} < \left(\frac{\mu}{\sqrt{2}}\right)^{-\sigma} \frac{\exp[-\pi(K+1)/\kappa\sqrt{2}+\Re(i\pi p_{-K-1}^2)]\,|p_{-K-1}^{-it}|}{\big(1-\exp[-\sqrt{2}\pi(K+1)/\kappa]\big)(1-\exp[-\pi/\kappa\sqrt{2}])}$$

⟦57⟧ $$< \frac{2^{\frac{1}{2}\sigma} u_{-K-1}}{\big(1-\exp[-\sqrt{2}\pi(K+1)/\kappa]\big)^2(1-\exp[-\pi/\kappa\sqrt{2}])}.$$

Similarly, if (as is always the case if $K' \geqslant 0$ and $a \leqslant 1$) $\Im(p_{K'+1})$ is greater than the value of y for which the maximum occurs,

⟦58⟧ $$\sum_{K'+1}^{\infty} u_k < \frac{u_{K'+1}}{\big(1-\exp[-\sqrt{2}\pi(K'+1)/\kappa]\big)^2(1-\exp[-\pi/\kappa\sqrt{2}])},$$

so that

$$|R^*| = \left|\left(\sum_{-\infty}^{\infty} - \sum_{-K}^{K'}\right)\frac{\epsilon}{\kappa} h(p_k)\right|$$

$$< \frac{2^{\frac{1}{2}\sigma}|h(p_{-K-1})|+|h(p_{K'+1})|}{\kappa\big(1-\exp[-\sqrt{2}\pi(K^*+1)/\kappa]\big)^2(1-\exp[-\pi/\kappa\sqrt{2}])},$$

where $K^* = \min(K, K')$.

In the case in which κ is small compared with $\tau^{\frac{1}{2}}$ we can easily obtain a rough estimate of the number of terms required for a given accuracy.
⟦59⟧ For in this case K, K' are such that $p_{-K}/\mu-1$ and $p_{K'}/\mu-1$ are small,

⟦38⟧

and u_K is approximately $\exp[-2\pi(K+1)^2/\kappa^2]$. If the remainders R and ⟦60⟧
R^* are of the same order of magnitude, then we have approximately $2\pi(K+1)^2/\kappa^2 = \frac{1}{2}\pi\kappa^2$, *i.e.*, $K+1 = \frac{1}{2}\kappa^2$. The number of terms T that we take is $2K+1$, *i.e.*, approximately κ^2, and the total error is of the order of magnitude of $e^{-\frac{1}{2}\pi T}$. If this statement is to be put into an exact form we must say that if μ and κ are suitably chosen as functions of t, η, and σ lies in the interval $0 \leqslant \sigma \leqslant 1$, and if η tends to 0 and t to infinity in such a way that κ^{-1}, $\kappa\tau^{-\frac{1}{2}}$ also tend to 0, then the error does not exceed η and $T^{-1}\log\eta^{-1}$ tends to $\frac{1}{2}\pi$; this does not hold for any number larger than $\frac{1}{2}\pi$.

When $\kappa\tau^{-\frac{1}{2}}$ is of the order of magnitude of 1 we cannot get so simple an estimate of the number of terms needed, but we can obtain an estimate in the limiting case $\kappa\tau^{-\frac{1}{2}} \to \infty$. We may then neglect all the factors in u_h
except $|e^{i\pi} p_k^2|$. Putting $\mu + \epsilon v = p_{-K}$, we have approximately ⟦61⟧

$$|e^{i\pi(\mu+\epsilon v)^2}| = e^{-\frac{1}{2}\pi\kappa^2}$$

if R and R^* are of the same order of magnitude; *i.e.*,

$$\sqrt{2}\pi\mu v + \pi v^2 = \tfrac{1}{2}\pi\kappa^2.$$

But for large values of μ we have approximately $\mu = \kappa/2\sqrt{2}$, and therefore approximately

$$2v^2 + \kappa v - \kappa^2 = 0.$$

The two roots of this equation are $\bar{\epsilon}(P_{-K}-\mu)$ and $\bar{\epsilon}(P_{K'}-\mu)$ (approximately); the difference of the roots is $\frac{3}{2}\kappa$ and therefore $T = \frac{3}{2}\kappa^2$ approxi- ⟦62⟧
mately, and the error is of the order of magnitude of $e^{-\frac{1}{3}\pi T}$.

It is possible that this might be improved by taking μ to be something different from μ_0; for if we take μ closer to $\tau^{\frac{1}{2}}$ the remainder R^* is made smaller for any given value of T. Such an improvement would necessarily be at the expense of the remainder R; I do not think that any appreciable improvement really can be made along these lines.

In the case in which $T = 3$ we may put $\kappa = 1{\cdot}6\sqrt{2}$, and then, if $\sigma = \frac{1}{2}$,
$\mu = \tau^{\frac{1}{2}}$ and $t > 350$, and if the factor $\left(1-\exp[2\pi\kappa\epsilon(r-\mu)]\right)^{-1}$ is replaced ⟦63⟧
by 0 or 1 except for two terms of the main series, the error from all sources does not exceed $0{\cdot}0044\tau^{-\frac{3}{4}}$. ⟦64⟧

7. *A similar method.* There is an alternative, and better known, integral representation of the zeta-function on which we may base our calculations, viz.,

$$\zeta(s) = \sum_{r=1}^{m} r^{-s} + 2(2\pi)^{s-1}\sin\tfrac{1}{2}\pi s\,\Gamma(1-s)\sum_{r=1}^{m'} r^{-s} + \frac{1}{1+e^{i\pi s}}\int_{Q_m}\frac{e^{2m'i\pi z}z^{-s}\,dz}{1-e^{-2\pi i z}}.$$

Here Q_m is a curve coming from infinity in the first quadrant, crossing the real axis between m and $m+1$ and again between $-m$ and $-m-1$, and going on to infinity in the second quadrant; z^{-s} is defined as in § 2. If we choose $m = m' = [\tau^{\frac{1}{2}}]$, and let the part of Q_m in the neighbourhood of the positive real axis be a straight line cutting the negatively directed real axis at μ and at an angle of $+\frac{1}{4}\pi$, then, for large t, the only appreciable contribution to the integral comes from the neighbourhood of the positive real axis. We can approximate to this integral in the same way as before, the resulting approximate value for $\zeta(s)$ being

$$\sum_{r=1}^{\infty} r^{-s}\left\{1-(1+e^{i\pi s})^{-1}\left(1-\exp\left[-2\pi\kappa\epsilon(r-\mu)\right]\right)^{-1}\right\}$$
$$+2(2\pi)^{s-1}\sin\tfrac{1}{2}\pi s\,\Gamma(1-s)\sum_{r=1}^{m} r^{-s}$$
$$-\frac{\epsilon}{\kappa(1+e^{i\pi s})}\sum_{k=-K}^{K}\frac{(\mu+\epsilon k/\kappa)^{-s}\exp\left[2\pi i m(\mu+\epsilon k/\kappa)\right]}{1-\exp\left[-2\pi i(\mu+\epsilon k/\kappa)\right]}.$$

The integer K must not be chosen too large; $K < \frac{1}{2}\tau^{\frac{1}{2}}\kappa$ is usually sufficiently small. This method has the advantage that for points not on the critical line only two real integrals have to be evaluated, and not four. This may be of value for calculation of zeros not on the critical line. For this purpose it will not matter that the method is only applicable for large values of t; it is, however, possible to remove this restriction by integrating along a parabola, *e.g.*, the parabola

$$x^2 = 2\mu y+\mu^2 \quad ([\tau^{\frac{1}{2}}] \leqslant \mu \leqslant [\tau^{\frac{1}{2}}]+1).$$

The conformal mapping $u^2 = z$ transforms this parabola into a straight line, so that

$$\int_{Q_m}\frac{e^{2\pi i m z}z^{-s}\,dz}{1-e^{-2\pi i z}} = -2\int_{m^{\frac{1}{2}}\nearrow(m+1)^{\frac{1}{2}}}\frac{e^{2\pi i m u^2}u^{-2s+1}\,du}{1-e^{-2\pi i u^2}}.$$

The line of integration cuts the imaginary axis between $-im^{\frac{1}{2}}$ and $-i(m+1)^{\frac{1}{2}}$.

King's College,
Cambridge.

ROUNDING-OFF ERRORS IN MATRIX PROCESSES

By A. M. TURING

(*National Physical Laboratory, Teddington, Middlesex*)

[Received 4 November 1947]

SUMMARY

A number of methods of solving sets of linear equations and inverting matrices are discussed. The theory of the rounding-off errors involved is investigated for some of the methods. In all cases examined, including the well-known 'Gauss elimination process', it is found that the errors are normally quite moderate: no exponential build-up need occur.

Included amongst the methods considered is a generalization of Choleski's method which appears to have advantages over other known methods both as regards accuracy and convenience. This method may also be regarded as a rearrangement of the elimination process.

THIS paper contains descriptions of a number of methods for solving sets of linear simultaneous equations and for inverting matrices, but its main concern is with the theoretical limits of accuracy that may be obtained in the application of these methods, due to rounding-off errors.

The best known method for the solution of linear equations is Gauss's elimination method. This is the method almost universally taught in schools. It has, unfortunately, recently come into disrepute on the ground that rounding off will give rise to very large errors. It has, for instance, been argued by Hotelling (ref. **5**) that in solving a set of n equations we should keep $n \log_{10} 4$ extra or 'guarding' figures. Actually, although examples can be constructed where as many as $n \log_{10} 2$ extra figures would be required, these are exceptional. In the present paper the magnitude of the error is described in terms of quantities not considered in Hotelling's analysis; from the inequalities proved here it can immediately be seen that in all normal cases the Hotelling estimate is far too pessimistic.

The belief that the elimination method and other 'direct' methods of solution lead to large errors has been responsible for a recent search for other methods which would be free from this weakness. These were mainly methods of successive approximation and considerably more laborious than the direct ones. There now appears to be no real advantage in the indirect methods, except in connexion with matrices having special properties, for example, where the vast majority of the coefficients are very small, but there is at least one large one in each row.

The writer was prompted to carry out this research largely by the practical work of L. Fox in applying the elimination method (ref. **2**). Fox found that no exponential build-up of errors such as that envisaged by Hotelling actually occurred. In the meantime another theoretical investigation was being carried out by J. v. Neumann, who reached conclusions similar to those of this paper for the case of positive definite matrices, and communicated them to the writer at Princeton in January 1947 before the proofs given here were complete. These results are now published (ref. **6**).

1. Measure of work in a process

It is convenient to have a measure of the amount of work involved in a computing process, even though it be a very crude one. We may count up the number of times that various elementary operations are applied in the whole process and then give them various weights. We might, for instance, count the number of additions, subtractions, multiplications, divisions, recordings of numbers, and extractions of figures from tables. In the case of computing with matrices most of the work consists of multiplications and writing down numbers, and we shall therefore only attempt to count the number of multiplications and recordings. For this purpose a reciprocation will count as a multiplication. This is purely formal. A division will then count as two multiplications; this seems a little too much, and there may be other anomalies, but on the whole substantial justice should be done.

2. Solution of equations versus inversion

⟦1⟧ Let us suppose we are given a set of linear equations $\mathbf{Ax} = \mathbf{b}$ to solve. Here $\mathbf{A}$ represents a square matrix of the nth order and $\mathbf{x}$ and $\mathbf{b}$ vectors of the nth order. We may either treat this problem as it stands and attempt to find $\mathbf{x}$, or we may solve the more general problem of finding the inverse of the matrix $\mathbf{A}$, and then allow it to operate on $\mathbf{b}$ giving the required solution of the equations as $\mathbf{x} = \mathbf{A}^{-1}\mathbf{b}$. If we are quite certain that we only require the solution to the one set of equations, the former approach has the advantage of involving less work (about one-third the number of multiplications by almost all methods). If, however, we wish to solve a number of sets of equations with the same matrix $\mathbf{A}$ it is more convenient to work out the inverse and apply it to each of the vectors $\mathbf{b}$. This involves, in addition, n^2 multiplications and n recordings for each vector, compared with a total of about $\frac{1}{3}n^3$ multiplications in an independent solution. There are other advantages in having an inverse. From the coefficients of the inverse we can see at once how sensitive the solution is to small changes in the coefficients of $\mathbf{A}$ and of $\mathbf{b}$. We have, in fact,

⟦2⟧ $$\frac{\partial x_i}{\partial b_j} = (\mathbf{A}^{-1})_{ij}, \qquad \frac{\partial x_i}{\partial a_{jk}} = -(\mathbf{A}^{-1})_{ij}\, x_k.$$

This enables us to estimate the accuracy of the solution if we can judge the accuracy of the data, that is, of the matrix $\mathbf{A}$ and the vector $\mathbf{b}$, and also enables us to correct for any small changes which we may wish to make in these data.

It seems probable that with the advent of electronic computers it will become standard practice to find the inverse. This time has, however, not yet arrived and some consideration is therefore given in this paper to solutions without inversion. A form of compromise involving less work than inversion, but including some of the advantages, is also considered.

3. Triangular resolution of a matrix

A number of the methods for the solution of equations and, more particularly, for the inversion of matrices, depend on the resolution of a matrix into the product of two triangular matrices. Let us describe a matrix which has zeros above the diagonal as 'lower triangular' and one which has zeros below as 'upper triangular'. If in addition the coefficients on the diagonal are unity the expressions 'unit upper triangular' and 'unit lower triangular' may be used. The resolution is essentially unique, in fact we have the following

THEOREM ON TRIANGULAR RESOLUTION. *If the principal minors of the* ⟦3⟧
matrix $\mathbf{A}$ *are non-singular, then there is a unique unit lower triangular matrix* $\mathbf{L}$, *a unique diagonal matrix* $\mathbf{D}$, *with non-zero diagonal elements, and a unique unit upper triangular matrix* $\mathbf{U}$ *such that* $\mathbf{A} = \mathbf{LDU}$. *Similarly there are unique* $\mathbf{L}'$, $\mathbf{D}'$, $\mathbf{U}'$ *such that* $\mathbf{A} = \mathbf{U}'\mathbf{D}'\mathbf{L}'$.

The kth diagonal element of $\mathbf{D}$ will be denoted by d_k. The $1k$ coefficient ⟦4⟧
of the equation $\mathbf{A} = \mathbf{LDU}$ gives us $l_{11} d_1 u_{1k} = a_{1k}$ and since $l_{11} = u_{11} = 1$ this determines d_1 to be a_{11} and u_{1k} to be a_{1k}/d_1; these choices satisfy the equations in question. Suppose now that we have found values of l_{ij}, u_{jk} ⟦5⟧
with $j < i_0$ (that is, we have found the first $i_0 - 1$ rows of $\mathbf{L}$ and columns of $\mathbf{U}$) and the first $i_0 - 1$ diagonal elements d_k, so that the equations arising from the first $i_0 - 1$ rows of the equation $\mathbf{A} = \mathbf{LDU}$ are satisfied; and suppose further that these choices are unique and $d_k \neq 0$. It will be shown how the next row of $\mathbf{L}$ and the next column of $\mathbf{U}$, and the next diagonal element $d_{i_0} \neq 0$ are to be chosen so as to satisfy the equations arising from the next row of $\mathbf{A} = \mathbf{LDU}$, and that the choice is unique. The equations to be satisfied in fact state

$$l_{i_0 i_0} d_{i_0} u_{i_0 k} = a_{i_0 k} - \sum_{j<i_0} l_{i_0 j} d_j u_{jk} \quad (k \geqslant i_0),$$

$$l_{i_0 k} d_k u_{kk} = a_{i_0 k} - \sum_{j<k} l_{i_0 j} d_j u_{jk} \quad (k < i_0).$$

The right-hand sides of these equations are entirely in terms of quantities already determined. When $k = i_0$ the first equation is satisfied and can only be satisfied by putting d_{i_0} = right-hand side, determining d_{i_0}. The equations for $k > i_0$ can then be satisfied by one and only one set of values of u_{i_0k}, provided $d_{i_0} \neq 0$. The equations for $k < i_0$ can also be satisfied by one and only one set of values of l_{i_0k}, since each d_k is different from 0 The new diagonal element d_{i_0} is not 0 because the i_0th principal minor of $\mathbf{A}$ is equal to the product of the first i_0 diagonal elements d_k.

4. The elimination method

Suppose that we wish to solve the equations $\mathbf{Ax} = \mathbf{b}$ by the elimination method. The procedure is as follows. We first add such multiples of the first equation to the others that the coefficient of x_1 is reduced to zero in all of them (excepting the first). We then add multiples of the second equation to the later ones until the coefficient of x_2 is reduced to zero. After $n-1$ steps of this nature we shall be left with a set of equations of the form $\sum_{i \leqslant j} v_{ij} x_j = c_i$. From the equation $v_{nn} x_n = c_n$ the unknown x_n can then be found immediately, and by substituting it in the equation $v_{n-1,n-1} x_{n-1} + v_{n-1,n} x_n = c_{n-1}$ we then find x_{n-1}, and so on until by repeated back-substitution we have found all the coefficients of the (originally) unknown vector $\mathbf{x}$. This description of the elimination process is all that is required in order to apply it. We shall find it instructive, however, to look at it further from a number of points of view.

(1) The process of replacing the rows of a matrix by linear combinations of other rows may be regarded as left-multiplication of the matrix by another matrix, this second matrix having coefficients which describe the linear combinations required. Each stage of the above-described elimination process is of this nature, so that we first convert the equations $\mathbf{Ax} = \mathbf{b}$ into $\mathbf{J}_1 \mathbf{Ax} = \mathbf{J}_1 \mathbf{b}$ and record $\mathbf{J}_1 \mathbf{A}$ and $\mathbf{J}_1 \mathbf{b}$. We then convert them into $\mathbf{J}_2 \mathbf{J}_1 \mathbf{Ax} = \mathbf{J}_2 \mathbf{J}_1 \mathbf{b}$, and so on, until we finally have $\mathbf{J}_{n-1} \ldots \mathbf{J}_1 \mathbf{Ax} = \mathbf{J}_{n-1} \ldots \mathbf{J}_1 \mathbf{b}$. In accordance with the theorem on triangular resolution we may write $\mathbf{J}_{n-1} \ldots \mathbf{J}_1 = \mathbf{L}^{-1}$ and $\mathbf{J}_{n-1} \ldots \mathbf{J}_1 \mathbf{A} = \mathbf{DU}$. The matrix $\mathbf{DU}$ is upper triangular, that is, it has no coefficients other than zeros below the diagonal. The matrix $\mathbf{L}^{-1}$ and its inverse $\mathbf{L}$ are lower triangular.

(2) The matrix $\mathbf{L}$ can be very easily obtained from the matrices
⟦6⟧ $\mathbf{J}_1, \ldots, \mathbf{J}_{n-1}$. We have in fact $\mathbf{L} = 1 + \sum_{r=1}^{n-1} (1 - \mathbf{J}_r)$. The proof of this will be left to the reader.

(3) There is no need for us to take either the equations or the unknowns in the order in which they are given. In other words, if $\mathbf{P}$, $\mathbf{Q}$ represent permutations we may solve instead $\mathbf{A'x'} = \mathbf{b'}$, where $\mathbf{A'} = \mathbf{PAQ}$, $\mathbf{b'} = \mathbf{Pb}$, $\mathbf{x} = \mathbf{Qx'}$. The permutations $\mathbf{P}$, $\mathbf{Q}$ may be chosen bit by bit as we carry the process through. One popular method is to let $\mathbf{Q}$ be the identity, that is,

to take the variables in the order given, and to choose **P** so that the coefficients in the matrices $\mathbf{J}_r$ do not exceed unity in absolute magnitude. This is always possible, and for almost all matrices gives a unique **P**. Alternatively, this variation of the method may be described by saying that **P** is chosen so that d_1 shall have the largest possible value, and subject to this, d_2 to be as large as possible, and so on. This procedure is called 'taking the largest coefficient in the column as pivot'. The diagonal elements $d_1, d_2,..., d_n$ are known as the first, second,..., last pivots. There seems to be a definite advantage in using the largest pivot in the column as it is likely to have smaller proportionate errors than other possible pivots, and saves us from the embarrassment of getting a pivot which is little different from zero. It is possible that there is also a further advantage in choosing the largest coefficient in the matrix as pivot.

(4) The leading terms of the work involved in solving a set of n equations ⟦7⟧ by the elimination method are as follows: $\frac{1}{3}n^3+O(n^2)$ multiplications and recordings of which $\frac{1}{2}n^2+O(n)$ recordings involve the vector **b**.

(5) If, after we have solved one set of equations $\mathbf{Ax} = \mathbf{b}$, we are asked to solve a second set $\mathbf{Ax}' = \mathbf{b}'$ with the same matrix **A**, we have only to operate on **b** with the matrices $\mathbf{J}_1,..., \mathbf{J}_{n-1}$ the values of which may be supposed to have been kept for reference, and then solve $\mathbf{DUx} = \mathbf{J}_{n-1}...\mathbf{J}_1\mathbf{b}$. In other words, if the matrices $\mathbf{J}_1,..., \mathbf{J}_{n-1}$ have been kept (amounting to $\frac{1}{2}n(n-1)$ numbers) the work involved in solving a second set with the same **A** is that part of the original work which involved **b**, namely, $\frac{1}{2}n^2+O(n)$ multiplications and n recordings.

This process may also be expressed in another form, which appears to be quite different, but actually is an identical calculation. As mentioned in (2), the triangle **L** in the resolution $\mathbf{A} = \mathbf{LDU}$ may be obtained immediately from the matrices $\mathbf{J}_1,..., \mathbf{J}_{n-1}$. If we put $\mathbf{DUx}' = \mathbf{y}'$ we shall then have $\mathbf{Ly}' = \mathbf{b}'$. The equations $\mathbf{Ly}' = \mathbf{b}'$ may be solved for $\mathbf{y}'$ by one back-substitution process and then the equation $\mathbf{DUx}' = \mathbf{y}'$ solved by a second back-substitution.

(6) As we have described it, the matrices $\mathbf{J}_1\mathbf{A}$, $\mathbf{J}_2\mathbf{J}_1\mathbf{A}$,..., are all written down in full. Actually, however, we are not really interested in all the coefficients of all these matrices. All we need in the end are $\mathbf{J}_1,..., \mathbf{J}_{n-1}$ and $\mathbf{J}_{n-1}...\mathbf{J}_1\mathbf{A}$. It is sufficient, therefore, to calculate all coefficients of $\mathbf{J}_{n-1}...\mathbf{J}_1\mathbf{A}$, and those coefficients of $\mathbf{J}_r...\mathbf{J}_1\mathbf{A}$ which are required for the determination of $\mathbf{J}_{r+1}$. If we write $\mathbf{A}^{(r)}$ for $\mathbf{J}_r...\mathbf{J}_1\mathbf{A}$ we have

$$\mathbf{A}_{ij}^{(r)} = \mathbf{A}_{ij}^{(r-1)}+(\mathbf{J}_r)_{ir}\mathbf{A}_{rj}^{(r-1)} \quad (i > r),$$ ⟦8⟧

where $$(\mathbf{J}_r)_{ir} = -\frac{\mathbf{A}_{ir}^{(r-1)}}{\mathbf{A}_{rr}^{(r-1)}},$$

and by addition

$$\mathbf{A}^{(r)}_{ij} = \mathbf{A}_{ij} + \sum_{s=1}^{r} (\mathbf{J}_s)_{is} \mathbf{A}^{(s-1)}_{sj}.$$

If $i \leqslant r$ we have $\mathbf{A}^{(r)}_{ij} = \mathbf{A}^{(r-1)}_{ij}$ and so

$$\mathbf{A}^{(n)}_{ij} = \mathbf{A}_{ij} + \sum_{s=1}^{n-1} (\mathbf{J}_s)_{is} \mathbf{A}^{(n)}_{sj},$$

$$(\mathbf{J}_r)_{ir} = -\frac{\mathbf{A}_{ir} + \sum_{s=1}^{r-1} (\mathbf{J}_s)_{is} \mathbf{A}^{(s-1)}_{sr}}{\mathbf{A}_{rr} + \sum_{s=1}^{r-1} (\mathbf{J}_s)_{rs} \mathbf{A}^{(s-1)}_{sr}}.$$

Thus we can obtain the numbers actually required ($\mathbf{A}^{(n)}_{ij}$, $(\mathbf{J}_r)_{ir}$) without recording intermediate quantities. This variation of the elimination method will be seen to be identical with the method (1) of § 6 (the 'unsymmetrical Choleski method').

This form of the elimination method is to be preferred to the original form in every way. The recording involved in the work on the matrix is reduced from $\frac{1}{3}n^3 + O(n^2)$ to $n^2 + O(n)$, and the rounding off is at the same time made correspondingly less frequent.

(7) The elimination method may be used to invert a matrix. One method is to solve a succession of sets of equations $\mathbf{A}\mathbf{x}^{(r)} = \mathbf{b}^{(r)}$, where $\mathbf{b}^{(r)} = \delta_{ir}$. The total work involved in the inversion is then $n^3 + O(n^2)$ multiplications. Alternatively, we may invert the matrices $\mathbf{L}$ and $\mathbf{DU}$ separately by back-substitution and then multiply them together. The work is still $n^3 + O(n^2)$ multiplications.

(8) When the matrix $\mathbf{A}$ is symmetric, the matrices $\mathbf{L}$ and $\mathbf{U}$ are transposes, and it is therefore unnecessary to calculate both of them. The best arrangement is probably to proceed as with an unsymmetrical matrix, but to ignore all the coefficients below the diagonal in the matrices $\mathbf{A}^{(r)}$. These coefficients are all either zero or equal to the corresponding elements of the transpose. This fact enables us to find the appropriate matrices $\mathbf{J}_r$ at each stage.

(9) The elimination method can be described in another, superficially quite unrelated form. We may combine multiplication of rows and addition to other rows with multiplication of columns and adding to other columns. In other words, we may form a product $\mathbf{J}_{n-1} \ldots \mathbf{J}_1 \mathbf{A} \mathbf{K}_1 \ldots \mathbf{K}_{n-1}$, and try to arrange that it shall be diagonal. The matrix $\mathbf{J}_r$ is to differ from unity only in the rth column below the diagonal, and $\mathbf{K}_r$ is to differ from unity only in the rth row above the diagonal. If we carry out the multiplications by $\mathbf{J}_1, \ldots, \mathbf{J}_{n-1}$ before the multiplications by $\mathbf{K}_1, \ldots, \mathbf{K}_{n-1}$, then it is clear that we have only the elimination method, for in either case we form $\mathbf{J}_1 \mathbf{A}$,

$\mathbf{J}_2\mathbf{J}_1\mathbf{A}$,... and the multiplications by $\mathbf{K}_1,\ldots,\mathbf{K}_{n-1}$ which come after actually involve no computation; they merely result in replacing certain coefficients in the matrix $\mathbf{J}_{n-1}\ldots\mathbf{J}_1\mathbf{A}$ by zeros (compare note (2)). It is not quite so clear in the case where the order of calculation is $\mathbf{A}$, $\mathbf{J}_1\mathbf{A}$, $\mathbf{J}_1\mathbf{A}\mathbf{K}_1$, $\mathbf{J}_2\mathbf{J}_1\mathbf{A}\mathbf{K}_1,\ldots$. In this case, however, the right-multiplications do not alter that part of the matrix which will be required later; in fact, they again do nothing but replace certain coefficients by zeros. So far as the subsequent work is concerned, we may consider that these right-multiplications were omitted, and that we formed $\mathbf{J}_{n-1}\ldots\mathbf{J}_1\mathbf{A}$ as in the elimination method.

When this method is used and we choose the largest pivot in the matrix, it is clear that all the coefficients of $\mathbf{J}_r$ and of $\mathbf{K}_r$ do not exceed unity. This provides one proof that when the largest pivot in the matrix is chosen the coefficients of $\mathbf{L}$, $\mathbf{U}$ do not exceed unity (in absolute magnitude).

5. Jordan's method for inversion

In § 4 (1) we mentioned that the elimination process could be regarded as the reduction of a matrix to triangular form by left-multiplication of it by a sequence of matrices $\mathbf{J}_1,\ldots,\mathbf{J}_{n-1}$. In the Jordan method we left-multiply the matrix $\mathbf{A}$ by a similar sequence of matrices. The difference is that with the Jordan method we aim at reducing $\mathbf{A}$ to a diagonal,† or preferably to the unit matrix, instead of merely to a triangle.†

The process consists in forming the successive matrices $\mathbf{J}_1\mathbf{A}$, $\mathbf{J}_2\mathbf{J}_1\mathbf{A},\ldots$, where $\mathbf{J}_r$ differs from the unit matrix only in the rth column, and where $\mathbf{J}_r\ldots\mathbf{J}_1\mathbf{A}$ differs from a diagonal matrix only in the columns after the rth. Let us put

$$\mathbf{A}^{(r)} = \mathbf{J}_r\ldots\mathbf{J}_1\mathbf{A}, \qquad \mathbf{X}^{(r)} = \mathbf{J}_r\ldots\mathbf{J}_1, \tag{9}$$

we shall then have

$$\mathbf{A}^{(r)}_{ij} = \mathbf{A}^{(r-1)}_{ij}+(\mathbf{J}_r)_{ir}\mathbf{A}^{(r-1)}_{rj} \quad (i \neq r),$$

$$(\mathbf{J}_r)_{ir} = -\frac{\mathbf{A}^{(r-1)}_{ir}}{\mathbf{A}^{(r-1)}_{rr}} \quad (i \neq r)$$

(so that $\mathbf{A}^{(r)}_{ir} = 0$ if $i \neq r$),

$$\mathbf{A}^{(r)}_{rj} = (\mathbf{J}_r)_{rr}\mathbf{A}^{(r-1)}_{rj},$$

$$\mathbf{X}^{(0)}_{ij} = \delta_{ij},$$

$$\mathbf{X}^{(r)}_{ij} = \mathbf{X}^{(r-1)}_{ij}+(\mathbf{J}_r)_{ir}\mathbf{X}^{(r-1)}_{rj}. \tag{10}$$

The particular diagonal to which $\mathbf{A}$ is reduced is at our disposal. Possible choices include the following. The diagonal may be the unit matrix. Or we may arrange that the diagonal elements of the $\mathbf{J}_r$ are all unity and tolerate the non-unit diagonal elements in $\mathbf{J}_n\ldots\mathbf{J}_1\mathbf{A}$. A third alternative is to arrange that the diagonal elements in $\mathbf{J}_n\ldots\mathbf{J}_1\mathbf{A}$ shall be between 0·1 and 1 and that the diagonal elements in $\mathbf{J}_r$ shall be powers of 10.

† Hereafter 'triangle' and 'diagonal' will be written for 'triangular matrix' and 'diagonal matrix'.

Jordan's method is probably the most straightforward one for inversion. Although it can be used for the solution of equations, it is not very economical for that purpose. For hand work it has the serious disadvantage that the recording is very heavy and cannot be avoided by methods such as that suggested in connexion with the elimination method. It may be the best method for use with electronic computing machinery.

6. Other methods involving the triangular resolution

There are several ways of obtaining the triangular resolution. When it has been obtained, it can be used for the solution of sets of equations, or for the inversion of the matrix as has been described under the elimination method. Possible methods of resolution are described below.

(1) We may use the formulae given in the proof of the theorem on triangular resolution. This involves $\frac{1}{3}n^3+O(n^2)$ multiplications, $n^2+O(n)$ recordings. This method is closely related to Choleski's method for symmetrical matrices ((7) below), and we may therefore describe it as the 'unsymmetrical Choleski method'.

(2) We may apply the elimination method, regarded as a means of obtaining the triangular resolution; see notes (1), (2), (6) on the elimination method.

(3) We may obtain simultaneously, and bit by bit, the four triangles $\mathbf{L}$, $\mathbf{L}^{-1}$, $\mathbf{U}$, $\mathbf{U}^{-1}$ and the diagonal $\mathbf{D}$. The method makes use of the following simple facts about triangles:

(*a*) If we wish to invert a triangle, but only know the values in a subtriangle, we can obtain the coefficients of the inverse in the corresponding subtriangle: for example, if we know the first 5 rows of a lower triangle $\mathbf{L}$, then we can obtain the first 5 rows of $\mathbf{L}^{-1}$.

(*b*) If we know the first r columns of a unit lower triangle then we know its first $r+1$ rows: likewise, if we know the first r rows of a unit upper triangle we know also its first $r+1$ columns.

Let us suppose that we have carried the process to the point of knowing the first r rows of $\mathbf{L}$, the first $r-2$ of $\mathbf{L}^{-1}$ and $r-1$ of $\mathbf{U}$ and $\mathbf{U}^{-1}$. We carry on the inversion of $\mathbf{L}$ to obtain the $(r-1)$th and rth rows of $\mathbf{L}^{-1}$, and then multiply these rows into $\mathbf{A}$ to obtain the rth and $(r-1)$th rows of $\mathbf{L}^{-1}\mathbf{A}$, i.e. of $\mathbf{DU}$. From this we obtain at once the rth and $(r-1)$th rows of $\mathbf{D}$, and dividing obtain the rth and $(r-1)$th rows of $\mathbf{U}$. By (*b*) we have the rth and $(r+1)$th columns of $\mathbf{U}$ and by (*a*) obtain those of $\mathbf{U}^{-1}$. Multiplying we obtain the rth and $(r+1)$th columns of $\mathbf{AU}^{-1}$, i.e. of $\mathbf{LD}$, and from this the rth and $(r+1)$th elements of $\mathbf{D}$ and columns of $\mathbf{L}$. By (*b*) we have the $(r+1)$th and $(r+2)$th rows of $\mathbf{L}$.

We can, of course, arrange to increase r by 1 instead of 2 at each stage.

This is essentially Morris's escalator method (ref. **4**), so called because by breaking off the work at any stage we obtain the solution for one of the principal minors of $\mathbf{A}$; the order of the minor increases in steps. Morris's method differs in one small point. The diagonal elements $\mathbf{D}$ are not obtained as the diagonal of $\mathbf{L}^{-1}\mathbf{A}$ or of $\mathbf{AU}^{-1}$, but by using the identity $d_k = a_{kk} - \sum_{i<k} (\mathbf{AU}^{-1})_{ki} d_i^{-1} (\mathbf{L}^{-1}\mathbf{A})_{ik}$, which follows from the (kk) coefficient of the matrix equation $\mathbf{A} = (\mathbf{AU}^{-1})\mathbf{D}^{-1}(\mathbf{L}^{-1}\mathbf{A})$.

If Morris's method is used for the inversion of a matrix the work involved consists of $\frac{5}{3}n^3+O(n^2)$ multiplications (two triangle inversions each $\frac{1}{6}n^3+O(n^2)$, two multiplications of a triangle by $\mathbf{A}$, each $\frac{1}{2}n^3+O(n^2)$, and one multiplication of two triangles of opposite type, $\frac{1}{3}n^3+O(n^2)$), and $3n^3+O(n^2)$ recordings (this can be slightly reduced). It does not appear to be especially satisfactory in either respect.

To relate the above account to Morris's put

$q_k = d_k, \quad x_i = (U^{-1})_{1i}, \quad y_i = (U^{-1})_{2i}, \ldots, \quad x_i' = (L^{-1})_{i1}, \quad y_i' = (L^{-1})_{i2}, \ldots.$

(4) We may look for an upper triangular matrix $\mathbf{M}$ such that

$$\mathbf{M^*A^*AM} = 1,$$ ⟦11⟧

that is, so that $\mathbf{AM}$ is orthogonal. From the first r rows of $\mathbf{M}$ (which are also the first r columns of $\mathbf{M}^*$) we can obtain the first r rows of $\mathbf{M}^*$ because ⟦12⟧
of its triangular character, and hence the corresponding rows of $\mathbf{M^*A^*}$ and $\mathbf{M^*A^*A}$. The equation $\mathbf{M^*A^*A.M} = 1$ is then applied, using the first r columns in the $(r+1)$th row of the product. This determines the ratios of the coefficients of $\mathbf{M}$ in the $(r+1)$th row. The $(r+1)$th diagonal element of the equation then determines the multiplying factor. Having found $\mathbf{M}$ and $\mathbf{AM}$ we obtain the inverse as $\mathbf{M(AM)^*}$, or we may solve $\mathbf{Ax} = \mathbf{b}$ by forming $\mathbf{(AM)^*b}$ and then $\mathbf{M(AM)^*b}$. In the terminology of orthogonal vectors, as described below, the formation of $\mathbf{(AM)^*b}$ would be 'expressing $\mathbf{b}$ in terms of the base of orthogonal vectors'.

This method is the orthogonalization process described in ref. (**3**), p. 9. It is closely related to the Morris method for symmetrical matrices (see (5) below). We may apply Morris's method by forming $\mathbf{A^*A}$ and then looking for the upper triangular matrix $\mathbf{M}$ to satisfy $\mathbf{M^*A^*AM} = 1$. This would only involve $\mathbf{A}$ through the formation of $\mathbf{A^*A}$ and hence of $\mathbf{MA^*A}$. Thus Morris's method applied to the normalized matrix $\mathbf{A^*A}$ differs from the orthogonalization process only in that $\mathbf{M^*A^*A}$ is obtained as $\mathbf{M^*(A^*A)}$ instead of as $\mathbf{(M^*A^*)A}$.

We now come to methods for symmetrical matrices. These can all be made to provide methods for unsymmetrical matrices by normalizing the given matrix, that is, forming $\mathbf{AA^*}$ from $\mathbf{A}$. For instance, if we wish to solve $\mathbf{Ax} = \mathbf{b}$, we may form $\mathbf{A^*A}$ and $\mathbf{A^*b}$, and then solve $\mathbf{A^*Ax} = \mathbf{A^*b}$ by one of these methods. This normalizing technique is, however, of doubtful value. The formation of $\mathbf{A}^*\mathbf{A}$ involves $\frac{1}{2}n^3+O(n^2)$ multiplications, so that the work involved is greater with normalization than without, in the case of solving equations, and is no less for the case of inversion. Moreover, normalizing tends to make equations more 'ill-conditioned' (see § 8 below).

(5) A scheme mentioned in note (8) under the elimination method.

(6) We may apply the method (1), but we shall only need to find $\mathbf{L}$ and $\mathbf{D}$, since $\mathbf{U} = \mathbf{L}^*$. As a slight variation we may find $\mathbf{LD}$.

(7) Another variation on (6) is to find $\mathbf{LD}^{\frac{1}{2}}$. This method is due to Choleski (ref. **1**). The matrix $\mathbf{LD}^{\frac{1}{2}}$ may involve some pure imaginary numbers, but no strictly complex ones.

(8) Morris's method simplifies considerably for symmetric matrices. From the first r rows of $\mathbf{L}$ we can obtain the first r columns of $\mathbf{L}^{*-1}$, i.e. $\mathbf{U}^{-1}$, by inverting. Left-multiplication by $\mathbf{A}$ gives the first r columns of $\mathbf{AU}^{-1}$, i.e. of $\mathbf{LD}$, and from this we obtain the first $(r+1)$ rows of $\mathbf{L}$. Again Morris obtains $\mathbf{D}$ differently.

This method is identical with a variation of the orthogonalization method, applicable to symmetric matrices and due to L. Fox (ref. **2**). Fox regards two vectors $\mathbf{b}$ and $\mathbf{c}$ as 'orthogonal' relative to $\mathbf{A}$ if $(\mathbf{c}, \mathbf{Ab}) = 0$ (scalar product). Fox finds a set of vectors $\mathbf{v}_1, \mathbf{v}_2,..., \mathbf{v}_n$ which are orthogonal in this sense. The vectors $\mathbf{Av}_r$ may be used as a base for other vectors: we have in fact

$$\mathbf{b} = \sum_r \frac{(\mathbf{b}, \mathbf{v}_r)}{(\mathbf{v}_r, \mathbf{Av}_r)} \mathbf{Av}_r.$$

The solution of equations is effected by means of the formula

$$\mathbf{A}^{-1}\mathbf{b} = \sum_r \frac{(\mathbf{b}, \mathbf{v}_r)}{(\mathbf{v}_r, \mathbf{Av}_r)} \mathbf{v}_r.$$

It is best to obtain $\mathbf{v}_1, \mathbf{v}_2,..., \mathbf{v}_n$ by orthogonalizing the unit coordinate-axis vectors, that is, besides the vectors being orthogonal, $\mathbf{v}_r$ is restricted to be a linear combination of $\mathbf{e}_1, \mathbf{e}_2,..., \mathbf{e}_n$, or in other words, to have all coefficients after the rth equal to 0. In this case the vectors $\mathbf{v}_r$ are the rows of $\mathbf{L}^{-1}$, and the orthogonality relation is $\mathbf{L}^{-1}(\mathbf{AL}^{-1*}) = \mathbf{D}$. The orthogonalization process by which $\mathbf{L}^{-1}$ is found is identical with the inversion of $\mathbf{AL}^{-1*}\mathbf{D}^{-1}$.

7. Measure of the magnitude of a matrix

There are a number of ways in which the magnitude of a matrix may be measured by a real number. They include:

The norm. The norm $N(\mathbf{A})$ of the matrix $\mathbf{A}$ is given by

$$N(\mathbf{A}) = (\text{trace } \mathbf{A}^*\mathbf{A})^{\frac{1}{2}} = \Big(\sum_{i,j} a_{ij}^2\Big)^{\frac{1}{2}}.$$

The maximum expansion $B(\mathbf{A})$. This is given by

$$B(\mathbf{A}) = \max_{\mathbf{x}} \frac{|\mathbf{Ax}|}{|\mathbf{x}|} = \max_{\mathbf{x}} \frac{(\mathbf{Ax}, \mathbf{Ax})^{\frac{1}{2}}}{(\mathbf{x}, \mathbf{x})^{\frac{1}{2}}}.$$

The maximum coefficient $M(\mathbf{A})$. This is the largest coefficient in the matrix:

$$M(\mathbf{A}) = \max_{i,j} |a_{ij}|.$$

Of these measures one of the first two above is probably of greatest theoretical significance. In this paper we deal chiefly with the maximum coefficient, since it is the most easily computed.

A number of inequalities relating these are listed below.

$$M(\mathbf{X}+\mathbf{Y}) \leqslant M(\mathbf{X})+M(\mathbf{Y}) \tag{7.1}$$

$$M(\mathbf{XY}) \leqslant nM(\mathbf{X})M(\mathbf{Y}) \tag{7.2}$$

$$B(\mathbf{X}+\mathbf{Y}) \leqslant B(\mathbf{X})+B(\mathbf{Y}) \tag{7.3}$$

$$B(\mathbf{XY}) \leqslant B(\mathbf{X})B(\mathbf{Y}) \tag{7.4}$$

$$N(\mathbf{X}+\mathbf{Y}) \leqslant N(\mathbf{X})+N(\mathbf{Y}) \tag{7.5}$$

$$N(\mathbf{XY}) \leqslant N(\mathbf{X})N(\mathbf{Y}) \tag{7.6}$$ [[13]]

$$N(\mathbf{X}) \leqslant nM(\mathbf{X}) \tag{7.7}$$

$$M(\mathbf{X}) \leqslant N(\mathbf{X}) \tag{7.8}$$

$$M(\mathbf{X}) \leqslant B(\mathbf{X}) \tag{7.9}$$

$$B(\mathbf{X}) \leqslant n^{\frac{1}{2}}M(\mathbf{X}) \tag{7.10}$$ [[14]]

$$B(\mathbf{X}) \leqslant N(\mathbf{X}) \tag{7.11}$$

$$N(\mathbf{X}) \leqslant n^{\frac{1}{2}}B(\mathbf{X}) \tag{7.12}$$

8. Ill-conditioned matrices and equations

When we come to make estimates of errors in matrix processes we shall find that the chief factor limiting the accuracy that can be obtained is 'ill-conditioning' of the matrices involved. The expression 'ill-conditioned' is sometimes used merely as a term of abuse applicable to matrices or equations, but it seems most often to carry a meaning somewhat similar to that defined below.

Consider the equations

$$\left.\begin{aligned} 1{\cdot}4x+0{\cdot}9y &= 2{\cdot}7 \\ -0{\cdot}8x+1{\cdot}7y &= -1{\cdot}2 \end{aligned}\right\} \tag{8.1}$$

and form from them another set by adding one-hundredth of the first to the second, to give a new equation replacing the first

$$\left.\begin{aligned} -0{\cdot}786x+1{\cdot}709y &= -1{\cdot}173 \\ -0{\cdot}800x+1{\cdot}700y &= -1{\cdot}200 \end{aligned}\right\}. \tag{8.2}$$

The set of equations (8.2) is fully equivalent to (8.1), but clearly if we attempt to solve (8.2) by numerical methods involving rounding-off errors

we are almost certain to get much less accuracy than if we worked with equations (8.1). We should describe the equations (8.2) as an *ill-conditioned* set, or, at any rate, as ill-conditioned compared with (8.1). It is characteristic of ill-conditioned sets of equations that small percentage errors in the coefficients given may lead to large percentage errors in the solution. If we are required to solve the equations $\mathbf{Ax} = \mathbf{b}$, but the coefficients used are those of $\mathbf{A}-\mathbf{S}$ instead of those of $\mathbf{A}$, $\mathbf{S}$ being a small matrix, then, to
⟦15⟧ first order in $\mathbf{S}$, the solution obtained will be $\mathbf{x}_0+\mathbf{A}^{-1}\mathbf{S}\mathbf{x}_0$, where $\mathbf{x}_0$ is the correct solution. We may average the effect of this over a random population of matrices $\mathbf{S}$, and over the coefficients in the solution and matrix, and we shall find the

⟦16⟧
$$\frac{\text{R.M.S. error of coefficients of solution}}{\text{R.M.S. coefficient of solution}} = \frac{1}{n}N(\mathbf{A})N(\mathbf{A}^{-1})\frac{\text{R.M.S. error of coefficients of } \mathbf{A}}{\text{R.M.S. coefficient of } \mathbf{A}}.$$

This equation suggests that we might take either $N(\mathbf{A})N(\mathbf{A}^{-1})$ or $\frac{1}{n}N(\mathbf{A})N(\mathbf{A}^{-1})$ as a measure of the degree of ill-conditioning in a matrix. We will adopt the latter and call $\frac{1}{n}N(\mathbf{A})N(\mathbf{A}^{-1})$ *the N-condition number of* $\mathbf{A}$. We will also use $nM(\mathbf{A})M(\mathbf{A}^{-1})$ as another measure of ill-conditioning and call it *the M-condition number of* $\mathbf{A}$. There is substantial agreement between the two measures, though the M-number tends to be the larger, especially with diagonal or nearly diagonal matrices.

It should be noted that if all the coefficients of a matrix are multiplied by the same factor the condition numbers are unaltered, but that if a row or column is multiplied by a very large or a very small number the condition numbers are usually increased. For instance, the matrices

$$\begin{pmatrix} 0{\cdot}8 & 0{\cdot}6 \\ -0{\cdot}6 & 0{\cdot}8 \end{pmatrix} \quad (8.3) \qquad \text{and} \qquad \begin{pmatrix} 0{\cdot}008 & 0{\cdot}006 \\ -0{\cdot}6 & 0{\cdot}8 \end{pmatrix} \quad (8.4)$$

have the M-condition numbers 1·28 and 128 respectively and N-condition numbers 1 and 50·005. This may be considered quite a satisfactory example of the application of the definition. In practice one will tend to work with the same number of figures throughout a matrix, and the small values in the first row of (8.4) will prejudice the accuracy obtainable, because of the number of significant figures available. It is certainly true that a trivial modification improves the conditioning, but we should consider that until the possibility of this modification has been observed and action taken, the matrix remains ill-conditioned.

It is often stated that ill-conditioned matrices are ones which have small determinants, that is, small considering the magnitudes of the coefficients. This statement contains a certain amount of truth. It is certainly the case that bad conditioning and small determinants tend to go together. However, the determinant may differ very greatly from the above-defined condition numbers as a measure of conditioning. This may be illustrated by the cases of the matrices

$$\begin{pmatrix} 1 & 0 & 0 \\ 0 & 0{\cdot}1 & 0 \\ 0 & 0 & 0{\cdot}1 \end{pmatrix}; \begin{pmatrix} 1 & 0 & 0 \\ 0 & 1 & 1 \\ 0 & 0 & 0{\cdot}01 \end{pmatrix}; \begin{pmatrix} 1 & 1 & 1 \\ 1 & 1{\cdot}1 & 1 \\ 1 & 1 & 1{\cdot}1 \end{pmatrix}; \begin{pmatrix} 1 & 1 & 1 \\ 1 & 2 & 1 \\ 1 & 1 & 1{\cdot}01 \end{pmatrix}$$

all of which have the determinant 0·01, and which have the M-condition numbers 30, 300, 69·3, 612, respectively, and N-condition numbers 4·77, 47·1, 33·0, 232.

The best conditioned matrices are the orthogonal ones, which have N-condition numbers of 1. Their M-condition numbers are mostly of the order of magnitude of $\log n$ (for large order n). If the coefficients of a matrix are chosen at random from a normal population we shall get N-condition numbers of the order of $n^{\frac{1}{2}}$ and M-condition numbers about $\log n$ times greater. Thus random matrices are only slightly ill-conditioned.

The matrices which occur in practical problems are by no means random in this sense. There is a very large class of problems which naturally give rise to highly ill-conditioned equations. Suppose, for example, that we have reason to believe that some function of position in two dimensions can be represented by a polynomial of the fourth degree and that we wish to determine the coefficients. To this end we measure the values of the function at 25 points, and so obtain 25 linear equations for the desired coefficients. It may well happen that we are only able to make the measurements within a small region, and this will certainly mean that the equations are ill-conditioned. In such a case the equations might be improved by a differencing procedure, but this will not necessarily be the case with all problems. Preconditioning of equations in this way will always require considerable liaison between the experimenter and the computer, and this will limit its applicability.

9. The classical iterative method

Suppose that $\mathbf{B}$ is an approximate inverse of $\mathbf{A}$. Then we can obtain from it a better inverse $\mathbf{B}_2$ by the formula $\mathbf{B}_2 = 2\mathbf{B}-\mathbf{BAB}$. If we write $\mathbf{E} = 1-\mathbf{AB}$, $\mathbf{E}_2 = 1-\mathbf{AB}_2$, so that $\mathbf{E}$ and $\mathbf{E}_2$ give a measure of the incorrectness of the two inverses: we have $\mathbf{E}_2 = \mathbf{E}^2$, so that at each application of this process the error is essentially squared.

The work involved in applying this method is considerable, since it involves $2n^3$ multiplications at each stage. It may be useful in cases where a good approximate inverse is already available, and $1-\mathbf{AB}$ has already been calculated, but found to be a little larger than can be tolerated. We may then calculate $\mathbf{B}_2$ but carry the process no farther. This involves n^3 multiplications, but since we may write $\mathbf{B}_2 = \mathbf{B}+\mathbf{BE}$, the number of figures in one of the factors (viz. in $\mathbf{E}$) may be kept small.

A somewhat similar type of method applies for the improvement of solutions of sets of equations. Suppose, for example, we have to solve the equations $\mathbf{Ax} = \mathbf{b}$ and that we have obtained a resolution $\mathbf{A} = \mathbf{L}.\mathbf{DU}$ (say), somewhat inaccurately. By double back-substitution we obtain a solution $\mathbf{x}_1$ of $\mathbf{L}.\mathbf{DUx} = \mathbf{b}$, which is an inaccurate solution of $\mathbf{Ax} = \mathbf{b}$. We may further test this solution by forming the 'residual' vector $\mathbf{b}_1 = \mathbf{b}-\mathbf{Ax}_1$, and if this is too large we solve $\mathbf{Ax} = \mathbf{b}_1$ to obtain a correction. In this process we do not obtain 'quadratic convergence' but only convergence in geometric progression. On the other hand, the method is very practical because the work involved per stage is only $2n^2$ multiplications.

10. General remarks on error estimates. The error in a reputed inverse

Error estimates can be of two kinds. We may wish to know how accurate a certain result is, and be willing to do some additional computation to find out. A different kind of estimate is required if we are planning calculations and wish to know whether a given method will lead to accurate results. In the former case we do not care what quantities the error is expressed in terms of, provided they are reasonably easily computed. With these estimates we wish to be absolutely sure that the error is within the range stated, but at the same time not to state a range which is very much larger than necessary. With the second type of estimate, the error is preferably expressed in terms of quantities whose meaning is sufficiently familiar that the general run of values involved may at least be guessed at. We are also as much interested in the statistical behaviour of the errors as in the maximum possible value.

This paper is mainly concerned with estimates of the second kind, since those of the first kind can be quickly dismissed. Let $\mathbf{B}$ be a reputed inverse of $\mathbf{A}$. To determine its accuracy we form $\mathbf{E} = 1-\mathbf{AB}$. Then in view of the inequalities (7.1), (7.2), and the equation

$$\mathbf{A}^{-1}-\mathbf{B} = \mathbf{B}(\mathbf{E}+\mathbf{E}^2+\ldots)$$

we have

$$M(\mathbf{B}-\mathbf{A}^{-1}) \leqslant \sum_{r=1}^{\infty} M(\mathbf{BE}^r) \leqslant \sum_{r=1}^{\infty} n^r M(\mathbf{B})\{M(\mathbf{E})\}^r = \frac{nM(\mathbf{B})M(\mathbf{E})}{1-nM(\mathbf{E})}, \quad [17]$$

which is the required error estimate. In order to apply this inequality it is necessary to carry out the matrix multiplication **BA**, involving n^3 multiplications. However, if it is intended to apply the classical iteration method for improving the inverse at least once, we shall have to calculate E in doing so, and we shall have $1-\mathbf{AB}_2 = \mathbf{E}_2 = \mathbf{E}^2$ and therefore

$$M(\mathbf{B}_2-\mathbf{A}^{-1}) \leqslant \frac{nM(\mathbf{B}_2)M(\mathbf{E}_2)}{1-nM(\mathbf{E}_2)} \leqslant \frac{n^2M(\mathbf{B}_2)\{M(\mathbf{E})\}^2}{1-n^2\{M(\mathbf{E})\}^2}.$$

It should be observed that this inequality is only applicable to the inversion of a matrix, and not to the solution of equations. It is difficult to determine the accuracy of the solution of a set of equations without inverting the matrix. This is another reason why it is preferable to treat inversion rather than solution of equations as a standard process.

When making estimates of the effects of rounding-off errors we need the process under examination to be rather minutely described. If, for instance, a product abc is to be formed, we need to know whether it is obtained as $ab.c$ or as $a.bc$. If it is obtained as $ab.c$ we shall need to know how many figures are kept in ab. This may be either a definite number of decimal or binary places, or a definite number of significant figures, or the number of figures kept may be made to depend on the results of previous calculations. Usually, however, by a trivial modification of the quantity recorded, these latter cases can be reduced to one of the former.

The variety of possible detailed calculation procedures is, of course, vastly greater than the list of methods which we have considered, for these can be subdivided into numerous alternatives which appear only trivially different at first sight, but which may differ very seriously from the point of view of error estimates. We cannot here carry out the analysis for more than a very few of the procedures. These have been chosen so as to give bounds of error which are both reasonably small and also fairly simple in their analytical form. We have concentrated particularly on error estimates which can be expressed in terms of the matrix **A** and its inverse. In practical work the details of the procedure must be determined by other considerations. With any particular procedure it will usually be found possible to obtain some estimate of the type proved in this paper, but usually quantities such as $M(\mathbf{L})$, $M(\mathbf{D}^{-1})$, etc., will be involved. These can be obtained conveniently as a by-product in the calculation. Alternatively, one may find bounds of error by calculating $1-\mathbf{AB}$ as above. In this case the importance of the analysis which follows is to show that it is probable that the error obtained will be reasonably small if a process is used which is somewhat similar to one of those here considered, and that these methods are therefore reasonable ones to use. Our main purpose in this paper is to establish that the exponential build-up of errors need not occur, and this will be proved when we have found one method of inversion where it is absent.

11. Rounding-off errors in Jordan's method

The Jordan method was described in § 5, but we have now to specify the details of the rounding-off and the diagonal. We shall consider the case where $\mathbf{A}$ is reduced to a unit matrix. We assume that in the calculation of each quantity

⟦18⟧ $$\mathbf{A}_{ij}^{(r-1)} - \frac{\mathbf{A}_{rj}^{(r-1)}\mathbf{A}_{ir}^{(r-1)}}{\mathbf{A}_{rr}^{(r-1)}},$$

an error of at most ϵ is made. How this is to be secured need not be specified, but it is clear that the number of figures to be retained in $\mathbf{A}_{ir}^{(r-1)}/\mathbf{A}_{rr}^{(r-1)}$ will have to depend on the values of the $\mathbf{A}_{rj}^{(r-1)}$. Likewise, we assume that in the calculation of

$$\mathbf{X}_{ij}^{(r-1)} - \frac{\mathbf{X}_{rj}^{(r-1)}\mathbf{A}_{ir}^{(r-1)}}{\mathbf{A}_{rr}^{(r-1)}}$$

an error of at most ϵ' is made. It is convenient to think of these errors as quantities deliberately added after the accurate calculation has been made. If the quantities added after the calculation of $\mathbf{A}^{(r)}$, $\mathbf{X}^{(r)}$ are the matrices $\mathbf{S}_r$, $\mathbf{S}'_r$ we shall have

$$\begin{aligned} \mathbf{J}_n[...\{\mathbf{J}_2(\mathbf{J}_1\mathbf{A}+\mathbf{S}_1)+\mathbf{S}_2\}...]+\mathbf{S}_n &= 1, \\ \mathbf{J}_n[...\{\mathbf{J}_2(\mathbf{J}_1+\mathbf{S}'_1)+\mathbf{S}'_2\}...]+\mathbf{S}'_n &= \Xi, \end{aligned} \tag{11.1}$$

where Ξ represents the actual matrix obtained at the end of the calculation as the value of $\mathbf{A}^{-1}$.

The equations (11.1) give us

$$\begin{aligned} \mathbf{A}+\sum \mathbf{X}_r^{-1}\mathbf{S}_r &= \mathbf{X}_n^{-1}, \\ 1+\sum \mathbf{X}_r^{-1}\mathbf{S}'_r &= \mathbf{X}_n^{-1}\Xi \end{aligned} \tag{11.2}$$

⟦19⟧ and hence $$\Xi = \Big(1+\mathbf{A}^{-1}\sum_r \mathbf{X}_r^{-1}\mathbf{S}_r\Big)\mathbf{A}^{-1}\Big(1+\sum_r \mathbf{X}_r^{-1}\mathbf{S}'_r\Big). \tag{11.3}$$

The matrix $\mathbf{X}_r\mathbf{A}$ is the result of the first r stages of the reduction of $\mathbf{A}$ and agrees with $\mathbf{D}$ in the first r columns. This fact may be expressed in the equation

$$(\mathbf{X}_r\mathbf{A}-1)\mathbf{I}_r = 0, \tag{11.4}$$

where $\mathbf{I}_r$ is that matrix which agrees with the unit matrix in the first r
⟦20⟧ columns and with the zero matrix elsewhere. It is also clear that $\mathbf{X}_r$ differs

from the unit matrix only in the first r columns; this fact may be expressed in the equation

$$(\mathbf{X}_r-1)(1-\mathbf{I}_r) = 0. \tag{11.5}$$

From (11.4) and (11.5) we now find $\mathbf{X}_r^{-1}$;

$$\mathbf{X}_r^{-1} = \mathbf{A}\mathbf{I}_r+1-\mathbf{I}_r. \tag{11.6}$$ [[21]]

When we ignore the second-order terms in the rounding-off errors (11.3), (11.6) give us

$$\Xi-\mathbf{A}^{-1} = -\mathbf{A}^{-1}\Big(\sum_r \mathbf{X}_r^{-1}\mathbf{S}_r\Big)\mathbf{A}^{-1}+\mathbf{A}^{-1}\sum_r \mathbf{X}_r^{-1}\mathbf{S}_r'$$ [[22]]

$$= \sum_r \{\mathbf{I}_r+\mathbf{A}^{-1}(1-\mathbf{I}_r)\}(\mathbf{S}_r\mathbf{A}^{-1}-\mathbf{S}_r'). \tag{11.7}$$

Let us now assume that each coefficient $\mathbf{S}_r$ is at most ϵ and each coefficient of $\mathbf{S}_r'$ at most ϵ'. From (11.7) we can estimate the error in M-measure

$$M(\Xi-\mathbf{A}^{-1}) \leqslant \sum_r n\{1+M(\mathbf{A}^{-1})\}M(\mathbf{S}_r\mathbf{A}^{-1}-\mathbf{S}_r')$$ [[23]]

$$\leqslant \sum_r n\{1+M(\mathbf{A}^{-1})\}\{n\epsilon M(\mathbf{A}^{-1})+\epsilon'\}$$

$$\leqslant n^2\{1+M(\mathbf{A}^{-1})\}\{\epsilon'+n\epsilon M(\mathbf{A}^{-1})\}, \tag{11.8}$$

or in B-measure,

$$B(\Xi-\mathbf{A}^{-1}) \leqslant \sum_r B\{\mathbf{I}_r+\mathbf{A}^{-1}(1-\mathbf{I}_r)\}\{B(\mathbf{S}_r\mathbf{A}^{-1})+B(\mathbf{S}_r')\}$$

$$\leqslant \sum_r \{1+B(\mathbf{A}^{-1})\}\{\epsilon n^{\frac{1}{2}}B(\mathbf{A}^{-1})+\epsilon' n^{\frac{1}{2}}\}$$ [[24]]

$$\leqslant n^{\frac{3}{2}}\{1+B(\mathbf{A}^{-1})\}\{\epsilon B(\mathbf{A}^{-1})+\epsilon'\}, \tag{11.9}$$

or in N-measure,

$$N(\Xi-\mathbf{A}^{-1}) \leqslant \sum_r N\{\mathbf{I}_r+\mathbf{A}^{-1}(1-\mathbf{I}_r)\}\{N(\mathbf{S}_r\mathbf{A}^{-1})+N(\mathbf{S}_r')\}$$

$$\leqslant \sum_r \{r^{\frac{1}{2}}+(1-r)^{\frac{1}{2}}N(\mathbf{A}^{-1})\}\{n\epsilon N(\mathbf{A}^{-1})+n\epsilon'\}$$ [[25]]

$$\leqslant \tfrac{2}{3}(n+1)^{\frac{3}{2}}\{1+N(\mathbf{A}^{-1})\}\{\epsilon'+\epsilon N(\mathbf{A}^{-1})\}. \tag{11.10}$$ [[26]]

If we use the relations $\mathbf{S}_r\mathbf{I}_r = \mathbf{S}_r'(1-\mathbf{I}_r) = 0$, which follow from the restrictions on the coefficients which can suffer rounding-off errors, (11.8) may be improved to

$$M(\Xi-\mathbf{A}^{-1}) \leqslant n\epsilon'+\frac{n(n-1)}{2}M(\mathbf{A}^{-1})\Big\{\epsilon+\epsilon'+\frac{2n-1}{3}\epsilon M(\mathbf{A}^{-1})\Big\}. \tag{11.11}$$ [[27]]

This result is best possible in the sense that given ϵ, ϵ', M we can find $\mathbf{S}_r$, $\mathbf{S}_r'$, $\mathbf{A}$ so that $M(\mathbf{S}_r) \leqslant \epsilon$, $M(\mathbf{S}_r') \leqslant \epsilon'$, $M(\mathbf{A}^{-1}) = M$ and the error $M(\Xi-\mathbf{A}^{-1})$, still ignoring second-order terms, is exactly

$$n\epsilon'+\frac{n(n-1)}{2}M\Big(\epsilon+\epsilon'+\frac{2n-1}{3}\epsilon M\Big).$$

We may also use (11.7) to give us an estimate of the statistical error. Let the coefficients of the matrices $\mathbf{S}_1,\ldots, \mathbf{S}_n$ which are not obliged to be

0 be $s_1, \ldots, s_K$ in some order, and likewise let the coefficients of $\mathbf{S}'_1, \ldots, \mathbf{S}'_n$ which are not necessarily zero be $s_{K+1}, \ldots, s_P$. The equation (11.7) may then be put in the form

$$(\mathbf{\Xi}-\mathbf{A}^{-1})_{ij} = \sum_{u=1}^{P} c_{iju} s_u,$$

where c_{iju} depends only on the coefficients of $\mathbf{A}^{-1}$. Suppose that the rounding-off errors s_u are independent and have standard deviation σ_u and zero mean, then the mean square value of $(\mathbf{\Xi}-\mathbf{A}^{-1})_{ij}$ is $\sum_{u=1}^{P} c_{iju}^2 \sigma_u^2$. Let us put $\sigma_u = \eta$ for $u \leqslant K$, $\sigma_u = \eta'$ for $u > K$ and the mean square error in $\mathbf{A}_{ij}^{-1}$ becomes $\eta^2 \sum_{u=1}^{K} c_{iju}^2 + \eta'^2 \sum_{u=K+1}^{P} c_{iju}^2$. When we substitute in the correct values for c_{iju} we obtain:

mean square error in $(\mathbf{A}^{-1})_{ij}$

⟦28⟧
$$= \eta^2 \sum_{m,K} (\mathbf{A}^{-1})_{im}^2 (\mathbf{A}^{-1})_{Kj}^2 \min(K, i-1) + \eta^2 \sum_{K>i} (\mathbf{A}^{-1})_{Kj}^2 (K-i) + \\ + \eta'^2 \left[\sum_{m} (\mathbf{A}^{-1})_{im}^2 \min(j, m-1) + \frac{(n-1)(n-i+1)}{2} \right],$$

where η is the standard deviation and zero the mean of each coefficient of $\mathbf{S}_r$, and η' is the standard deviation and zero the mean of each coefficient of $\mathbf{S}'_r$.

Also

mean square error in $(\mathbf{A}^{-1})_{ij}$

$$\leqslant \eta^2 \left[\{M(\mathbf{A}^{-1})\}^4 \frac{n(n+1)(n-\frac{1}{2})}{3} + \{M(\mathbf{A}^{-1})\}^2 \frac{(n-1)(n-i+1)}{2} \right] + \\ + \eta'^2 \left[\{M(\mathbf{A}^{-1})\}^2 (n-\tfrac{1}{2}-\tfrac{1}{2}j) + \frac{(n-i)(n-i+1)}{2} \right].$$

The leading term in the R.M.S. error in $(\mathbf{A}^{-1})_{ij}$ is therefore at most

$$\eta \{M(\mathbf{A}^{-1})\}^2 \frac{n^{\frac{3}{2}}}{\sqrt{3}}.$$

The assumptions $M(\mathbf{S}_r) < \epsilon$, $M(\mathbf{S}'_r) < \epsilon'$ in the above analysis state in effect that we are working to a fixed number of decimal places both in the reduction of the original matrix to unity and in the building up of the inverse. It is not easy to obtain corresponding results for the case where a definite number of *significant* figures are kept, but we may make some qualitative suggestions.

⟦29⟧ The error when working with a fixed number of decimal places arose almost entirely from the reduction of the original matrix, and very little from the building up of the inverse. This, at any rate, applies for the inversion of ill-conditioned matrices with coefficients of moderate size.

However, the coefficients of the inverse are larger than those of the original matrix, so that if we work to the same number of significant figures in both we may expect the discrepancy to disappear. The general idea of this may be expressed by putting

$$M(\mathbf{S}_r) < \delta M(\mathbf{A}), \qquad M(\mathbf{S}'_r) < \delta' M(\mathbf{A}^{-1}),$$

so that $$\frac{M(\boldsymbol{\Xi}-\mathbf{A}^{-1})}{M(\mathbf{A}^{-1})} < n^3 M(\mathbf{A}) M(\mathbf{A}^{-1}) \left(1+\frac{1}{M(\mathbf{A}^{-1})}\right)\left(\delta+\frac{\delta'}{M(\mathbf{A})}\right).$$ ⟦30⟧

There still remains the factor $\frac{1}{M(\mathbf{A})}$ multiplying δ'. This could be removed by arranging to reduce $\mathbf{A}$, not to the unit matrix, $\mathbf{1}$, but to $M(\mathbf{A}).\mathbf{1}$. This would be a reasonable procedure in any case, though it would be more convenient to choose the nearest power of 10 to take the place of $M(\mathbf{A})$. We see now that it is the M-condition number $nM(\mathbf{A})M(\mathbf{A}^{-1})$ which determines the magnitude of the errors when we work to a definite number of figures.

In the case of positive definite, symmetric matrices it is possible to give more definite estimates for the case where calculation is limited to a specific number of significant figures. Results of this nature have been obtained by J. v. Neumann and H. H. Goldstine (ref. **6**).

It is instructive to compare the estimates of error given above with the errors liable to arise from the inaccuracy of the original matrix. If we desire the inverse of $\mathbf{A}$, but the figures given to us are not those of $\mathbf{A}$ but of $\mathbf{A}-\mathbf{S}$, then if we invert perfectly correctly we shall get $(\mathbf{A}-\mathbf{S})^{-1}$ instead of $\mathbf{A}^{-1}$, that is, we shall make an error of $(\mathbf{A}-\mathbf{S})^{-1}-\mathbf{A}^{-1}$, i.e. of

$$(1-\mathbf{A}^{-1}\mathbf{S})^{-1}\mathbf{A}^{-1}\mathbf{S}\mathbf{A}^{-1}.$$

If we ignore the second-order terms this is $\mathbf{A}^{-1}\mathbf{S}\mathbf{A}^{-1}$. The leading terms in the error in the Jordan method were $\mathbf{A}^{-1}\left(\sum_r (1-\mathbf{I}_r)\mathbf{S}_r\right)\mathbf{A}^{-1}$ so that we ⟦31⟧ might say that the greater part of the error is equal to that error which would have been produced by an original error in the matrix of $\sum_r (1-\mathbf{I}_r)\mathbf{S}_r$.

It is possible to give error estimates also for several others amongst the methods suggested elsewhere in this paper. This is, for instance, the case for the elimination method.

The elimination method in its first phase proceeds similarly to the Jordan process, but we only attempt to reduce $\mathbf{A}$ to a triangle and not to a diagonal: also the matrix representing the complete operation in this first phase is triangular.

12. Errors in the Gauss elimination process

We will consider the errors in the Gauss elimination process as consisting of two parts, one arising from the reduction of the matrix to the

triangular form, and the other from the back-substitution. Of these we are mainly interested in the error arising from the reduction, since this is the part of the process which has been most criticized. We adopt the description of the process given in § 4, note (1), and observe that apart from a slight difference in the form of the matrices $\mathbf{J}_r$, the reduction is similar to the Jordan process. As in the Jordan process, we shall assume that we make matrix errors $\mathbf{S}_1, \mathbf{S}_2, \ldots, \mathbf{S}_n$ in the various stages of the
⟦32⟧ reduction of $\mathbf{A}$, and vector errors $\mathbf{s}_1, \mathbf{s}_2, \ldots, \mathbf{s}_n$ in the operations on $\mathbf{b}$. Assuming there are no back-substitution errors, and ignoring the second-order terms in the errors we should have:

⟦33⟧
$$\text{error in } \mathbf{x} = \mathbf{U}^{-1}\mathbf{X}_n \sum_{r=1}^{n} \mathbf{X}_r^{-1}(\mathbf{s}_r' - \mathbf{S}_r \mathbf{U}^{-1}\mathbf{X}_n \mathbf{b}),$$

where $\mathbf{X}_r = \mathbf{J}_r \ldots \mathbf{J}_1$. Now, assuming that the process has been done with the largest pivot chosen from each column, we shall have $M(\mathbf{X}_r^{-1}) = 1$, for
⟦34⟧ $\mathbf{X}_r^{-1} = 1 + \sum_{s \leqslant r} (1 - \mathbf{J}_s)$ as mentioned in § 4 (2). Then

⟦35⟧
$$|\text{error in } \mathbf{x}_m| = |(\mathbf{A}^{-1} \sum \mathbf{X}_r^{-1}(\mathbf{s}_r' - \mathbf{S}_r \mathbf{A}^{-1}\mathbf{b}))_m|$$
$$= \left| \sum_{\substack{j,k,r \\ j \geqslant k}} (\mathbf{A}^{-1})_{mj} (\mathbf{X}_r^{-1})_{jk} (\mathbf{s}_r')_k - \sum_{l,p} (\mathbf{S}_r)_{kl} (\mathbf{A}^{-1})_{lp} \mathbf{b}_p \right|$$

⟦36⟧
$$\leqslant \frac{n^2(n+1)}{2} M(\mathbf{A}^{-1})\epsilon' + \frac{n^4(n+1)}{2} \{M(\mathbf{A}^{-1})\}^2 M(\mathbf{b})\epsilon,$$

where $M(\mathbf{s}_r') \leqslant \epsilon'$, $M(\mathbf{S}_r) \leqslant \epsilon$.

To these errors we have to add those which arise from the back-substitution. This consists in solving the equations $\mathbf{DUx} = \mathbf{L}^{-1}\mathbf{b}$, where $\mathbf{U}$ is unit upper triangular and $\mathbf{D}$ diagonal. We obtain x_n first and then x_r in order
⟦37⟧ of decreasing r by means of the formula $x_r = \mathbf{d}_r^{-1}(\mathbf{L}^{-1}\mathbf{b})_r - \sum_{i>r} (\mathbf{DU})_{ri} x_i$. Now if we make an error of t_r in the calculation of x_r from the previously obtained coefficients of $\mathbf{x}$, then we shall have solved accurately the
⟦38⟧ equations $\mathbf{DUx} = \mathbf{L}^{-1}\mathbf{b} + \mathbf{Dt}$, that is, we shall have introduced an error of $\mathbf{U}^{-1}\mathbf{t}$, or, since $\mathbf{A} = \mathbf{LDU}$, of $\mathbf{A}^{-1}\mathbf{LDt}$. If we arrange that $M|t_r| \leqslant \epsilon d_r^{-1}$,
⟦39⟧ the greatest error in any coefficient from this source is $n^2 M(\mathbf{A}^{-1})\epsilon$, and normally much smaller than the error arising from the first part of the process. Furthermore, d_r will normally tend to be less than 1.

It is interesting to note the value of the error in the last pivot, that is, the error in the (nn) coefficient of $\mathbf{J}_n \ldots \mathbf{J}_1 \mathbf{A}$. The matrix error in $\mathbf{J}_n \ldots \mathbf{J}_1 \mathbf{A}$ is $\mathbf{X}_n \sum_r \mathbf{X}_r^{-1}\mathbf{S}_r$, that is, since $\mathbf{X}_n \mathbf{L}^{-1} = \mathbf{DUA}^{-1}$, it is $\mathbf{DUA}^{-1} \sum_r \mathbf{X}_r^{-1}\mathbf{S}_r$. The
⟦40⟧ (nn) coefficient is $d_n \sum_r a_{nj}(\mathbf{X}_r^{-1}\mathbf{S}_r)_{jn}$ and since $M(\mathbf{X}_r^{-1}) = 1$, $M(\mathbf{S}_r) \leqslant \epsilon$ it does not exceed $n^2 d_n M(\mathbf{A}^{-1})\epsilon$ in absolute magnitude, that is, the proportionate error in the last pivot is at most $n^2 M(\mathbf{A}^{-1})\epsilon$. This cannot be very large

unless the matrix is ill-conditioned. With worst possible conditioning we find an error somewhat similar to Hotelling's estimate. The matrix error in $\mathbf{J}_n \ldots \mathbf{J}_1 \mathbf{A}$ may be written $\mathbf{L}^{-1} \sum_r \mathbf{X}_r^{-1} \mathbf{S}_r$, from which we find that the error in the last pivot cannot exceed $n^2 \epsilon M(\mathbf{L}^{-1})$. But since $M(\mathbf{L}) = 1$ we find $M(\mathbf{L}^{-1}) \leqslant 2^{n-1}$ (and equality can be attained): that this error may [[41]]
actually be as great as $2^{n-2}\epsilon$ may be seen by considering the inversion of a matrix differing only slightly from

$$\begin{bmatrix} 1 & 0 & 0 & . & . & . & 0 \\ -1 & 1 & 0 & . & . & . & 0 \\ -1 & -1 & 1 & . & . & . & 0 \\ . & . & . & . & . & . & . \\ -1 & -1 & -1 & . & . & . & 1 \end{bmatrix}$$

It appears then that the error in the last pivot can only be large if $\mathbf{L}^{-1}$ is large, and that this can only happen with ill-conditioned equations. Actually even then we may consider ourselves very unlucky if $\mathbf{L}^{-1}$ is large. Normally, even with ill-conditioned equations we may expect the off-diagonal coefficients of $\mathbf{L}$ to be distributed fairly uniformly between -1 and 1, possibly with a tendency to be near 0. Only when there is a strong tendency for negative values will we find a large $\mathbf{L}^{-1}$.

13. Errors in the unsymmetrical Choleski method

When obtaining the triangular resolution of a matrix by the method of the theorem (§ 3) it is convenient to think of the process as follows. We are given a matrix $\mathbf{A}$ and the matrices $\mathbf{L}$ and $\mathbf{DU}$ ($= \mathbf{W}$, say). We form the product $\mathbf{LW}$ coefficient by coefficient. When calculating any one of the coefficients of $\mathbf{LW}$, we always find that the data are incomplete to the extent of one number, and we therefore choose this number so as to give the required coefficient in $\mathbf{A}$. The unknown quantity when forming a_{ij} is always either l_{ij} or w_{ij}. Regarding the process in this way suggests the following rule for deciding the number of figures to be retained. We always retain sufficient figures to give us an error of not more than ϵ in the coefficient of $\mathbf{A}$ under consideration. In actual hand computation this rule is extremely simple to apply. Suppose, for example, that ϵ is $\frac{1}{2}10^{-7}$ and that we are forming the product $(\mathbf{LW})_{94}$, i.e. $\sum_{j=1}^{4} l_{9j} w_{j4}$. We first form $\sum_{j=1}^{3} l_{9j} w_{j4}$ accumulating the products in the machine. All the relevant quantities should be available at this stage. We then set up the multiplicand w_{44} which should also be known and 'turn the handle' until the quantity in the product register, rounded off to seven figures, first agrees with the given value of a_{94} (which is assumed to have zeros in the eighth and later figures). All the figures in the multiplier register are then written down as the value of l_{94}.

[[42]] The theory of the errors in this method is peculiarly simple. The triangular resolution obtained is an exact resolution of a matrix $\mathbf{A}-\mathbf{S}$, where $M(\mathbf{S}) < \epsilon$, and the resultant error in the inverse is $\mathbf{A}^{-1}\mathbf{S}\mathbf{A}^{-1}$, and in any coefficient at most $n^2\{M(\mathbf{A}^{-1})\}^2\epsilon$. A similar procedure is appropriate in the inversion of the triangles $\mathbf{L}$ and $\mathbf{W}$. When inverting $\mathbf{W}$ (say) we can arrange, by an exactly similar computing procedure, that its product with its reputed inverse differs from unity by at most ϵ' in each coefficient, i.e. $\mathbf{LK} = 1-\mathbf{S}'$, where $M(\mathbf{S}') < \epsilon$ and $\mathbf{K}$ is the reputed inverse. Note the order in the product which is significant. Likewise we find a reputed inverse $\mathbf{V}$ for $\mathbf{DU}$ such that $\mathbf{V}.\mathbf{DU} = 1-\mathbf{S}''$ and $M(\mathbf{S}'') < \epsilon'$. The error arising from using these reputed inverses is $-(1-\mathbf{S}'')^{-1}\mathbf{VK}(1-\mathbf{S}')+\mathbf{VK}$, or neglecting second-order terms, $\mathbf{S}''\mathbf{A}^{-1}+\mathbf{A}^{-1}\mathbf{S}'$. Finally, there is a possible source of error due to rounding off in the actual formation of the product $\mathbf{VK}$. If this does not exceed ϵ'' in any coefficient, the error in any coefficient of the reputed inverse of $\mathbf{A}$ is in all at most

$$n^2\epsilon\{M(\mathbf{A}^{-1})\}^2+2n\epsilon' M(\mathbf{A}^{-1})+\epsilon''.$$

This paper is published with the permission of the Director of the National Physical Laboratory.

REFERENCES

1. Commandant Bénoit, 'Note sur une méthode, etc.' (Procédé du Commandant Cholesky), *Bull. Géod.* (Toulouse), 1924, No. 2, 5–77.
2. L. Fox, H. D. Huskey, and J. H. Wilkinson, 'Notes on the solution of algebraic linear simultaneous equations', see above, pp. 149–73.
3. J. v. Neumann, V. Bargmann, and D. Montgomery, *Solution of Linear Systems of High Order*, lithographed, Princeton (1946).
4. J. Morris, 'An escalator method for the solution of linear simultaneous equations', *Phil. Mag.*, series 7, **37** (1946), 106.
5. H. Hotelling, 'Some new methods in matrix calculation', *Ann. Math. Stat.* **14** (1943), 34.
6. J. v. Neumann and H. H. Goldstine, 'Numerical inverting of matrices of high order', *Bull. Amer. Math. Soc.* **53** (1947), 1021–99.

ANNALS OF MATHEMATICS
Vol. 52, No. 2, September, 1950

THE WORD PROBLEM IN SEMI-GROUPS WITH CANCELLATION

BY A. M. TURING

(Received August 13, 1949)

It will be shown that the word problem in semi-groups with cancellation is not solvable. The method depends on reducing the unsolvability of the problem in question to a known unsolvable problem connected with the logical computing machines introduced by Post (Post, [1]) and the author (Turing, [1]). In this we follow Post (Post, [2]) who reduced the problem of Thue to this same unsolvable problem.

1. Semi-groups with cancellation

By a semi-group with cancellation we understand a set $\mathfrak{S}$ within which is defined a function $f(a, b)$ described as a product and satisfying

(i) The associative law $f(f(a, b), c) = f(a, f(b, c))$

(ii) The cancellation laws $f(a, b) = f(a, c) \supset b = c$

$$f(b, a) = f(c, a) \supset b = c$$

for any a, b, c in $\mathfrak{S}$.

In view of the associative law we naturally write *abc* for both $f(f(a, b), c)$ and $f(a, f(b, c))$, and similarly for strings of letters of any length. If A and B are two such strings (e.g. *arghm* and *gog*) then AB represents the result of writing one after the other (i.e. *arghmgog*). We have here used the convention (to be followed throughout) that capital letters are variables for strings of letters.

2. The word problem

One may construct a semi-group with cancellation by means of generators and relations as follows. We first define a certain finite set $\mathfrak{G}$ of letters to be the *generators*. Strings of such letters are described as *words*. A class $\mathfrak{F}$ of pairs of words is chosen and described as the *fundamental relations* (abbreviated sometimes to F.R.). The words are understood to represent members of a semi-group, and pairs forming a fundamental relation to represent equal members, but no pairs of words to represent the same member unless obliged to by this condition. The object of the 'word problem' is to find a method for determining whether a given pair of words do or do not represent the same semi-group element. Such a pair of words will be described as a *relation* of the semi-group.

Treating the above explanation of the meaning of a relation as relatively informal and intuitive, we may also give a more formal definition which is equivalent to the first. This greater formality would in many applications tend to dryness and obscurity, but for our present purpose it gives greater clarity, since we are not concerned so much with obtaining relations as with showing that certain relations cannot be obtained. This requires a very unambiguous form of definition.

Immediate deducibility and assertibility

We say that

(i) Any pair in $\mathfrak{F}$ is immediately assertible.

(ii) Any pair of form (A, A) is immediately assertible.

(iii) (A, B) is immediately deducible from (B, A).

(iv) (A, B) is immediately deducible from (A, C) and (C, B).

(v) (AG, BG) is immediately deducible from (A, B) for any generator G.

(vi) (GA, GB) is immediately deducible from (A, B) for any generator G.

(vii) (A, B) is immediately deducible from (AG, BG) for any generator G.

(viii) (A, B) is immediately deducible from (GA, GB) for any generator G.

Note that we use a capital for G. If g is a generator it is a quite definite one.

Immediate deducibility and assertibility apply in no other cases but those listed above.

Semi-group relations

A pair (A, B) is a relation of the semi-group $\mathfrak{S}$ arising from the fundamental relations $\mathfrak{F}$ if there is a 'proof' of (A, B) consisting of a sequence of pairs of which each is either immediately assertible or immediately deducible from previous pairs of the sequence, and the last pair is (A, B).

We wish to find out whether the word problem for semi-groups is solvable or not. The possibilities may be divided into three alternatives.

(a) There might be a general method applicable for all sets $\mathfrak{F}$ of fundamental relations, and all pairs (A, B).

(b) For any given set of fundamental relations there might be a special method which could be used, but no general method applicable for all sets.

(c) There may be some particular set of F.R. such that no method will apply with it: i.e., any reputed method will give the wrong answer for some pair of words.

We shall show that (c) is the correct alternative.

3. Computing machines

As in Turing [1], Post [1], we identify the existence of a 'general method for determining . . . ' with that of a 'computing machine constructed to determine . . . '. Numerous alternative definitions have been given (Church [1], Kleene [1], see also Hilbert and Bernays, Appendix to vol. II) and proved equivalent (Kleene [2], Turing [2]).

A computing machine is imagined as a mechanical device capable of a finite number of states or *internal configurations* (I.C.) $q_1, q_2, \cdots, q_r$, working on a (both ways) infinite tape divided into squares on which symbols are printed. These symbols are drawn from the set $s_0, s_1, s_2, \cdots, s_n$. At any moment one particular square is especially closely connected with the machine. It is called the *scanned square*. The behavior of the machine at this moment is determined by its own I.C. and the symbol on the scanned square. The behavior consists in altering this symbol and possibly also shifting the scanned square one place to right or left. The state of the whole system at any moment is called its *complete configuration* (C.C.). The C.C. could be described by giving the sequence of symbols on the tape, starting with the first which is not blank and ending with the last, and interposing the name of the I.C. on the left of the symbol on the scanned square. In Turing [1] the expression *configuration* (unqualified) was used to describe the combination of I.C. with scanned symbol, so that the behavior of the machine is determined by the configuration at each moment.

For our present purpose many of the conventions of Turing [1] are not ideally chosen, but we are to some extent bound by them since we have to use the results of that paper. We shall describe a C.C. differently. We shall take the form described above and flank it with s_4 at either end. Thus for instance $s_4s_1s_3q_2s_1s_4$ represents a C.C. in which the I.C. is q_2 and the tape bears $s_1s_3s_1$ the last of these symbols being on the scanned square. This device is due to Post (who uses h for s_4) and greatly facilitates the connection with the word problem and Thue systems. We shall divide the internal configurations into two classes, the left-facing and the right-facing. The left-facing I.C.'s will be l_0 , l_1 , l_2 , $\cdots$, l_R and the right facing will be r_1 , r_2 , $\cdots$, r_R (some otiose ones may be included in one of the classes to equalize numbers). If the I.C. is an r_i the new convention requires the machine's behavior to be determined by it and the symbol next on its right, but in the case of an l_i it is determined by l_i and the symbol on its left. It is easy to modify a machine constructed according to the original conventions to give one with similar properties but satisfying the new conventions. We leave the details to the reader. The new convention is not of course considered desirable in general, but only where semi-groups (and perhaps groups) are concerned.

In accordance with these conventions we make the following definitions.

A *computing machine* is determined by giving a (finite) set of symbols, comprising *tape symbols, left-facing internal configurations*, and *right-facing internal configurations*, and also a *table*.

A *complete configuration* is a word made up of tape symbols together with one internal configuration. A table consists of a number of *entries*, each of which is a pair of words of one of the following eight forms

$$\begin{array}{ll} (r_is_k\ ,\ s_{k'}r_{i'}) & (s_0l_i\ ,\ s_0s_0l_{i'}) \\ (s_kl_i\ ,\ l_{i'}s_{k'}) & (s_0l_i\ ,\ l_{i'}) \\ (s_kl_i\ ,\ s_{k'}r_{i'}) & (r_is_0\ ,\ r_{i'}\ s_0s_0) \\ (r_is_k\ ,\ l_{i'}s_{k'}) & (r_is_0\ ,\ r_{i'}) \end{array}$$

where the left-facing internal configurations are l_0 , l_1 , l_2 , $\cdots$, l_R and the right-facing are $r_1\, r_2$, $\cdots$, r_R . The tape symbols are s_0 , s_1 , $\cdots$, s_N

A table must be so constructed that no C.C. has two subwords which are first members of different entries. As a consequence of this condition we may define the (unique if existent) *successor* of a C.C. as follows. If (U, V) is an entry of a table then the successor of $X\ U\ Y$ is $X\ V\ Y$. We say that B is a descendant of A if it is possible to connect A to B by a finite sequence of words each of which is the successor (in our technical sense) of the one which immediately precedes it in the sequence.

We might restrict the definition of a C.C. by insisting that a C.C. must begin and end with s_4. But it will appear shortly that it will make no difference for our purpose whether we do so or not.

Note that if B is a descendant of A in a machine $\mathfrak{M}$ then (A, B) is a relation of the semi-group whose F.R. are the entries of $\mathfrak{M}$.

The symbol s_0 represents 'blank'. If a machine following the new convention is made to correspond in behavior with one following the old it is necessary that the entries $(s_0 l_i, s_0 s_0 l_{i'})$, $(s_0 l_i, l_{i'})$, $(r_i s_0, r_{i'} s_0 s_0)$, $(r_i s_0, r_{i'})$ should only come into play when the machine is scanning a blank square in one of the two continuous infinite sequences of blank squares at the two ends of the tape. This would be ensured by appropriate entries in the machine's table, and would be made possible by the presence of the 'end symbols' s_4. However so long as we are satisfied that the behavior of the machines in Turing [1] can be modelled by machines following the present conventions we need not consider these details any further.

There is no general method which can be used to determine, given a machine and a C.C., whether that C.C. has any descendants in which a given generator occurs. The equivalent of this was stated in Turing [1] (p. 248), but Post has exposed a weakness in the proof which was given there. Post pointed out that certain conventions which should only have been used in the construction of particular machines had actually been assumed to apply to arbitrary machines. This convention required partial answers never to be erased, and to have a space order corresponding to the time order of their printing. Such a restriction applied to machines in general would be most undesirable. Post suggests overcoming the difficulty by considering the 'answer' to be given by the time sequence of printings, and ignoring the space sequence altogether. Another possibility is to show that the objectionable convention is in some sense inessential. This could be done by making use of the fact that the 'universal machine' itself obeys the convention. However there is no intention to enter into these questions in detail here, and we shall assume the theorem proved in the form in which it is stated at the beginning of this paragraph.

If we were content to prove the unsolvability of the word problem in the weak form ((b) or (c)), it would be natural to attempt to reduce it to the unsolvable problem just mentioned. For the strong form (c) however we prefer a result about a particular machine which may be translated into a result about a particular semi-group. This is made possible by the existence of the 'universal machine' $\mathfrak{U}$. This machine has the property that, given the table of any other machine $\mathfrak{M}$ and a C.C. K of $\mathfrak{M}$ we can find a C.C. of $\mathfrak{U}$, $c(\mathfrak{M}, K)$ say, such that a certain easily distinguishable subsequence of the descendants of $c(\mathfrak{M}, K)$ corresponds in a certain simple way to the descendants of K in the motion of $\mathfrak{M}$. The details of $\mathfrak{U}$ are to a very large extent arbitrary, and so are the details of the above-mentioned correspondences. One possible convention is that the special subsequence of descendants of $c(\mathfrak{M}, K)$ consists of those whose penultimate symbol is s_5 , and that of the corresponding descendant of K is obtained by omitting all symbols which are not symbols for $\mathfrak{M}$. (This of course requires that there are certain symbols which are sacred to the use of $\mathfrak{U}$). However in spite of the arbitrariness we shall speak of $\mathfrak{U}$ as though it were a quite definite machine whose table was before us. In order actually to obtain such a table we should have to take the description of the universal machine in Turing [1] and modify it in several ways. Firstly it would be necessary to expand all the abbreviations used: secondly it must be modified to accord with the conventions of the present paper, and finally various matters of detail noted by Post (Post [2], p. 7, footnote) must be corrected. As a consequence of the properties of $\mathfrak{U}$ and the above mentioned unsolvable problem concerning machines in general *there is a machine $\mathfrak{B}$ such that there is no general method by which given a C.C. of $\mathfrak{B}$ we can determine whether $s_4 l_0 s_4$ is one of its descendants or not. $\mathfrak{B}$ is such that any C.C. containing l_0 has no successor.* Here $\mathfrak{B}$ is a machine obtained by a small modification of $\mathfrak{U}$. $\mathfrak{B}$ behaves like $\mathfrak{U}$ in its earlier steps, but as soon as it reaches a C.C. in which s_3 appears it behaves differently, proceeding quickly to the C.C. $s_4 l_0 s_4$ by erasures, etc.   [[1]]

In what follows we shall not be concerned with any other machines than $\mathfrak{B}$, apart from some parenthetical remarks about a machine $\mathfrak{B}'$. Our interest in machines in general, and the conventions by which they are restricted, therefore fades. All we need to know is that $\mathfrak{B}$ satisfies the above italicized statement and that its table follows our new conventions.

4. The semi-group $\mathfrak{S}_0$.

The semi-group $\mathfrak{S}_0$ is obtained from the table for $\mathfrak{B}$ as follows. Let the tape symbols for $\mathfrak{B}$ be s_0 , s_1 , $\cdots$, s_N , the left-facing internal configurations l_1 , l_2 , $\cdots$, l_R and the right-facing ones r_1 , r_2 , $\cdots$, r_R : let the entries of the table for $\mathfrak{B}$ be E_1 , $\cdots$, E_M . Then the generators of $\mathfrak{S}_0$ are to be y, σ_m , τ_m , j_i , k_i , n_i , p_i , u_k , v_k , w_k , z_k $(m = 1, 2, \cdots, M; i = 0, 1, \cdots, N; k = 1, 2, \cdots, R)$. The F.R. of $\mathfrak{S}_0$ will be classified into three kinds, *principal relations*, *commutation relations*, and *change relations*.

Principal relations. For each entry of $\mathfrak{B}$ we have two principal relations, called *first phase* and *second phase* relations. A table is given below of the forms that these relations take for the entry E_m according to the eight different forms that this entry can have

Entry E_m	First phase relation	Second phase relation	
$(r_i s_h , s_h' r_{i'})$	$(v_i k_h , \sigma_m j_h , v_i , \tau_m)$	$(\sigma_m n_h , z_i , \tau_m , z_i p_h)$	[[2]]
$(s_h l_i , l_{i'} s_h')$	$(j_h u_i , \sigma_m u_i , k_{h'} \tau_m)$	$(\sigma_m w_i , p_{h'} \tau_m , n_h w_i)$	
$(s_h l_i , s_h' r_{i'})$	$(j_h u_i , \sigma_m j_{h'} v_{i'} \tau_m)$	$(\sigma_m n_{h'} z_{i'} \tau_m , n_h w_i)$	
$(r_i s_h , l_{i'} s_h')$	$(v_i k_h , \sigma_m u_{i'} k_{h'} \tau_m)$	$(\sigma_m w_{i'} p_{h'} \tau_m , z_i p_h)$	
$(s_0 l_i , s_0 s_0 l_{i'})$	$(j_0 u_i , \sigma_m j_0 j_0 u_{i'} \tau_m)$	$(\sigma_m n_0 n_0 w_{i'} \tau_m , n_0 w_i)$	
$(s_0 l_i , l_{i'})$	$(j_0 u_i , \sigma_m u_{i'} \tau_m)$	$(\sigma_m w_{i'} \tau_m , n_0 w_i)$	
$(r_i s_0 , r_{i'} s_0 s_0)$	$(v_i k_0 , \sigma_m v_i , k_0 k_0 \tau_m)$	$(\sigma_m z_{i'} p_0 p_0 \tau_m , z_i k_0)$	
$(r_i s_0 , r_{i'})$	$(v_i k_0 , \sigma_m v_{i'} \tau_m)$	$(\sigma_m z_{i'} \tau_m , z_i k_0).$	

Commutation relations

$$\left.\begin{array}{l}(\sigma_m u_i\,,\ u_i\tau_m)\\(v_i\tau_m\,,\ \sigma_m v_i)\\(w_i\tau_m\,,\ \sigma_m w_i)\\(\sigma_m z_i\,,\ z_i\tau_m)\end{array}\right\}\begin{array}{l}(i = 1, \cdots, R\\ \ \ m = 1, \cdots, M).\end{array}$$

Change relations

$$\begin{array}{ll}(j_4u_0\,,\ n_4y)&\\(yk_4\,,\ w_0p_4)&\\(y\tau_m\,,\ \sigma_m y)&\qquad (m = 1, \cdots, M).\end{array}$$

The symbols which occur in first phase relations may be called *first phase symbols*. Words made up of such symbols may be called first phase words. Second phase symbols and words may be defined similarly.

The symbols u_i, v_i, w_i, z_i (which correspond to internal configurations) will
[[3]] be called *active* symbols. The symbols j_h, h_h, σ_m which always appear to the left of the active symbols in the F.R., are called *left symbols*, and k_h, p_h, τ_m are called right symbols.

A word is described as *normal* if it contains exactly one active symbol and all the symbols to the left of this active symbol are left symbols and all to the right of it are right symbols. Both members of an F.R. are evidently normal.

It will be seen that the first phase relations are obtained from the corresponding entries by

(a) Substituting v_i for r_i and u_i for l_i. Also substituting j_h for s_h when it occurs on the left of the I.C. but substituting k_h for it when it occurs on the right.

(b) Inserting σ_m on the left of the second term and τ_m on its right.

The second phase entries are obtained by

(c) Substituting z_i for r_i and w_i for l_i. Also substituting n_h for s_h when it occurs on the left of the I.C. but substituting p_h for it when it occurs on the right.

(d) As (b).

(e) Interchanging the two terms.

The complete set of relations for $\mathfrak{S}_0$ may be regarded as forming the table for a certain machine $\mathfrak{B}'$. The movements of $\mathfrak{B}$ correspond to the earlier steps in the movement of $\mathfrak{B}'$, which may be called the first phase of the movement of $\mathfrak{B}'$. The C.C.'s of the two machines correspond if the generators σ_m, τ_m are ignored and the substitutions (a) applied. This correspondence applies throughout if $\mathfrak{B}$ never reaches the C.C. $s_4l_0s_4$. However if it does so the motion of $\mathfrak{B}'$ continues further. After the first phase comes an 'intermediate phase' in which the C.C. $\Sigma j_4u_0T_1K_4T_2$ is changed to $\Sigma n_4w_0T_1p_4T_2$. Here Σ is a word formed entirely from the generators $\sigma_1, \sigma_2, \cdots, \sigma_M$ and T_1, T_2 are formed entirely from $\tau_1, \tau_2, \ldots, \tau_M$. Finally we have the second phase in which the motions of the first phase are carried out again, but in reverse order and with different tape symbols and internal configurations. At the end of this we return to the original C.C. with certain substitutions applied. We intend to show that the necessary and sufficient condition that $s_4 l_0 s_4$ be a descendant of C is that $(\varphi_1(C), \varphi_2(C))$ be a relation of $\mathfrak{S}_0$ where φ_1 , φ_2 represent the substitutions (a) and (c) above. The proof of the necessity is easy and consists essentially in establishing the above properties of $\mathfrak{B}'$. This argument is in effect reproduced in Lemmas 1–4 below, but without explicit reference to the machine $\mathfrak{B}'$. Indeed the machine $\mathfrak{B}'$ is wholly fictitious since it does not obey either the old or the new conventions. It has the power of "manufacturing tape" on which to write the symbols σ_m , τ_m . Such machines are certainly worthy of consideration, but are not recognized here. The sufficiency is more difficult, for we are permitted to do certain operations in semi-groups which do not correspond to motions of the machines, e.g., cancellation. The generators σ_m , τ_m have been introduced into the F.R. to ensure that any results which can be obtained by the application of these operations could also be obtained without. This means that the permission to cancel becomes a dead letter. The introduction of different sets of symbols for use on the right and the left of the active symbol has a similar purpose. It ensures that relations which contain more than one active symbol in each term can be split up into ones containing only one: this enables these relations to be interpreted in terms of machines. More picturesquely we can say that when the left and right symbols are distinguishable we can let two or more machines work on the same tape without danger of their interfering with one another.

We shall eventually prove

THEOREM 1. *The necessary and sufficient condition that* $(\varphi_1(C), \varphi_2(C))$ *be a relation of* $\mathfrak{S}_0$ *is that* $s_4 l_0 s_4$ *be a descendant of* C *in the machine* $\mathfrak{B}$.

As we have mentioned φ_1 , φ_2 represent the substitutions (a) and (c). An equivalent definition is given below. Combining Theorem 1 with the principle enunciated at the end of §3 we have

THEOREM 2. *There is no general method by which we can determine whether a given pair of words is or is not a relation of* $\mathfrak{S}_0$. *For if there were such a method we could apply it to pairs of the form* $(\varphi_1(C), \varphi_2(C))$ *and so determine whether* $s_4 l_0 s_4$ *is a descendant of* C *in* $\mathfrak{B}$.

Since we shall be concerned with no other semi-groups than $\mathfrak{S}_0$ we shall use the word 'relation' to mean 'relation of $\mathfrak{S}_0$'.

5. Sufficiency in Theorem 1

We begin by defining certain functions of words and C.C.'s. The null word is denoted by Λ. The functions φ_1 , φ_2 (already mentioned above) are defined for all C.C.'s of $\mathfrak{B}$ by the equations

$$\begin{aligned} &\varphi_1(l_i) = u_i , \quad \varphi_1(r_i) = v_i , \quad \varphi_1(s_h X) = j_h \varphi_1(X), \quad \varphi_1(X s_h) = \varphi_1(X) k_h \\ &\varphi_1(l_i) = w_i , \quad \varphi_2(r_i) = z_i , \quad \varphi_2(s_h X) = n_h \varphi_2(X), \quad \varphi_2(X s_h) = \varphi_2(X) p_h . \end{aligned}$$ ⟦4⟧

The functions ψ_1 , ψ_2 are defined for first and second phase words respectively by equations

⟦5⟧ $\psi_1(XG) = \psi_1(X)\psi_1(G), \quad \psi_1(u_i) = l_i, \quad \psi_1(v_i) = v_i, \quad \psi_1(\Lambda) = \Lambda$

$\psi_1(k_h) = \psi_1(j_h) = s_h, \quad \psi_1(\sigma_m) = \psi_1(\tau_m) = \Lambda$

$\psi_2(XG) = \psi_2(X)\psi_2(G), \quad \psi_2(w_i) = l_i, \quad \psi_2(z_i) = v_i, \quad \psi_2(\Lambda) = \Lambda$

⟦6⟧ $\psi_2(r_h) = \psi_2(p_h) = s_h, \quad \psi_2(\sigma_m) = \psi_2(\tau_m) = \Lambda.$

The function χ is defined for first phase words by the equations

$$\chi(XG) = \chi(X)\chi(G), \quad \chi(u_i) = w_i, \quad \chi(v_i) = z_i, \quad \chi(\Lambda) = \Lambda$$
$$\chi(j_h) = r_h, \quad \chi(k_h) = p_h, \quad \chi(\sigma_m) = \sigma_m, \quad \chi(\tau_m) = \tau_m.$$

The function τ is defined for words containing only $\sigma_1, \sigma_2, \cdots, \sigma_M$ and σ for words containing only $\tau_1 \tau_2, \cdots, \tau_M$ by the equations

$$\tau(\Lambda) = \Lambda, \quad \tau(X\sigma_m) = \tau(X)\tau_m$$
$$\sigma(\Lambda) = \Lambda, \quad \sigma(X\tau_m) = \sigma(X)\sigma_m.$$

In these definitions X is any word and G any generator.

LEMMA 1. *For all arguments for which the functions are defined we have*

$$\psi_1(XY) = \psi_1(X)\psi_1(Y), \qquad \psi_2(XY) = \psi_2(X)\psi_2(Y)$$
$$\chi(\psi_1(X)) = \psi_2(X) \qquad \varphi_2(\psi_1(X)) = \chi(X)$$
$$\psi_1(\varphi_1(V)) = V$$
$$\psi_2(\varphi_2(V)) = V.$$

These results are all trivial.

LEMMA 2. *The fundamental relations corresponding to the mth entry (U, V) of the table for $\mathfrak{B}$ are $(\varphi_1(U), \sigma_m\varphi_1(V)\tau_m)$ (first phase) and $(\sigma_m\varphi_2(V)\tau_m, \varphi_2(U))$ (second phase).*

This merely requires verification in each of the eight cases.

LEMMA 3. *If $C = C_0, C_1, C_2, \cdots$ are the descendants of C in $\mathfrak{B}$ (the sequence possibly terminating), then for each Cr we can find a normal word Hr such that (H_0, Hr) and $(\chi(Hr), \chi(H_0))$ are relations and $H_0 = \varphi_1(C_0)$, $\psi(Hr) = Cr$.*

In the case $r = 0$ the value $\varphi_1(C_0)$ evidently satisfies all the requirements. Suppose H_{r-1} has been defined and satisfies the conditions. We may write
⟦7⟧ C_{r-1} as AUB and C_r as AVB where (U, V) is one of the entries of the machine. Now if $H_{r-1} = PQR$ then $AUB = \psi_1(P)\psi_1(Q)\psi_1(R)$. If P and R are sufficiently short, $\psi_1(P)$ and $\psi_1(R)$ are respectively contained in A and B. If we take them as long as can be done consistently with this, we shall have $\psi_1(P) = A$ and $\psi_1(R) = B$, for each increase of P or R by one symbol does not increase the length of $\psi_1(P)$ or $\psi_1(R)$ by more than one symbol. Hence $\psi_1(Q) = U$. We now define H_r to be $P\sigma_m\varphi_1(V)\tau_mR$ where (U, V) is the entry E_m. Then

$$\psi_1(H_r) = \psi_1(P)\psi_1(\sigma_m)\psi_1(\varphi_1(V))\psi_1(\tau_m)\psi_1(R) = AVB = Cr$$

The relation (H_{r-1}, H_r) then follows from $(\varphi_1(U), \sigma_m(\varphi_1(V)\tau_m)$ which is an F.R. by Lemma 2. The F.R. $(\sigma_m\varphi_2(V)\tau_m, \varphi_2(U))$ can also be written as

$$(\chi(\psi_1(\sigma_m)\psi_1(\varphi_1(V))\psi_1(\tau_m)), \ \chi(Q))$$

in virtue of the equation $\psi_1(Q) = U$ and Lemma 1. From this F.R. follows the relation $(\chi(H_r), \chi(H_{r-1}))$.

LEMMA 4. *If* $s_4l_0s_4$ *is a descendant of* C *in* $\mathfrak{B}$ *then* $(\varphi_1(C), \varphi_2(C))$ *is a relation.*

Using the notation of Lemma 3 suppose that C_R is $s_4l_0s_4$. Then $(\varphi_r(C), H_r)$ and $(\chi(H_R), \varphi_2(C))$ are relations and it remains to obtain the relation $(H_R, \chi(H_R))$. Since $\psi_1(H_R) = s_4l_0s_4$ by Lemma 3 we can write the relation $(H_R, \chi(H_R))$ as $(\Sigma_1j_4\Sigma_2u_0T_1K_4T_2, \Sigma_1n_4\Sigma_2w_0T_1P_4T_2)$ where $\Sigma_1\Sigma_2$ contain only $\sigma_1, \sigma_2, \cdots, \sigma_M$, and T_1T_2 contain only $\tau_1, \tau_2, \cdots, \tau_M$. We obtain $(\Sigma_2u_0, u_0\tau(\Sigma_2))$ and $(\sigma(T_1)w_0, w_0T_1)$ from the commutation relations and (j_4u_0, n_4y), $(y\tau(\Sigma_2)T_1, \Sigma_2\sigma(T_1)y)$, (yK_4, w_0P_4) from the change relations. From these the required relation follows.

6. The necessity in Theorem 1 ⟦8⟧

This completes the sufficiency part of Theorem 1, and it remains to prove the necessity. We are no longer interested in the detailed form of $\varphi_2(C)$, but only in the fact that it is not a first phase word, for this is in itself enough to ensure that $\mathfrak{B}$ reaches $s_4l_0s_4$. We proceed by an induction over the steps of the proof of the semi-group relation. We shall show that this proof need not contain any cancellations, and need only involve normal words.

Length of proof. Suppose that a proof contains a applications of (vii) or (viii), b applications of (i), c of (ii), (iii) or (iv) and d of (v) or (vi). We describe the length of the proof by the polynomial $a\omega^2 + b\omega + c + 2d$. We regard one polynomial as greater than another (and the corresponding proof as longer) if its values are greater for all sufficiently large positive values of ω. Alternatively we may regard $a\omega^2 + b\omega + c + 2d$ as an ordinal, but this may be misleading since our addition will be the addition of polynomials rather than the addition of ordinals. Thus for instance $(\omega + 1) + (\omega^2 + \omega)$ is $\omega^2 + 2\omega + 1$ and not $\omega^2 + \omega$.

We shall use the notation $[A, B]$ to mean "the length of the proof of the relation ⟦9⟧
(A, B)".

The proofs are "well-ordered" when arranged according to their length in this sense. This means that if we can show that a property which holds of all proofs shorter than the proof P must also hold of P, then the property will hold of all proofs. Or in other words there is always a shortest proof not having a given property if there is any not having it. This is the basis of the arguments below.

LEMMA 5. *If a relation* (A_1B) *satisfies* $n\omega \leqq [A, B] < (n + 1)\omega$, *then there is a* ⟦10⟧
chain $A = A_0, A_1, \cdots, A_n = B$ *of words such that each pair of consecutive members (or links) is either of form* (XUY, XVY) *or of form* (XVY, XUY) *where* (U, V) *is an* F.R. *In the former case the pair is described as a progressive joint, and in the latter as a retrogressive joint.*

The notation of this lemma is used throughout what follows. Consider the relation for which the lemma fails and for which the proof is shortest. The last step of the proof uses as premisses relations with shorter proofs and therefore ones for which the lemma holds. This last step is not by (vii) or (viii) for if it were the length would be at least ω^2. If the last step is by (i) then the length is ω, and the lemma holds with $n = 1$ by taking $U = A$, $V = B$ and x, y void. If the last step is by (ii) then the length is 1 and the lemma holds with $n = 0$. If the last step is by (iii) then the chain for (B, A) gives ones for (A, B) when reversed in order. If it is by (iv) the chains for (A, C) and for (C, B) give one for (A, B) when taken in combination. If it is by (v) and (F, H) has the chain $F_0, F_1, \cdots, F_n$ then (FG, HG) has the chain $F_0G, F_1G, \cdots, F_nG$. Similarly if it is by (vi). In each case there is a chain for the conclusion with the appropriate length, contrary to hypothesis. Hence the assumption that the lemma fails for some proof is false.

LEMMA 6. *If one link of a chain is nomal then every link is normal.*

It will suffice to show that the neighbors of a normal link are normal. Suppose XUY, XVY are two consecutive links and XUY is normal, and either (U, V) or (V, U) is an F.R. Since XUY is normal X consists of left symbols and Y of right symbols. Also V is normal being a term of an F.R. Hence XVY is normal.

Chains such as those concerned in Lemma 6 will be called *normal chains.*

If (XUY, XVY) is a joint of a chain and (U, V) or (V, U) is an F.R. then the joint is said to be *justified* by that F.R.

A two-generator word is described as a *barrier* if the proposition that the first generator is a left symbol is logically equivalent to the proposition that the second is a right symbol. A word is said to *have* a barrier if a barrier is a subword.

LEMMA 7. *No barrier occurs as a subword of either term of an* F.R.

This must be verified by detailed examination of the various F.R.

LEMMA 8. *A word without barriers is either normal or else consists entirely of left symbols or entirely of right symbols.*

This follows at once from the fact that two consecutive generators which do not form a barrier must form one of the four combinations left-left, left-active, active-right, right-right.

[[11]] LEMMA 9. *If one link of a chain has no barriers, then no link of the chain has a barrier.*

We have only to show that if XUY has no barriers and either (U, V) or (V, U) is an F.R. then XVY has no barriers. Any barrier which it has cannot be wholly contained in either X or Y, since it would then appear in XUY; nor can it be contained in V by Lemma 7. If it overlaps from X to V then its first generator is a left symbol, for it precedes a symbol in U which is not a right symbol. The first generator of V must therefore be a right symbol, which is impossible. Likewise the barrier cannot overlap from V to Y.

[[12]] LEMMA 10. *If* $[G_1A, G_2B] < \omega^2$ *and* GG_1 *is a barrier then* GG_2 *is a barrier. Likewise if* $[AG_1, BG_2] < \omega^2$ *and* G_1G *is a barrier then* G_2G *is a barrier.*

We will consider only the first assertion. There is a chain for (G_1A, G_2B) by Lemma 5, and therefore one for (GG_1A, GG_2B), by preceding each link with G. The result follows by lemma 9.

[[13]] LEMMA 11. *Let* AG_1G_2F *and* BG_3G_4H *be abbreviated to* P *and* Q, *and let* $[P, Q] < \omega^2$. *Suppose also that* G_1G_2 *and* G_3G_4 *are barriers and either* AG_1 *and* BG_3 *contain no barriers or* G_2F *and* G_4H *contain no barriers. Suppose further that* P, Q *are not identical. Then* $[AG_1, BG_3] + [G_2F, G_4H] + 1 \leqq [P, Q]$.

If the lemma is false we suppose that (P, Q) is the relation with the shortest proof consistent with denying the lemma. We deal only with the case where AG_1 and BG_3 contain no barriers, the case where G_2F and G_4H contain none being similar. Let us consider the last step in the proof of (P, Q). The premisses to this last step must obey the lemma, if they are of appropriate form, i.e. if both terms have a barrier and they are not identical. This last step could not be by (i) since an F.R. has no barriers, nor by (ii) since P, Q are not identical. If it is by (iii) the proof could be reduced in length by omitting the last step and adopting a modified conclusion, while still denying the lemma. Now take the case where the last step is by (iv). We may suppose the premisses were (P, D) and (D, Q). We may suppose that D is not identical with either P or Q for if it were we could omit the last step and the identical premiss. If D contains a barrier it is of form EG_5G_6I where G_5G_6 is a barrier and E contains no barriers. The lemma then applies to both premises, i.e., to (AG_1G_2F, EG_5G_6I) and (EG_5G_6I, BG_3G_6H) so that ⟦14⟧

$$[AG_1, EG_5] + [G_2F, G_6I] + 1 \leqq [P, D].$$
$$[EG_5, BG_3] + [G_6I, G_4H] + 1 \leqq [D, Q].$$

But

$$[AG_1, EG_5] + [EG_5, BG_3] + 1 \geqq [AG_1, BG_3]$$
$$[G_2F, G_6I] + [G_6I, G_4H] + 1 \geqq [G_2F_1G_4H]$$
$$[P, D] + [D, Q] + 1 = [P, Q]$$

and so

$$[AG_1, BG_3] + [G_2F, G_4H] + 1 \leqq [P, Q].$$

Hence D cannot contain a barrier. But this contradicts Lemma 9. If the last step is by (v) we may write $F'G$ and $H'G$ for F and H. Then

$$[G_2F, G_4H] \leqq [G_2F', G_4H'] + 1$$
$$[AG_1, BG_3] + [G_2F, G_4H] + 1 \leqq [AG_1, BG_3] + [G_2F', G_4H'] + 2$$
$$\leqq [AG_1\, G_2\, F', BG_3G_4H'] + 1 = [P, Q].$$

If the last step is by (vi) then (P, Q) is of form (GG_5P', GG_6Q') where $G_1G_5G_6$ are generators and P', Q' may possibly be void. If GG_5 is a barrier then so is GG_6 by Lemma 10 and we see that A, B are void. (AG_1, BG_3) reduces to (G, G) whose proof is of length 1. The premiss of the last step is (G_2F, G_4H) and ⟦15⟧

$$[G_2F, G_4H] + 2 = [P, Q]. \text{ Hence } [AG_1, BG_3] + [G_2F, G_4H] + 1 = [P, Q].$$

If GG_5 is not a barrier then GG_6 is not either and we see that A, B are not void. Put $A = GA'$, $B = GB'$. A' and B' may be void. The premiss of the last step is $(A'G_1G_2F, B'G_3G_4H)$ and so since G_1G_2, G_3G_4 are barriers and $A'G_1$, $B'G_3$ have no barriers and $A'G_1G_2F$, $B'G_3G_4H$ are not identical

$$[A'G_1, B'G_3] + [G_2F, G_4H] + 3 \leqq [A'G_1G_2F', B'G_3G_4H] + 2 = [P, Q].$$

But

$$[A'G_1, B'G_3] + 2 \geqq [AG_1, BG_3]$$

so that

⟦16⟧ $$[AG_1, BG_3] + [G_1F, G_4H] + 1 \leqq [P, Q].$$

⟦17⟧ Lemma 12. *No normal first phase word can be expressed in more than one way in the form XUY where U is the first term of an* F.R. *nor in more than one way in the form XVY where V is the second term of an* F.R.

If $XUY = X'U'Y'$ then $\psi(X)\psi(U)\psi(Y) = \psi(X')\psi(U')\psi(Y')$. But then

$$\varphi_1^{-1}(\psi(X))\varphi_1^{-1}(\psi(U))\varphi_1^{-1}(\psi(Y)) = \phi_1^{-1}(\psi(X'))\varphi_1^{-1}(\psi(U'))\varphi_1^{-1}(\psi(Y'))$$

and $\varphi_1^{-1}(\psi(U))$, $\varphi_1^{-1}(\psi(U'))$ are first terms in entries of the table for $\mathfrak{B}$, and $\varphi_1^{-1}(\psi(XUY))$, $\varphi_1^{-1}(\psi(X'U'Y'))$ are C.C.'s. Then by the restrictions to which all tables are subject we have $\varphi_1^{-1}(\psi(U)) = \varphi_1^{-1}(\psi(U'))$. Also U and U' have no generators σ_m (being first terms) and therefore $U = \psi(U)$, $U' = \psi(U')$. Hence $U = U'$. If $XVY = X'V'Y'$ then since both sides are normal V and V' must overlap, and since they both terminate with a σ_m or τ_m they must have a σ_m or τ_m in common. Hence the value of m is the same for both, i.e., V and V' are the same relation. Since the words are normal they contain at most one active symbol each. Hence in XVY, $X'V'Y'$ the active symbol of V must occur at the same distance from the beginning of both expressions, so that $X = X'$ and likewise $Y = Y'$.

Lemma 13. *In a normal chain without repetitions there is no change of direction, i.e. the joints are either all progressive or all retrogressive.*

Suppose there is a link which is flanked both by a progressive and by a retrogressive joint.

Then it is expressible in both the forms XUY and $X'U'Y'$ where (U, V) and (U', V') are F.R. or else in both the forms XVY and $X'V'Y'$ In the former case the flanking links are XVY and $X'V'Y'$, and in the latter they are XUY and $X'U'Y'$. In either case they are identical, by Lemma 12 together with the fact that an F.R. is uniquely determined by either of its terms. The chain therefore has a repetition contrary to hypothesis.

Lemma 14. *If the end links of a chain are first phase words, and there is no change of direction, then all links are first phase.*

We will suppose the chain progressive. Consider the last link which is not first phase, and suppose that the joint following it is justified by (U, V). Then V is first phase but U is not. But looking through the fundamental relations we see that there is no such (U, V).

Lemma 15. *A normal chain without change of direction is completely determined by its first link.*

This is an immediate consequence of Lemma 12 and the definition of a chain.

⟦18⟧ Lemma 16. *Let Σ_1, Σ_2 be words formed from $\sigma_1\sigma_2, \cdots, \sigma_M$, and let G be a first phase symbol, G' a second phase symbol. Then $\Sigma_1G'X$ cannot precede Σ_2GY in a progressive chain.*

Every link can be written in the form ΣFZ where Σ (and also Σ_3, Σ_4 etc.) contains only $\sigma_1\sigma_2, \cdots, \sigma_M$, and the generator F is not a σ_m. If $\Sigma_3F'X'$, Σ_4FY' are two consecutive links of the chain then the corresponding joint must be justified by an F.R. of the form $\Sigma_5F'X''$, Σ_6FY''. If $\Sigma_3F'X'$ is the last link for which F' is second phase, then F' is second phase but F is not. Clearly such an F.R. is neither first phase nor second phase, and no such relation can be found among the commutation relations or the change relations.

LEMMA 17. *If G is a generator and there is a normal chain for* (GA, GB) *or for* (AG, BG) *then there is one for* (A, B). [[19]]

Consider the case where the given chain is one for (GA, GB), and suppose that there is no chain for (A, B). Let the chain for (GA, GB) be $F_0, F_1, \cdots, F_n$ Not all the links of the chain begin with G, for if they did we could obtain a chain for (A, B) simply by omitting G from each link. Let the first and last links which do not begin with G be F_r and F_s. We will assume the chain progressive. Then (F_{r-1}, F_r) is justified by some F.R. of form (GU, V) and (F_s, F_{s+1}) by one of form (U', GV') where V and U' do not begin with G. Thus G is both the leftmost symbol of the first member of a relation and the leftmost symbol of a second member. The only symbols which can satisfy this condition are σ_m. The first symbol of V must be of form z_i, u_i, or n_h. If it is z_i or u_i then it is the active symbol in V, and the link (F_r, F_{r+1}) must be justified by a relation whose first member begins with that symbol. But there is no such relation, so that V begins with n_h. Likewise U' begins with an j_h. But there is a progressive chain from V to U' contradicting Lemma 16. This completes the proof for the case (GA, GB). The argument for the case (AG, BG) is similar; although there is not complete symmetry in the change relations between left and right, it does not affect the argument at any point.

LEMMA 18. *If one term of a relation contains no active symbol and the proof of the relation does not apply* (vii) *or* (viii), *then the relation is an identity.*

We consider the last step of the shortest proof of such a relation which is not an identity. It is not by (i) since every F.R. has an active symbol in each term, nor by (ii) since it is not an identity. If it is by (iii) we could omit the last step and obtain a shorter proof denying the lemma. If it is by (iv) one of the premisses has a term which contains no active symbol. This premiss has a shorter proof than the relation in question and therefore is an identity, and the last step may be omitted. If it is by (v) or (vi) the premiss must be an identity by a similar argument, and the conclusion is therefore also an identity. A step by (vii) or (viii) has been excluded by the conditions of the lemma. Hence there is no such relation which is not an identity.

LEMMA 19. *The shortest proof of a normal relation does not use* (vii) *or* (viii) *or words which are not normal.* [[20]]

We consider the relation (P, Q) which denies the lemma and, consistently with this, has the shortest possible proof. If there is an application of (vii) or (viii) consider the first one, and suppose it is applied to (GA, GB) to obtain (A, B). Then either (A, B) is (P, Q) or else A, B are not normal for in any other case (A, B) would deny the lemma and have a shorter proof than (P, Q). If (A, B) is (P, Q) then (GA, GB) cannot be normal, for if it were its proof would use only normal words, being shorter than that of (P, Q). There would therefore be a normal chain for (GA, GB) by Lemmas 5, 6 and one for (A, B) (i.e., for (P, Q)) by Lemma 17. Then (P, Q) satisfies the lemma contrary to hypothesis. Equally if (A, B) is not normal (GA, GB) will not be normal. Then by Lemma 8 there are either barriers in GA, GB or one of them contains no active symbol. In the latter case since (vii) and (viii) are not used in the proof of (GA, GB) this relation must be an identity (Lemma 18). But in this case we can shorten the proof of (A, B) since it is an identity. Hence there are barriers in GA and GB. Applying Lemma 11 we see that either the barriers are right at the beginning and $[G, G] + [A, B] + 1 \leqq [GA, GB]$ or else (A, B) is of form $(A'G_1G_2F, B'G_3G_4H)$ where G_1G_2 and G_3G_4 are barriers and $GA'G_1$, $GB'G_3$ without barriers and

$$[GA'G_1, GB'G_3] + [G_2F, G_4H] + 1 \leqq [GA, GB] < [P, Q].$$

We may apply Lemma 17 to $(GA'G, GB'G_3)$: this shows that there is a chain for $(A'G_1, B'G_3)$. Combining this with the proof of (G_2F, G_4H) gives us a proof of (P, Q) which does not apply (vii) or (viii) and is therefore shorter than the original one. The same is true if $[G, G] + [A, B] + 1 \leqq [GA, GB]$. Hence there is not any application of (viii) and likewise none of (vii). The whole proof is therefore of length less than ω^2, so that by Lemma 5 there is a chain for (P, Q). But then (P, Q) does not deny the lemma, contrary to hypothesis. This establishes the lemma.

⟦21⟧ LEMMA 20. *If for some W which is not first phase $(\varphi_1(C), W)$ is a relation, then $s_4l_0s_4$ is a descendant of C.*

By Lemmas 19, 15 there is a chain for $(\varphi_1(C), W)$. Let the last first phase link XUY and the joint connecting it to the next link be ustified by (U, V). Then U is first phase but V is not. This identifies (U, V) as (j_4u_0, u_4y). Now consider the words H_r described in Lemma 3. As each H_r consists only of first phase symbols the same is true of each link of its chain (Lemma 14). This chain therefore is wholly contained in the chain for $(\varphi_1(C), XUY)$, which we will suppose is of length n. Then $r \leqq n$ so that C has only a finite number of descendants. Let the last be C_s, and consider the chain leading from H_s to XUY. Since it consists entirely of first phase symbols the fundamental relations involved must be either first phase relations or commutation relations. I assert that they are all commutation relations. For if one is a first phase relation then C_s has a successor. Since these steps are all by commutation relations, H_s contains u_0 and therefore C_s contains l_0, and is therefore $s_4l_0s_4$ by the properties of $\mathfrak{B}$.

Theorem 1 follows at once from Lemmas 4, 20 and the fact that $\varphi_2(C)$ is not first phase.

VICTORIA UNIVERSITY
MANCHESTER, ENGLAND

REFERENCES

ALONZO CHURCH, [1]. *An unsolvable problem of elementary number theory*, Amer. J. Math., 58 (1936), 345–363.

S. C. KLEENE, [1]. *General recursive functions of natural numbers*, Math. Ann., 112 (1935–6), 727–742.

S. C. KLEENE, [2]. *λ-definability and recursiveness*, Duke Math. J., 2 (1936), 340–353.

E. L. POST, [1]. *Finite combinatory processes—formulation 1*, J. Symbolic Logic, 1 (1936), 103–105.

E. L. POST, [2]. *Recursive unsolvability of a problem of Thue*, J. Symbolic Logic, 12 (1947), 1–11.

A. M. TURING, [1]. *On computable numbers, with an application to the Entscheidungsproblem*, Proc. London Math. Soc. (2), 42 (1937), 230–265. A correction to this paper has appeared in the same periodical, 43 (1937), 544–546.

A. M. TURING, [2]. *Computability and λ-definability*, J. Symbolic Logic, 2 (1937), 153–163.

D. HILBERT and P. BERNAYS, Grundlagen der Mathematik, 2 vols., Berlin, (1940).

SOME CALCULATIONS OF THE RIEMANN ZETA-FUNCTION

By A. M. TURING

[Received 29 February 1952.—Read 20 March 1952]

Introduction

IN June 1950 the Manchester University Mark 1 Electronic Computer was used to do some calculations concerned with the distribution of the zeros of the Riemann zeta-function. It was intended in fact to determine whether there are any zeros not on the critical line in certain particular intervals. The calculations had been planned some time in advance, but had in fact to be carried out in great haste. If it had not been for the fact that the computer remained in serviceable condition for an unusually long period from 3 p.m. one afternoon to 8 a.m. the following morning it is probable that the calculations would never have been done at all. As it was, the interval $2\pi.63^2 < t < 2\pi.64^2$ was investigated during that period, and very little more was accomplished.

The calculations were done in an optimistic hope that a zero would be found off the critical line, and the calculations were directed more towards finding such zeros than proving that none existed. The procedure was such that if it had been accurately followed, and if the machine made no errors in the period, then one could be sure that there were no zeros off the critical line in the interval in question. In practice only a few of the results were checked by repeating the calculation, so that the machine might well have made an error.

If more time had been available it was intended to do some more calculations in an altogether different spirit. There is no reason in principle why computation should not be carried through with the rigour usual in mathematical analysis. If definite rules are laid down as to how the computation is to be done one can predict bounds for the errors throughout. When the computations are done by hand there are serious practical difficulties about this. The computer will probably have his own ideas as to how certain steps should be done. When certain steps may be omitted without serious loss of accuracy he will wish to do so. Furthermore he will probably not see the point of the 'rigorous' computation and will probably say 'If you want more certainty about the accuracy why not just take more figures?' an argument difficult to counter. However, if the calculations are being done by an automatic computer one can feel sure that this kind of indiscipline does not occur. Even with the automatic computer this rigour can be rather tiresome to achieve, but in connexion with such a subject as the analytical theory of numbers, where rigour is the essence, it seems worth while. Unfortunately, although the details were all worked out, practically nothing was done on these lines. The interval $1414 < t < 1608$ was investigated and checked, but unfortunately at this point the machine broke down and no further work was done. Furthermore this interval was subsequently found to have been run with a wrong error value, and the most that can consequently be asserted with certainty is that the zeros lie on the critical line up to $t = 1540$, Titchmarsh having investigated as far as 1468 (Titchmarsh (**5**)).

This paper is divided into two parts. The first part is devoted to the analysis connected with the problem. All the results obtained in this part are likely to be applicable to any further calculations to the same end, whether carried out on the Manchester Computer or by any other means. The second part is concerned with the means by which the results were achieved on the Manchester Computer.

Part I. General

1. The Θ notation

In analysis it is customary to use the notation $O\{f(x)\}$ to indicate 'some function whose ratio to $f(x)$ is bounded'. In the theory of a computation one needs a similar notation, but one is interested in the value of the bound concerned. We therefore use the notation $\Theta(\alpha)$ to indicate 'some number not greater in modulus than α'. The symbol Θ has been chosen partly because of a typographical similarity to 0, partly because of the relation with the use of ϑ to indicate 'a number less than 1'.

2. The approximate functional equation

We shall use throughout the notation of Ingham (**1**) and Titchmarsh (**3**) without special definition. Our problem is to investigate the distribution of zeros of $\zeta(s)$ for large t. This will presumably depend on being able to calculate $\zeta(\sigma+it)$ or some closely associated function for large t, and σ not too far from $\frac{1}{2}$. We have to consider what formula to use and what associated function. For $\sigma > 1$ it is possible to use the defining series $\zeta(s) = \sum_{1}^{\infty} n^{-s}$, but this is too far from $\sigma = \frac{1}{2}$. For $0 < \sigma < 1$ there are other formulae which also involve calculating a number of terms of this series, but it is always necessary to take at least $t/2\pi$ terms.

Alternatively one can use the functional equation

$$\zeta(s)\Gamma(\tfrac{1}{2}s)\pi^{-\frac{1}{2}s} = \zeta(1-s)\Gamma(\tfrac{1}{2}-\tfrac{1}{2}s)\pi^{-\frac{1}{2}+\frac{1}{2}s}$$

and take $t/2\pi$ terms of the series $\zeta(1-s) = \sum_{1}^{\infty} n^{s-1}$. Another possible method which might suggest itself is to calculate at a number of points in the region $\sigma > 1$ and extrapolate, but this again involves much the same amount of work. However, if one considers an interpolation formula involving both values from the region $\sigma > 1$ and from the region $\sigma < 0$ one finds that it is possible to calculate the function by taking only about $\sqrt{(t/2\pi)}$ terms of the series $\sum n^{-s}$ and an equal number from $\sum n^{s-1}$. This result is embodied in

THEOREM 1. *Let m and ξ be respectively the integral and non-integral parts of $\tau^{\frac{1}{2}}$ and*

$$\tau \geqslant 64,$$

$$\kappa(\tau) = \frac{1}{4\pi i} \log \frac{\Gamma(\frac{1}{4}+\pi i\tau)}{\Gamma(\frac{1}{4}-\pi i\tau)} - \tfrac{1}{4}\tau \log \pi,$$ ⟦1⟧

$$Z(\tau) = \zeta(\tfrac{1}{2}+2\pi i\tau)e^{-2\pi i\kappa(\tau)},$$ ⟦2⟧

$$\kappa_1(\tau) = \tfrac{1}{2}(\tau \log \tau - \tau - \tfrac{1}{2}),$$ ⟦3⟧

$$h(\xi) = \frac{\cos 2\pi(\xi^2-\xi-\frac{1}{16})}{\cos 2\pi\xi}.$$

Then $Z(\tau)$ is real and

$$Z(\tau) = 2\sum_{n=1}^{m} n^{-\frac{1}{2}} \cos 2\pi\{\tau \log n - \kappa(\tau)\} + (-)^{m+1}\tau^{-\frac{1}{4}}h(\xi) + \Theta(1{\cdot}09\tau^{-\frac{3}{4}}),$$ ⟦4⟧

$$\kappa(\tau) = \kappa_1(\tau) + \Theta(0{\cdot}006\tau^{-1}).$$

It will be seen that $Z(\tau)$ may also be defined as being $\zeta(\frac{1}{2}+2\pi i\tau)$ for τ real, ⟦5⟧ $0 < \tau < 1$, and elsewhere by analytic continuation. The theorem could be proved by the argument outlined above, but is more conveniently proved by the method given as Theorem 22 of Titchmarsh (**3**). The numerical details are given in Titchmarsh (**4**). A more elaborate remainder is given there and is valid for $\tau \geqslant \delta$. The validity of the remainder given here follows trivially from it.

This formula can only give a limited accuracy, although it is nearly always adequate. If greater accuracy is required the formula given in Turing (**6**) may be applied. These agree with Titchmarsh's expression in ⟦6⟧ the sum of m terms, but $h(\xi)$ is replaced by another sum.

The function $h(\xi)$ is troublesome to calculate, largely because the numerator and denominator both vanish at $\xi = \frac{1}{4}$ and $\xi = \frac{3}{4}$, so that a special method would have to be applied for the neighbourhood of these points. The alternative of using a table and interpolation suggests itself. This possibility quickly leads to the suggestion of replacing the function by some polynomial which approximates it well enough in the region concerned.

In fact the polynomial $0{\cdot}373+2{\cdot}160(\xi-\frac{1}{2})^2$ is quite adequate, for we have

THEOREM 2. *If* $|\xi-\frac{1}{2}| < \frac{1}{2}$ *we have*

$$h(\xi) = 0{\cdot}373+2{\cdot}160(\xi-\tfrac{1}{2})^2+\Theta(0{\cdot}0153)$$

and if $|\xi-\frac{1}{2}| < 0{\cdot}53$ *we have* $h(\xi) = 0{\cdot}373+2{\cdot}160(\xi-\frac{1}{2})^2+\Theta(0{\cdot}0243)$.

This result is rather unexpectedly troublesome to prove. Its proof will be given in slightly more detail than it deserves, treating it as an example of 'rigorous computation'.

It may be said: 'As this is a purely numerical result surely it can be proved by straight computation.' This is in effect what is done, but it is not possible to avoid theory entirely. The function was calculated for the values $0, \frac{1}{30}, \frac{2}{30},\ldots, \frac{16}{30}, \frac{17}{30}$ of $\xi-\frac{1}{2}$ with an error $\Theta(10^{-4})$, and was found to satisfy the inequality with some margin. But nothing further can be deduced even if the differences are taken into account, unless something is known about the general behaviour of the function. An upper bound for the second derivative would be sufficient, but the labour of even the formal differentiation is discouraging, and the accidental singularities make the situation considerably worse. However, the function is integral, and it is therefore possible to obtain an inequality for any derivative by means of Cauchy's integral formula, taken in combination with an inequality for the function itself on a suitable contour. The method actually applied will be seen to be very similar to this. Instead of Cauchy's integral formula we use

$$f(\xi)-P(\xi) = \frac{(\xi-\xi_1)(\xi-\xi_2)(\xi-\xi_3)(\xi-\xi_4)}{2\pi i}\int\frac{f(u)\,du}{(u-\xi)(u-\xi_1)(u-\xi_2)(u-\xi_3)(u-\xi_4)},$$

where the function $f(\xi)$ is regular inside the anti-clockwise contour of integration, and $P(\xi)$ is the cubic polynomial agreeing with $f(\xi)$ at ξ_1, ξ_2, ξ_3, ξ_4. This equation follows from the fact that the right-hand side vanishes at the points ξ_1, ξ_2, ξ_3, ξ_4 and is of the form of $f(\xi)$ added to a cubic polynomial. We actually take the contour to be the square whose vertices are $\frac{1}{2}\pm i$, $\frac{1}{2}\pm 1$. One can prove without difficulty that $|h(\xi)| < \cosh\pi$ on this
[[7]] square and that if $f(\xi) = h(\xi)-0{\cdot}373-2{\cdot}160(\xi-\frac{1}{2})^2$ then $|f(\xi)| < 14{\cdot}3$ on the square. Taking ξ_1, ξ_2, ξ_3, ξ_4 to be of form $n/30$ and two of them to be on either side of ξ one easily deduces $|f(\xi)-P(\xi)| < 0{\cdot}0033$ if $|\xi-\frac{1}{2}| < 0{\cdot}053$, and a consideration of the values at the calculated points and the differences gives $|P(\xi)| < 0{\cdot}021$ if $|\xi-\frac{1}{2}| < 0{\cdot}53$ and $|P(\xi)| < 0{\cdot}012$ if $|\xi-\frac{1}{2}| < \frac{1}{2}$. It will be seen that the use of this approximation to $h(\xi)$ gives an extra error in $Z(\tau)$ of the order of $\tau^{-\frac{3}{4}}$ whereas Titchmarsh's formula has an error

of order only $\tau^{-\frac{3}{4}}$; but the errors are not equal until τ is over 2000, and both are then quite small. In the actual calculation described in Part II there were other errors of order as large as $\tau^{\frac{5}{4}}$.

Titchmarsh's formula as stated is valid only when the right value of m is used, i.e. if $\tau^{\frac{1}{2}} = m+\xi$ and $|\xi-\frac{1}{2}| \leqslant \frac{1}{2}$. This may be inconvenient as one may occasionally wish to go a little outside the range. One may justify doing so by means of

THEOREM 3. *Theorem 1 is valid with the error* $\Theta(1{\cdot}09\tau^{-\frac{3}{4}})$ *replaced by* $\Theta(1{\cdot}15m^{-\frac{3}{2}})$ *if the condition that* m *and* ξ *be the integral and non-integral parts of* $\tau^{\frac{1}{2}}$ *is replaced by the condition that* m *be an integer and*

$$\tau^{\frac{1}{2}} = m+\xi, \qquad |\xi-\tfrac{1}{2}| < 0{\cdot}53.$$

The new error introduced is

$$(-)^m\tau^{-\frac{1}{4}}\left[\frac{\cos 2\pi(\xi^2-\xi-\frac{1}{16})}{\cos 2\pi\xi}+\frac{\cos 2\pi\{(\xi-1)^2-(\xi-1)-\frac{1}{16}\}}{\cos 2\pi(\xi-1)}\right]- \qquad [\![8]\!]$$
$$-2(m+1)^{-\frac{1}{2}}\cos 2\pi[(m+\xi)^2\log(m+1)-\kappa\{(m+\xi)^2\}]$$

in the case that $1 < \xi < 1{\cdot}03$. But we have

$$\frac{\cos 2\pi(\xi^2-\xi-\frac{1}{16})}{\cos 2\pi\xi}+\frac{\cos 2\pi\{(\xi-1)^2-(\xi-1)-\frac{1}{16}\}}{\cos 2\pi(\xi-1)} = 2\cos 2\pi(\xi^2-2\xi-\tfrac{1}{16}).$$

Also if we put

$$j(\xi) = (m+\xi)^2\log(m+1)-\kappa_1\{(m+\xi)^2\}-\tfrac{1}{2}m^2-m+\tfrac{1}{2}+\xi^2-2\xi-\tfrac{1}{16},$$

then $j(\xi)$ and its first two derivatives vanish at $\xi = 1$ and $j'''(\xi) = \dfrac{-2}{m+\xi}$.

Hence by the mean value theorem $|j(\xi)| < \dfrac{(\xi-1)^3}{3(m+1)}$. Using also

$$|\kappa\{(m+\xi)^2\}-\kappa_1\{(m+\xi)^2\}| < \frac{0{\cdot}006}{(m+1)^2}$$

we see that the new error is at most [\![9]\!]

$$4\pi(m+1)^{-\frac{1}{4}}\left(\frac{(\xi-1)^3}{3(m+1)}+0{\cdot}006(m+1)^{-2}\right)+2|(m+\xi)^{-\frac{1}{4}}-(m+1)^{-\frac{1}{4}}|,$$

which is less than $0{\cdot}052(m+1)^{-\frac{3}{2}}$ since $m \geqslant 7$, $|\xi-1| < 0{\cdot}03$. A similar argument applies for the case $-0{\cdot}03 < \xi < 0$.

3. Principles of the calculations

We may now consider that with the aid of Theorems 1, 2, 3 we are in a position to calculate $Z(\tau)$ for any desired τ. How can we use this to obtain results about the distribution of the zeros? So long as the zeros are on the critical line the result is clearly applicable to enable us to find their position to an accuracy limited only by the accuracy to which we can find $Z(\tau)$. If there are zeros off the line we can find their position as follows. Suppose we have calculated $Z(\tau)$ for $\tau_1, \tau_2, \ldots, \tau_N$. Then we can approximate $Z(\tau)$ in the neighbourhood of these points by means of the polynomial $P(\tau)$ agreeing with $Z(\tau)$ at these points. The accuracy of the approximation may be determined as in Theorem 2. Suppose that in this way we find that $|Z(\tau)-P(\tau)| < \epsilon$ and $|P''(\tau)| < \epsilon'$ for $|\tau-\tau'| < \delta$ and that $|P(\tau')| < \epsilon''$ and $|P'(\tau')-a| < \epsilon'''$, then we see that

$$|Z(\tau)-a(\tau-\tau')| < \epsilon+\epsilon''+\tfrac{1}{2}\epsilon'\delta^2+\epsilon'''\delta$$

for $|\tau-\tau'| < \delta$, and we may conclude by Rouché's theorem that $Z(\tau)$ has a zero within this circle if $|a| > \epsilon'''+\frac{1}{2}\epsilon'\delta+\dfrac{\epsilon+\epsilon''}{\delta}$. This, however, is a tiresome procedure, and should be avoided unless we have good reason to believe that such a zero is really present. If there are any such zeros we may expect that the first ones to appear will be rather close to the critical line, and they will show themselves by the curve of $Z(\tau)$ approaching the zero line and receding without crossing it: in other words by behaving like a quadratic expression with complex zeros. In the absence of such behaviour we wish to prove that there are no complex zeros without using this interpolation procedure. Let us suppose that we have been investigating the range $T_0 < \tau < T_1$ and that we have found a certain number of real zeros in the interval. If by some means we can determine the total number of zeros in the rectangle $|\mathscr{I}\tau| < 2$, $T_0 < \mathscr{R}\tau < T_1$ (say) and find it to equal the number of changes of sign found, then we can be sure that there were no zeros off the critical line in this rectangle. This total number of zeros can be determined by calculating the function at various points round the rectangle. This might normally be expected to involve even more work than the calculations on the critical line. Fortunately, with the function concerned, the calculations on the lines $|\mathscr{I}\tau| = 2$ are not necessary. It is well known that the change in the argument of $Z(\tau)$ on these lines can be calculated to within $\frac{1}{2}\pi$ in terms of the gamma function. It remains to find the change on the lines $\mathscr{R}\tau = T_0$ and $\mathscr{R}\tau = T_1$. In principle this could be done by approximating $Z(\tau)$ with a polynomial, using an interpolation formula based on values calculated on the critical line. Since this interpolation procedure is necessary only at the ends of the interval investigated this would be a considerably smaller burden than the repeated application of it throughout the interval required by the method previously suggested. It will, however, be shown later on that even this application of the interpolation procedure is unnecessary, but for the sake of argument we will suppose for the moment that it is done. We may suppose then that the total number of zeros in the rectangle is known. If this differs from the number of changes of sign which have been found, then the deficit must be ascribed to a combination of four causes. Some may be due to pairs of complex zeros, some to pairs of changes of sign which were missed due to insufficiently many values $Z(\tau)$ being calculated, some to the accuracy of some of the values being inadequate to establish that changes of sign had occurred. Finally there may be some multiple zeros on the critical line. Each source accounts for an even number of zeros provided that the accuracy is sufficient for there to be no doubt about the signs of $Z(T_0)$ and $Z(T_1)$. By calculating further values and increasing the accuracy we can remove some of the discrepancies, but we cannot do anything about the multiple zeros by mere calculation. Assuming that there are no multiple zeros it is possible in principle to make sure that all the real zeros have been found by calculating $Z(\tau)$ at a sufficient number of real points, but the number of points would be many more than are required for finding all the real zeros. It is better to find the complex zeros in the manner already described.

To summarize. The method recommended is first to find the total number of zeros in the rectangle by methods to be described later. Then to calculate the function at sufficient points to account for all the zeros, either by changes of sign or as complex zeros determined by the use of Rouché's theorem. We know no way of dealing with multiple zeros, and simply hope that none are present.

4. Evaluation of $N(t)$

For reasons explained in the last section it is desirable to be able to determine the number of zeros of $Z(\tau)$ in a region $T_0 < \tau < T_1$. In practice this is best done by determining separately the numbers in the regions $0 < \mathscr{R}\tau < T_0$ and $0 < \mathscr{R}\tau < T_1$. If we write $\pi S(t)$ for the argument of $\zeta(\frac{1}{2}+it)$ obtained by continuation along a line parallel to the real axis from $\infty+it$, where the argument is defined to be zero, we have

$$N(T) = 2\kappa\left(\frac{T}{2\pi}\right)+1+S(T),$$

where $N(T)$ is the number of zeros of $\zeta(\sigma+it)$ in the region $0 < t < T$. The problem is thus reduced to the determination of $S(T)$. If the sign of $Z\left(\frac{T}{2\pi}\right)$ is known, the value of $S(T)$ is known modulo 2. It is not therefore ⟦10⟧
necessary to obtain $S(T)$ to any great accuracy. The principle of the method is that if $S_1(t) = \int_0^t S(u)\,du$ then $S_1(t)$ is known to be $O(\log t)$. If then the positions of the zeros are known in an interval of length L, $S(t)$ will be known

modulo 2 in this interval, the additive even integer being the same throughout. Hence $S_1(t_0+L)-S_1(t_0)$ will be known modulo $2L$, and if L is sufficiently large this will determine it exactly and thereby determine $S(t)$ throughout the interval. In order to complete the details of this argument it is necessary to replace the O result by a Θ result. It would also be desirable to try and arrange to manage with very limited knowledge of the positions of the zeros.

THEOREM 4. *If $t_2 > t_1 > 168\pi$, then*

$$S_1(t_2)-S_1(t_1) = \Theta\left(2{\cdot}30+0{\cdot}128\log\frac{t_2}{2\pi}\right).$$

The proof of this follows Theorem 40 of Titchmarsh (**3**). The essential step is

LEMMA 1. *If $t_2 > t_1 > 0$, then*

$$\pi\{S_1(t_2)-S_1(t_1)\} = \mathscr{R}\int_{\frac{1}{2}+it_2}^{\infty+it_2}\log\zeta(s)\,ds-\mathscr{R}\int_{\frac{1}{2}+it_1}^{\infty+it_1}\log\zeta(s)\,ds.$$

We apply Cauchy's theorem to $\log\zeta(s)$ and the rectangle with vertices $\frac{1}{2}+it_1$, $\frac{1}{2}+it_2$, $R+it_2$, $R+it_1$ and appropriate detours round the branch lines from zeros within the rectangle. The real part of the integral is

$$-\mathscr{R}\int_{\frac{1}{2}+it_2}^{R+it_2}\log\zeta(s)\,ds+\int_{\frac{1}{2}+it_1}^{\frac{1}{2}+it_2}\arg\zeta(s)(-i\,ds)+$$

[[11]] $$+\mathscr{R}\int_{\frac{1}{2}+it_1}^{R+it_1}\log\zeta(s)\,ds-\int_{R+it_1}^{R+it_2}\arg\zeta(s)(-i\,ds),$$

no contribution arising from the detours. The last of these integrals tends to 0 as $R\to\infty$ and the second is $\pi\{S_1(t_2)-S_1(t_1)\}$.

LEMMA 2. *If $\tau \geqslant 64$, we have*

$$|\zeta(\tfrac{1}{2}+2\pi i\tau)| < 4\tau^{\frac{1}{4}}.$$

[[12]] Since $|h(\zeta)| < 0{\cdot}95$ we have, by Theorem 1,

[[13]] $$|\tau(\tfrac{1}{2}+2\pi i\tau)| = |Z(\tau)| < 2\sum_{1\leqslant r\leqslant\tau^{\frac{1}{2}}} r^{-\frac{1}{2}}+1{\cdot}2\tau^{-\frac{3}{4}}+0{\cdot}95\tau^{-\frac{1}{4}}$$

and
$$\sum_{1\leqslant r\leqslant\tau^{\frac{1}{2}}} r^{-\frac{1}{2}} < 1+\int_1^{\tau^{\frac{1}{2}}} x^{-\frac{1}{2}}\,dx = 2\tau^{\frac{1}{4}}-1.$$

LEMMA 3.

$$|\zeta(1{\cdot}25+it)| < \zeta(1{\cdot}25) < 4{\cdot}6,$$

$$\left|\int_{1\cdot25+it}^{\infty} \log\zeta(s)\,ds\right| < \int_{1\cdot25}^{\infty} \log\zeta(\sigma)\,d\sigma < 1{\cdot}17,$$

$$\left|\frac{\zeta'}{\zeta}(1{\cdot}5+it)\right| < \frac{\zeta'}{\zeta}(1{\cdot}5) < 2{\cdot}62, \qquad [[14]]$$

$$\left|\int_{1\cdot5+it}^{2\cdot5+it} \log\zeta(s)\,ds\right| < \int_{1\cdot5}^{2\cdot5} \log\zeta(\sigma)\,d\sigma < 0{\cdot}548,$$

$$\left|\int_{1\cdot5+it}^{\infty} \log\zeta(s)\,ds\right| < \int_{1\cdot5}^{\infty} \log\zeta(\sigma)\,d\sigma < 0{\cdot}997,$$

$$\tfrac{1}{2}\log\pi > 0{\cdot}572.$$

These results are all based on the tables in Jahnke–Emde (**2**), p. 323. An error of two units in the last place is assumed. To the extent that we do not know how these tables were obtained we depart from the principles of the 'rigorous computation'.

LEMMA 4. *If* $\frac{1}{2} < \sigma < \frac{5}{4}$ *and* $t > 168\pi$, *then*

$$|\zeta(s)| < 4{\cdot}5t^{3/8-\frac{1}{4}\sigma}.$$

Consider $f(s)$:

$$f(s) = \zeta(s)\left(\frac{s-\frac{1}{2}}{i}\right)^{-3/8+\frac{1}{4}\sigma} \exp\left[\frac{-4\pi i}{s-\frac{1}{2}-127{\cdot}5\pi i}\right]. \qquad [[15]]$$

Now $$\left|\exp\left[\frac{-4\pi i}{s-\frac{1}{2}-127{\cdot}5\pi i}\right]\right| = \exp\left[\frac{-4\pi(t-127{\cdot}5\pi)}{(t-127{\cdot}5\pi)^2+(\sigma-\frac{1}{2})^2}\right].$$

Hence, by Lemma 3, $|f(s)| < 4$ on the line $\sigma = \frac{1}{2}$. Elsewhere, if $\frac{1}{2} < \sigma < \frac{5}{4}$, [[16]]
$t > 128\pi$, we have

$$\log\left|\left(\frac{s-\frac{1}{2}}{i}\right)^{-3/8+\frac{1}{4}s}\right| = \tfrac{1}{2}(-\tfrac{3}{8}+\tfrac{1}{4}\sigma)\log\{t^2+(\sigma-\tfrac{1}{2})^2\}+\tfrac{1}{4}t\tan^{-1}\frac{\sigma-\frac{1}{2}}{t}$$
$$\leqslant -\tfrac{1}{32}\log t^2+\tfrac{3}{16}. \qquad [[17]]$$

Hence on the line $\sigma = \frac{5}{4}$, $t \geqslant 128\pi$ we have

$$|f(s)| < \zeta(\tfrac{5}{4})e^{\frac{3}{16}}(128\pi)^{-\frac{1}{16}} < 4.$$

Finally on the line $t = 128\pi$, $\frac{1}{2} \leqslant \sigma \leqslant \frac{5}{4}$ we have

$$|f(s)| < |\zeta(s)|\exp\left[\frac{-2\pi^2}{\frac{1}{4}\pi^2+\frac{9}{16}}\right] \quad\text{and}\quad |\zeta(s)| < (128\pi)^{\frac{1}{4}} \qquad [[18]]$$

by equation (8) on p. 27 of Ingham (**1**). Hence $|f(s)| < 4$ on the whole boundary of the strip $t \geqslant 128\pi$, $\frac{1}{2} < \sigma < \frac{5}{4}$, and, since certainly $f(s) = O(t)$,

⟦19⟧ we have $|f(s)| < 4$ throughout the strip by the Phragmén–Lindelöf theorem. From this it follows that

⟦20⟧
$$|\zeta(s)| < 4e^{0\cdot 1}\left|\left(\frac{s-\frac{1}{2}}{i}\right)^{\frac{3}{8}-\frac{1}{4}s}\right|$$

⟦21⟧
$$< 4\cdot 5t^{\frac{3}{8}-\frac{1}{4}\sigma} \quad \text{for} \quad t > 168\pi.$$

The purpose of the factor $\exp\left[\dfrac{-4\pi i}{s-\frac{1}{2}-127\cdot 5\pi i}\right]$ is merely to enable us to do without accurate knowledge of $\zeta(s)$ over the end of the strip.

LEMMA 5. *If* $t > 168\pi$, *then*

$$\mathscr{R}\int\limits_{\frac{1}{2}+it}^{\infty+it} \log\zeta(s)\,ds \leqslant 2\cdot 30+0\cdot 12\log t.$$

For
$$\mathscr{R}\int\limits_{\frac{1}{2}+it}^{\infty+it} \log\zeta(s)\,ds \leqslant 1\cdot 17+\int\limits_{0\cdot 5}^{1\cdot 25} \log|\zeta(s)|\,ds, \text{ by Lemma 3,}$$
$$\leqslant 1\cdot 17+\int\limits_{0\cdot 5}^{1\cdot 25} \{\log 4\cdot 5+(\tfrac{3}{8}-\tfrac{1}{4}\sigma)\log t\}\,d\sigma,$$
by Lemma 4,
$$= 1\cdot 17+0\cdot 75\log 4\cdot 5+\tfrac{15}{128}\log t$$
$$< 2\cdot 30+0\cdot 12\log t.$$

It is certainly possible to improve the coefficient of $\log t$ in this result at the expense of the constant. The coefficient of $\log t$ could be reduced at any rate to $0\cdot 052$ using results stated on pp. 25, 26 of Titchmarsh (**3**).

LEMMA 6.

$$\frac{\zeta(s)\zeta(s+2)}{\{\zeta(s+1)\}^2} = \frac{s^2}{s^2-1}\frac{\{\Gamma(\frac{1}{2}s+\frac{3}{2})\}^2}{\Gamma(\frac{1}{2}s+1)\Gamma(\frac{1}{2}s+2)}\prod_{\rho}\frac{(s-\rho)(s-\rho+2)}{(s-\rho+1)^2},$$

where the product is over the non-trivial zeros of the zeta-function.

This is an immediate consequence of the Weierstrass product for the zeta-function.

⟦22⟧ LEMMA 7. *If* $k = 1\cdot 49$, $\mathscr{R}a \geqslant 0$, *then*

$$\mathscr{R}\left(\psi(a)+\frac{k}{a}\right) = \mathscr{R}\left[\int\limits_{a-1}^{a} \log\frac{z}{z+1}\,dz+\frac{k}{a}\right] \geqslant 0.$$

⟦23⟧ It is easily seen that if $\mathscr{R}a = 0$ then $\mathscr{R}\psi(a) = \mathscr{R}(k/a) = 0$. Also that $\mathscr{R}(k/a) \geqslant 0$ for $\mathscr{R}a \geqslant 0$ and that $\psi(a)$ is continuous at 0. $\psi(a)+(k/a) \to 0$ as $a \to \infty$. Hence applying the maximum modulus principle (or rather, the minimum real part principle) to $\psi(a)+(k/a)$ and various regions

$$\mathscr{R}a \geqslant 0, \qquad 0 < \epsilon < |a| < R,$$

we see by allowing $\epsilon \to 0$, $R \to \infty$ that the minimum real part must be achieved either on the boundary $\mathscr{R}a = 0$ or on the real axis (which may be a singularity). It only remains therefore to establish our inequality for the real axis. At any stationary point we must have

$$0 = \mathscr{R}\left(\psi'(a) - \frac{k}{a^2}\right) = -\log\left|1 - \frac{1}{a^2}\right| - \frac{k}{a^2}.$$

This equation only has solutions near to 0·91 and 1·2 both of which corre- ⟦24⟧
spond to minima of $\psi(a) + (k/a)$. There is no ordinary maximum separating them, but there is a singularity at $a = 1$. By computations near to these minima, and knowledge of an upper bound for the second derivative of the function in intervals enclosing them, one can show that the values at the minima are positive. The value at the lesser minimum (near 0·91) is about 0·0087. Hence $\psi(a) + (k/a) > 0$ on the real axis as required.

LEMMA 8. *If $\mathscr{R}z > 0$, then*

$$\frac{\Gamma'(z)}{\Gamma(z)} = \log z - \frac{1}{2z} + \Theta\left(\frac{2}{\pi^2 |(\mathscr{I}z)^2 - (\mathscr{R}z)^2|}\right).$$

We use the formula

$$\frac{\Gamma'(z)}{\Gamma(z)} = \log z - \frac{1}{2z} + 2\int_0^\infty \frac{u\,du}{(u^2+z^2)(e^{2\pi u}-1)}$$ ⟦25⟧

and take the line of integration to be $\mathscr{R}u = \mathscr{I}u = v > 0$. Then

$$\left|\frac{u}{e^{2\pi u}-1}\right| < \frac{e^{-\pi v}}{\pi\sqrt{2}}, \qquad |u^2+z^2| > |(\mathscr{I}z)^2 - (\mathscr{R}z)^2|.$$

No poles are encountered in the change of line of integration since $\mathscr{R}z > 0$. ⟦26⟧

LEMMA 9. *If $t > 50$, then*

$$-\mathscr{R}\int_{\frac{1}{2}+it}^{\infty} \log\zeta(s)\,ds < 4{\cdot}9 + 0{\cdot}245\log\frac{t}{2\pi}.$$

We have

$$\mathscr{R}\int_{\frac{1}{2}+it}^{\infty} \log\zeta(s)\,ds = \mathscr{R}\int_{\frac{1}{2}+it}^{\infty} \log\frac{\zeta(s)\zeta(s+2)}{\{\zeta(s+1)\}^2}\,ds +$$

$$+\mathscr{R}\int_{\frac{3}{2}+it}^{\infty} \log\zeta(s)\,ds + \mathscr{R}\int_{\frac{3}{2}+it}^{\frac{5}{2}+it} \log\zeta(s)\,ds$$

$$\geqslant \mathscr{R}\int_{\frac{1}{2}+it}^{\infty} \log\frac{\zeta(s)\zeta(s+2)}{\{\zeta(s+1)\}^2}\,ds - 1{\cdot}545,$$

by Lemma 3.

Also, by Lemma 6,

$$\int\limits_{\frac12+it}^{\infty} \log\frac{\zeta(s)\zeta(s+2)}{\{\zeta(s+1)\}^2}\,ds = \sum_{\rho}\int\limits_{\frac12+it}^{\frac32+it} \log\frac{s-\rho}{s-\rho+1}\,ds - $$

$$- \int\limits_{\frac12+it}^{\frac32+it} \log\frac{\Gamma(\frac12 s+1)}{\Gamma(\frac12 s+\frac32)}\,ds + \int\limits_{\frac12+it}^{\frac32+it} \log\frac{s}{s-1}\,ds.$$

Now if $\frac12 < \sigma < \frac32$, then $|s-1| < |s|$, and therefore

$$\mathscr{R}\int\limits_{\frac12+it}^{\frac32+it} \log\frac{s}{s-1}\,ds \geqslant 0.$$

Also $$\mathscr{R}\int\limits_{\frac12+it}^{\frac32+it} \log\frac{\Gamma(\frac12 s+1)}{\Gamma(\frac12 s+\frac32)}\,ds = -\tfrac12\mathscr{R}\frac{\Gamma'}{\Gamma}(\tfrac12 it+\sigma)$$

for some σ, $\frac54 < \sigma < \frac94$, by the mean value theorems,

⟦27⟧ $$\leqslant -\tfrac14\log(\tfrac14 t^2+\tfrac{25}{16}) - \frac{\frac52}{t^2+\frac{25}{4}} + \frac{2}{\pi^2(\frac14 t^2-\frac{81}{16})}$$

by Lemma 8,

⟦28⟧ $$< -\tfrac12\log\tfrac12 t \quad (\text{since } t > 50)$$

$$< -\tfrac12\log\frac{t}{2\pi} - 0{\cdot}572, \text{ by Lemma 3.}$$

Finally

⟦29⟧ $$\mathscr{R}\sum_{\rho}\int\limits_{\frac12+it}^{\frac32+it} \log\frac{s-\rho}{s-\rho+1}\,ds \geqslant -1{\cdot}49\mathscr{R}\sum_{\rho}\frac{1}{it-\rho+\frac32},$$

by Lemma 7,

$$= -1{\cdot}49\mathscr{R}\left[\frac{\zeta'}{\zeta}(it+\tfrac32) - \tfrac12\log\pi + \tfrac12\frac{\Gamma'}{\Gamma}(\tfrac12 it+\tfrac74)\right],$$

by the Mittag-Leffler series for $\dfrac{\zeta'}{\zeta}(s)$,

$$\geqslant -1{\cdot}49\left[\mathscr{R}\frac{\zeta'}{\zeta}(it+\tfrac32) - \tfrac12\log\pi + \tfrac14\log(\tfrac14 t^2+\tfrac{49}{16}) - \frac{7}{4t^2+49}\right],$$

by Lemma 8,

$$\geqslant -1{\cdot}49\left[\tfrac12\log\frac{t}{2\pi} + 2{\cdot}63\right],$$

using Lemma 3 and $t > 50$.

Combining these results gives the asserted inequality.

⟦90⟧

A variant of this method enables us to reduce the coefficient of τ to $\frac{1}{2}\log 2-\frac{1}{4}+\epsilon$, e.g. to 0·097, at the expense of the constant term.

Theorem 4 follows at once from Lemmas 1, 5, 9. ⟦30⟧

It is convenient to replace Theorem 4 by a similar result with $\kappa(\tau)$ or $\kappa_1(\tau)$ as the independent variable. This is because $\kappa(\tau)$ describes the 'expected' position of the zeros, and is therefore more informative than τ.

LEMMA 10. *If* $\tau_1 > 84$, *then*

$$\int_{\tau_1}^{\tau_2} S(2\pi\tau)\,d\kappa_1(\tau) = \Theta\{0{\cdot}184\log\tau_2+0{\cdot}0103(\log\tau_2)^2\}.$$

For

$$\int_{\tau_1}^{\tau_2} S(2\pi\tau)\,d\kappa_1(\tau) = \frac{1}{2\pi}\kappa_1'(\tau_1)\{S_1(2\pi\tau_2)-S_1(2\pi\tau_1)\}-$$

$$-\frac{1}{2\pi}\int_{\tau_1}^{\tau_2}\{S_1(2\pi\tau_2)-S_1(2\pi\tau)\}\kappa_1''(\tau)\,d\tau \qquad ⟦31⟧$$

$$= \Theta\left[\frac{2{\cdot}30+0{\cdot}128\log\tau_2}{2\pi}\left\{\kappa_1'(\tau_1)+\int_{\tau_1}^{\tau_2}|\kappa_1''(\tau)|\,d\tau\right\}\right]$$

$$= \Theta\left(\frac{2{\cdot}30+0{\cdot}128\log\tau_2}{4\pi}\log\tau_2\right).$$

THEOREM 5. *Let*

$$64 < \tau_{-R_1} < \tau_{1-R_1} < \ldots < \tau_0 < \ldots < \tau_{R_2-1} < \tau_{R_2}$$

and $\kappa(\tau_r) = c_r$, $\delta_r = c_r-c_0-\frac{1}{2}r$, $\delta_{R_2} = \delta_{-R_1} = 0$, *and* $Z(\tau_r)Z(\tau_{r+1}) < 0$ *if* $1-R_1 \leqslant r \leqslant R_2-2$, $\tau_{-R_1} > 84$. *Then* ⟦32⟧

$$-\frac{1}{2}+\frac{2}{R_1}\sum_{r=1-R_1}^{-1}\delta_r-\frac{0{\cdot}006}{\tau_{-R_1}}-\frac{2}{R_1}\{0{\cdot}184\log\tau_0+0{\cdot}0103(\log\tau_0)^2\}$$

$$\leqslant N(2\pi\tau_0)-2c_0-1$$

$$\leqslant \frac{1}{2}+\frac{2}{R_2}\sum_{r=1}^{R_2-1}\delta_r+\frac{0{\cdot}006}{\tau_0}+\frac{2}{R_1}\{0{\cdot}184\log\tau_{R_2}+0{\cdot}0103(\log\tau_{R_2})^2\}. \qquad ⟦33⟧$$

In the interval (τ_r, τ_{r+1}) we have $N(2\pi\tau) \geqslant N(2\pi\tau_0)+r$ if $0 \leqslant r \leqslant R_2-1$ ⟦34⟧ and therefore

$$\int_{\tau_0}^{\tau_R} N(2\pi\tau)\,d\kappa_1(\tau) \geqslant \sum_{r=0}^{R-1}(c_{r+1}-c_r)\{N(2\pi\tau_0)+r\} \qquad ⟦35⟧$$

$$= \tfrac{1}{2}R[N(2\pi\tau_0)+\tfrac{1}{2}(R-1)]-\sum_{r=1}^{R}\delta_r.$$

Also

⟦36⟧ $$\int_{\tau_0}^{\tau_R} \{2\kappa(\tau)+1\}\,d\kappa_1(\tau) = c_R - c_0 + (c_R^2 - c_0^2) + \Theta\left(\frac{0{\cdot}006(c_R - c_0)}{\tau_0}\right)$$

$$= \tfrac{1}{2}R(1+2c_0+\tfrac{1}{2}R) + \Theta\left(\frac{0{\cdot}003R}{\tau_0}\right).$$

The second inequality now follows since $S(2\pi\tau) = N(2\pi\tau) - 1 - 2\kappa(\tau)$ and the first may be proved similarly.

Example. It is known by computation that within distance 0·05 of each of the half-integers $547\frac{1}{2}$ to $554\frac{1}{2}$ there lie values of κ such that the corresponding value of Z has the same sign as $\cos 2\pi\kappa$. It is required to show that if τ_0 is that one of the points concerned which is within 0·05 of 551 then $N(2\pi\tau_0) = 1103$.

We take the values concerned to be $\tau_{-7}, \tau_{-6}, \ldots, \tau_7$ in Theorem 5,
⟦37⟧ and define τ_{-8}, τ_8 to satisfy $\delta_{-8} = \delta_8 = 0$. Then $|\delta_r| < 0{\cdot}1$ for each r, $-7 \leqslant r \leqslant 7$. The conditions of Theorem 5 are satisfied and it gives

⟦38⟧ $$-1{\cdot}0 \leqslant N(2\pi\tau_0) - 2c_0 - 1 \leqslant 1{\cdot}0.$$

⟦39⟧ $N(2\pi\tau_0)$ is odd since $Z(\tau_0)\cos 2\pi\kappa(\tau_0) > 0$ and we also have

$$|c_0 - 551| < 0{\cdot}05.$$

The required conclusion now follows.

PART II. THE COMPUTATIONS

Essentials of the Manchester Computer

It is not intended to give any detailed account of the Manchester Computer here, but a few facts must be mentioned if the strategy of the computation is to be understood. The storage of the machine is of two kinds, known as 'electronic' and 'magnetic' storage. The electronic storage consisted of four 'pages' each of thirty-two lines of forty binary digits. The magnetic storage consisted of a certain number of tracks each of two pages of similar capacity. Only about eight of these tracks were available for the zeta-function calculations. It was possible at any time to transfer one or both pages of a track to the electronic storage by an appropriate instruction. This operation takes about 60 ms. (milliseconds). Transfers to the magnetic store from the electronic were also possible, but were in fact only used for preparatory loading of the magnetic store. The course of the calculations is controlled by instructions each of twenty binary digits. These are normally magnetically stored, but must be transferred to the electronic store before they can be obeyed. In the initial state of the machine (with the magnetic store loaded) the electronic store is filled with zeros. A zero instruction, however, has a definite meaning, and in fact results in a transfer of instructions to the electronic store, thus initiating the calculation. Most instructions, such as transfer of 'lines' of forty digits, take 1·8 ms., but transfers to or from the magnetic store take longer, as has been mentioned, and multiplications take a time depending on the number of digits 1 in the multiplier, ranging from 3·6 ms. for a power of two to 39 ms. for $2^{40}-1$.

The results of the calculations are punched out on teleprint tape. This is a slow process in comparison with the calculations, taking about 150 ms. per character. The content of a tape may afterwards automatically be printed out with a typewriter if desired. The significance of what is printed out is determined by the 'programmer'. In the present case the output consisted mainly of numbers in the scale of 32 using the code

0	1	2	3	4	5	6	7	8	9	10	11	12	13	14	15	16	17	18	19
/	E	@	A	:	S	I	U	¼	D	R	J	N	F	C	K	T	Z	L	W

20	21	22	23	24	25	26	27	28	29	30	31
H	Y	P	Q	O	B	G	$\|$	M	X	V	£

and writing the most significant digit on the right. More conventionally the scale of 10 can be used, but this would require the storage of a conversion routine, and the writer was entirely content to see the results in the scale of 32, with which he is sufficiently familiar.

Outline of calculation method

The calculations had of course to be planned so that the total storage capacity used was within the capacity of the machine. So long as this was fulfilled it was desirable to make the time of calculation as short as possible without excessive trouble in programming. The most time-consuming part of the calculations is of course the computation of the terms

$$n^{-\frac{1}{2}} \cos 2\pi(\tau \log n - \kappa)$$

from given κ and τ. By storing tables of $\log n$ and $n^{-\frac{1}{2}}$ within the machine this was reduced essentially to two multiplications and the calculation of a cosine, together with arrangements for 'looking up' the logarithm and reciprocal square root. The cosines were obtained from a table giving $\cos(r\pi/128)$ for $0 \leqslant r \leqslant 64$ by linear interpolation and reducing to the first quadrant. This gives an error of less than 10^{-4}, which is quite sufficient accuracy for the purpose. Very much greater accuracy was of course required in the logarithms, for an error ϵ in $\log n$ gives rise to an error approaching $2\pi\tau\epsilon$ in the cosine, and $2\pi\tau$ may be very large, e.g. 25,000. These logarithms were calculated by the machine in a previous computation, and were given with an error not exceeding 2.10^{-10}. The reciprocal square roots were given with error not exceeding 10^{-5}. Both the logarithms and the reciprocal square roots were checked after loading into the magnetic store by automatic addition, the results obtained being compared with values based respectively on Stirling's formula and on the known value of $\zeta(\frac{1}{2})$. The table only went as far as $n = 63$. The tabular cosines were built up automatically from the values of $\cos(\pi/128)$ and $\sin(\pi/128)$ by using the addition formula. The values of $\cos(\pi/128)$ and $\sin(\pi/128)$ were calculated both automatically and manually. A hand-copying process was used in connexion with this table, but the final results when loaded were automatically thrice differenced and the results inspected.

The routine as a whole was checked (amongst other methods) by comparing the result given for a value of τ about 20,000 with an entirely different, slower, and simpler routine. This routine had itself been checked against a hand-computed value for $\tau = 16$ and against a value given by Titchmarsh (**5**) for $\tau = 201{\cdot}596$.

Since it was only necessary to calculate $\kappa(\tau)$ once for each value of τ this calculation did not have to be particularly quickly performed. It was considered sufficient to obtain the logarithm by means of a slow but simple routine taking about 1·2 sec. The time for each term $n^{-\frac{1}{2}} \cos 2\pi(\tau \log n - \kappa)$ was about 0·2 sec. With $m = 63$, and allowing for the calculation of $\kappa_1(\tau)$ this means about 14 secs. for each value of τ. The routine could be used for recording the results for given values of τ, a typical entry obtained in this way being:

ZETAFASTG/F@Q¼B£YNK@:ZSZ″XVMX///SA/////¼OTNR@O//.

This entry has to be divided into sequences of eight characters. In this case they are:

1. ZETAFAST. This occurs at the beginning of each entry. Its purpose is mainly to identify the document as referring to this zeta-function routine.
2. G/F@Q¼B£. This is a number useful in checking results and called the 'cumulant'. It appears in the scale of 32, with the most significant digit on the right. This is the standard method of representing numbers on documents connected with the Manchester Computer (a decimal method can also be used if desired).
3. YNK@:ZSZ. This is also in the scale of 32 and gives the residue of $2^{40}\kappa_1(\tau)$ modulo 2^{40}. Since Z is the symbol for 17 it will be seen that $\kappa_1(\tau)$ is near to $\frac{1}{2}$ mod 1.
4. ″XVMX///. This gives the value of $2^{17}\tau$; in this case τ is about 239·24.
5. SA/////¼. This was always included in the record due to a minor

difficulty in the programming. It did not seem worth while to take the trouble to eliminate it.

6. OTNR@O//. This is the value of $2^{30}Z(\tau)$ modulo 2^{40}. In this case $Z(\tau)$ is about 0·75.

The routine was not, however, used mostly for the calculation of values of $Z(\tau)$ with individually given τ. It was made to determine for itself appropriate values of τ, such as to give values of $2\kappa(\tau)$ near to successive integers. This was done by making each τ depend on the immediately previous one and on the previous κ by the formula $\tau' = \tau+(1-\delta)\alpha$, where τ', τ are the new and old values of τ respectively, δ is the difference of $2\kappa_1(\tau)$ from the nearest integer, and $(\alpha \log \tau)^{-1} = 1+\Theta(0{\cdot}1)$. This pro- [[40]]
cedure ensured that if the initial value of $\kappa(\tau)$ differs from an integer by less than 0·125, then the succeeding values will do likewise. It was decided not to record all the values of $Z(\tau)$, partly because the inspection and filing of the teleprint tape output would have a great burden to the experimenters. Values were only recorded when the unexpected sign occurred, i.e. when $Z(\tau)\cos 2\pi\kappa(\tau) < 0$. This reduced the amount of output data by about 90 per cent.

In order that there should be no doubt about the validity of the results it is necessary that one should also record all cases where the sign of $Z(\tau)$ is doubtful because of the limited accuracy of the computation. The criterion actually used was $Z(\tau)H(\kappa) > 0{\cdot}31E$, where [[41]]

$$H(\kappa) \equiv \kappa-\tfrac{1}{4}(\mathrm{mod}\,1), \qquad |H(\kappa)| < 0{\cdot}31.$$

The quantity $H(\kappa)$ arises very naturally with the computer. The condition $(\alpha \log \tau)^{-1} = 1+\Theta(0{\cdot}1)$ ensures that (except for one or two values at the [[42]]
beginning of a run), $|H(\kappa)| < 0{\cdot}31$. The actual errors involved in the calculation were:

Error arising from using Titchmarsh's formula (Theorem 3)	$1{\cdot}15m^{-\frac{3}{2}}$
Error due to replacing $\tau^{-\frac{1}{4}}h(\xi)$ by $m^{-\frac{1}{2}}h(\xi)$	$0{\cdot}47m^{-\frac{3}{2}}$
Error due to replacing ξ by $\frac{1}{2}\tau m^{-1}-\frac{1}{2}m$	$1{\cdot}08m^{-\frac{3}{2}}$
Error from using tabulated logarithms	$5{\cdot}1\times 10^{-10}m^{\frac{5}{2}}$
Error in replacing $\kappa(\tau)$ by $\kappa_1(\tau)$	$0{\cdot}15m^{-\frac{3}{2}}$
Error in calculating $\kappa_1(\tau)$	$2\times 10^{-10}m^{\frac{5}{2}}$
Error from using tabulated reciprocal square roots	$1{\cdot}3\times 10^{-4}m$
Error from using tabulated cosines and linear interpolations	$3{\cdot}2\times 10^{-4}m^{\frac{1}{2}}$
Error of Theorem 2	$0{\cdot}0243m^{-\frac{1}{2}}$

There are also numerous rounding off errors which are very small. These and all the 'cross terms' have been absorbed into the above errors so that we may put the whole error as not more than

$$E = 2{\cdot}85m^{-\frac{3}{2}}+0{\cdot}0243m^{-\frac{1}{2}}+3{\cdot}2\times 10^{-4}m^{\frac{1}{2}}+1{\cdot}3\times 10^{-4}m+7{\cdot}1\times 10^{-10}m^{\frac{5}{2}},$$

e.g. for $m = 15$

$$E < 0{\cdot}057,$$

and for $m = 65$

$$E < 0{\cdot}02.$$

The storage available was distributed as follows:

Magnetic store	
Logarithms routine (for κ)	1 page
Table of logarithms and reciprocal square roots	4 pages
Routine for calculating the terms $n^{-\frac{1}{2}}\cos 2\pi(\tau \log n - \kappa)$ and table of cosines	2 pages
Remainder of routine for calculating the function $Z(\tau)$	2 pages
Input routine	2 pages
Output routine	2 pages
Electronic store, as occupied during the greater part of the time	
Instructions and cosines	2 pages
Logarithms and reciprocal square roots	1 page
Miscellaneous data and working space	1 page

The principal investigation concerned the range $63^2 \leqslant \tau \leqslant 64^2$, i.e. the interval in which $m = 63$. Working at full efficiency it should have taken about 4 hours to calculate these values, the number of zeros concerned being about 1070. Full efficiency was not, however, achieved, and the calculation took about 9 hours. Only a small amount of this additional time was accounted for by duplicating the work. The special investigations in the neighbourhood of points where the unexpected sign occurred took a further 8 hours. The general reliability of the machine was checked from time to time by repeating small sections. The recorded cumulants were useful in this connexion. These cumulants were the totals of the values of $Z(\tau)$ computed since the last recorded value. If a calculation is repeated and there is agreement in cumulant value then there is a strong presumption that there is also agreement in all the individual values contributing to it. The result of this investigation, so far as it can be relied on, was that there are no complex zeros or multiple real zeros of $Z(\tau)$ in the region

$$63^2 \leqslant \tau \leqslant 64^2,$$

i.e. all zeros of $\zeta(s)$ in the region $2\pi.63^2 \leqslant t \leqslant 2\pi.64^2$ are simple zeros on the critical line.

Another investigation was also started with a view to extending the range of relatively small values of t for which the Riemann hypothesis holds. Titchmarsh has already proved that it holds up to $t = 1468$, i.e. to about $\tau = 231$. The new investigation started somewhat before $\tau = 225$ to allow a margin for the application of Theorem 5. It was intended to continue the work up to about $\tau = 500$, but an early breakdown resulted in its abandonment at $\tau = 256$. After applying Theorem 5 it would only be possible to assert the validity of the Riemann hypothesis up to about $\tau = 250$. All this part of the calculations was done twice, the unrecorded values being confirmed by means of the 'cumulants'.

Unfortunately $0{\cdot}31E$ was given the inappropriate value of $\frac{1}{128}$ and consequently we are only able to assert the validity of the Riemann hypothesis as far as $t = 1540$, a negligible advance.

REFERENCES

1. A. E. Ingham, *The distribution of prime numbers*, Cambridge Mathematical Tracts, No. 30 (1931).
2. Jahnke u. Emde (1), *Tafeln höherer Funktionen*, Leipzig, 1948.
3. E. C. Titchmarsh, *The zeta-function of Riemann*, Cambridge Mathematical Tracts, No. 26 (1930).
4. —— 'The zeros of the Riemann zeta-function', *Proc. Roy. Soc.* A, 151 (1935), 234–55.
5. —— 'The zeros of the Riemann zeta-function', ibid. 157 (1936), 261–3.
6. A. M. Turing, 'A method for the calculation of the zeta-function', *Proc. London Math. Soc.* (2) 48 (1943), 180–97.

The University,
Manchester.

SOLVABLE AND UNSOLVABLE PROBLEMS

A. M. TURING, F.R.S.

If one is given a puzzle to solve one will usually, if it proves to be difficult, ask the owner whether it can be done. Such a question should have a quite definite answer, yes or no, at any rate provided the rules describing what you are allowed to do are perfectly clear. Of course the owner of the puzzle may not know the answer. One might equally ask, 'How can one tell whether a puzzle is solvable?', but this cannot be answered so straightforwardly. The fact of the matter is that there is *no* systematic method of testing puzzles to see whether they are solvable or not. If by this one meant merely that nobody had ever yet found a test which could be applied to any puzzle, there would be nothing at all remarkable in the statement. It would have been a great achievement to have invented such a test, so we can hardly be surprised that it has never been done. But it is not merely that the test has never been found. It has been proved that no such test ever can be found.

Let us get away from generalities a little and consider a particular puzzle. One which has been on sale during the last few years and has probably been seen by most of the readers of this article illustrates a number of the points involved quite well. The puzzle consists of a large square within which are some smaller movable squares numbered 1 to 15, and one empty space, into which any of the neighbouring squares can be slid leaving a new empty space behind it. One may be asked to transform a given arrangement of the squares into another by a succession of such movements of a square into an empty space. For this puzzle there is a fairly simple and quite practicable rule by which one can tell whether the transformation required is possible or not. One first imagines the transformation carried out according to a different set of rules. As well as sliding the squares into the empty space one is allowed to make moves each consisting of two interchanges, each of one pair of squares. One would, for instance, be allowed as one move to interchange the squares numbered 4 and 7, and also the squares numbered 3 and 5. One is permitted to use the same number in both pairs. Thus one may replace 1 by 2, 2 by 3, and 3 by 1 as a move because this is the same as interchanging first (1, 2) and then (1, 3). The original puzzle is solvable by sliding if it is solvable according to the new rules. It is not solvable by sliding if the required position can be reached by the new rules, together with a 'cheat' consisting of *one single* interchange of a pair of squares.* Suppose, for instance, that one is asked to get back to the standard position –

1	2	3	4
5	6	7	8
9	10	11	12
13	14	15	

from the position

10	1	4	5
9	2	6	8
11	3		15
13	14	7	12

One may, according to the modified rules, first get the empty square into the correct position by moving the squares 15 and 12, and then get the squares 1, 2, 3, . . . successively into their correct positions by the interchanges (1, 10), (2, 10), (3, 4), (4, 5), (5, 9), (6, 10), (7, 10), (9, 11), (10, 11), (11, 15). The squares 8, 12, 13, 14, 15 are found to be already in their correct positions when their turns are reached. Since the number of interchanges required is

*It would take us too far from our main purpose to give the proof of this rule: the reader should have little difficulty in proving it by making use of the fact that an odd number of interchanges can never bring a set of objects back to the position it started from.

even, this transformation is possible by sliding.† If one were required after this to interchange say square 14 and 15 it could not be done.

This explanation of the theory of the puzzle can be regarded as entirely satisfactory. It gives one a simple rule for determining for any two positions whether one can get from one to the other or not. That the rule is so satisfactory depends very largely on the fact that it does not take very long to apply. No mathematical method can be useful for any problem if it involves much calculation. It is nevertheless sometimes interesting to consider whether something is possible at all or not, without worrying whether, in case it *is* possible, the amount of labour or calculation is economically prohibitive. These investigations that are not concerned with the amount of work involved are in some ways easier to carry out, and they certainly have a greater aesthetic appeal. The results are not altogether without value, for if one has proved that there is no method of doing something it follows *a fortiori* that there is no practicable method. On the other hand, if one method has been proved to exist by which the decision can be made, it gives some encouragement to anyone who wishes to find a workable method.

From this point of view, in which one is only interested in the question, 'Is there a systematic way of deciding whether puzzles of this kind are solvable?', the rules which have been described for the sliding-squares puzzle are much more special and detailed than is really necessary. It would be quite enough to say: 'Certainly one can find out whether one position can be reached from another by a systematic procedure. There are only a finite number of positions in which the numbered squares can be arranged (viz. 20922789888000) and only a finite number (2, 3, or 4) of moves in each position. By making a list of all the

†It can in fact be done by sliding successively the squares numbered 7, 14, 13, 11, 9, 10, 1, 2, 3, 7, 15, 8, 5, 4, 6, 3, 10, 1, 2, 6, 3, 10, 6, 2, 1, 6, 7, 15, 8, 5, 10, 8, 5, 10, 8, 7, 6, 9, 15, 5, 10, 8, 7, 6, 5, 15, 9, 5, 6, 7, 8, 12, 14, 13, 15, 10, 13, 15, 11, 9, 10, 11, 15, 13, 12, 14, 13, 15, 9, 10, 11, 12, 14, 13, 15, 14, 13, 15, 14, 13, 12, 11, 10, 9, 13, 14, 15, 12, 11, 10, 9, 13, 14, 15.

positions and working through all the moves, one can divide the positions into classes, such that sliding the squares allows one to get to any position which is in the same class as the one started from. By looking up which classes the two positions belong to one can tell whether one can get from one to the other or not.' This is all, of course, perfectly true, but one would hardly find such remarks helpful if they were made in reply to a request for an explanation of how the puzzle should be done. In fact they are so obvious that under such circumstances one might find them somehow rather insulting. But the fact of the matter is, that if one is interested in the question as put, 'Can one tell by a systematic method in which cases the puzzle is solvable?', this answer is entirely appropriate, because one wants to know if there is a systematic method, rather than to know of a good one.

The same kind of argument will apply for any puzzle where one is allowed to move certain 'pieces' around in a specified manner, provided that the total number of essentially different positions which the pieces can take up is finite. A slight variation on the argument is necessary in general to allow for the fact that in many puzzles some moves are allowed which one is not permitted to reverse. But one can still make a list of the positions, and list against these first the positions which can be reached from them in one move. One then adds the positions which are reached by two moves and so on until an increase in the number of moves does not give rise to any further entries. For instance, we can say at once that there is a method of deciding whether a patience can be got out with a given order of the cards in the pack: it is to be understood that there is only a finite number of places in which a card is ever to be placed on the table. It may be argued that one is permitted to put the cards down in a manner which is not perfectly regular, but one can still say that there is only a finite number of 'essentially different' positions. A more interesting example is provided by those puzzles made (apparently at least) of two or more pieces of very thick twisted wire which one is required to separate. It is understood that one is not allowed to bend the wires at all, and when one makes the right movement there is always plenty of room to get the pieces apart without them ever touching, if one wishes to do so. One may describe the positions of the pieces by saying where some three definite points of each piece are. Because of the spare space it is not necessary to give these positions quite exactly. It would be enough to give them to, say, a tenth of a millimetre. One does not need to take any notice of movements of the puzzle as a whole: in fact one could suppose one of the pieces quite fixed. The second piece can be supposed to be not very far away, for, if it is, the puzzle is already solved. These considerations enable us to reduce the number of 'essentially different' positions to a finite number, probably a few hundred millions, and the usual argument will then apply. There are some further complications, which we will not consider in detail, if we do not know how much clearance to allow for. It is necessary to repeat the process again and again allowing successively smaller and smaller clearances. Eventually one will find that either it can be solved, allowing a small clearance margin, or else it cannot be solved even allowing a small margin of 'cheating' (i.e. of 'forcing', or having the pieces slightly overlapping in space). It will, of course, be understood that this process of trying out the possible positions is not to be done with the physical puzzle itself, but on paper, with mathematical descriptions of the positions, and mathematical criteria for deciding whether in a given position the pieces overlap, etc.

These puzzles where one is asked to separate rigid bodies are in a way like the 'puzzle' of trying to undo a tangle, or more generally of trying to turn one knot into another without cutting the string. The difference is that one is allowed to bend the string, but not the wire forming the rigid bodies. In either case, if one wants to treat the problem seriously and systematically one has to replace the physical puzzle by a mathematical equivalent. The knot puzzle lends itself quite conveniently to this. A knot is just a closed curve in three dimensions nowhere crossing itself; but, for the purpose we are interested in, any knot can be given accurately enough as a series of segments in the directions of the three coordinate axes. Thus, for instance, the trefoil knot (Figure 1*a*) may be regarded as consisting of a number of segments joining the points given, in the usual (x, y, z) system of coordinates, as (1, 1, 1), (4, 1, 1,), (4, 2, 1), (4, 2, −1), (2, 2, −1), (2, 2, 2), (2, 0, 2), (3, 0, 2), (3, 0, 0), (3, 3, 0), (1, 3, 0), (1, 3, 1) and returning again with a twelfth segment to the starting point (1, 1, 1). This representation of the knot is shown in perspective in Figure 1*b*. There is no special virtue in the representation which has been chosen. If it is desired to follow the original curve more closely a greater number of segments must be used. Now let *a* and *d* represent unit steps in the positive and negative X-directions respectively, *b* and *e* in the Y-directions, and *c* and *f* in the Z-directions: then this knot may be described as *aaabffddccceeaffbbbddcee*. One can then, if one wishes, deal entirely with such sequences of letters. In order that such a sequence should represent a knot it is necessary and sufficient that the numbers of *a*'s and *d*'s should be equal, and likewise the number of *b*'s equal to the number of *e*'s and the number of *c*'s equal to the number of *f*'s, and it must not be possible to obtain another sequence of letters with these properties by omitting a number of consecutive letters at the beginning or the end or both. One can turn a knot into an equivalent one by operations of the following kinds—

(i) One may move a letter from one end of the row to the other.

(ii) One may interchange two consecutive letters provided this still gives a knot.

(iii) One may introduce a letter *a* in one place in the row, and *d* somewhere else, or *b* and *e*, or *c* and *f*, or take such pairs out, provided it still gives a knot.

(iv) One may replace *a* everywhere by *aa* and *d* by *dd* or replace each *b* and *e* by *bb* and *ee* or each *c* and *f* by *cc* and *ff*. One may also reverse any such operation.

[[1]] —and these are all the moves that are necessary.

It is also possible to give a similar symbolic equivalent for the problem of separating rigid bodies, but it is less straightforward than in the case of knots.

These knots provide an example of a puzzle where one cannot tell in advance how many arrangements of pieces may be involved (in this case the pieces are the letters *a, b, c, d. e, f*), so that

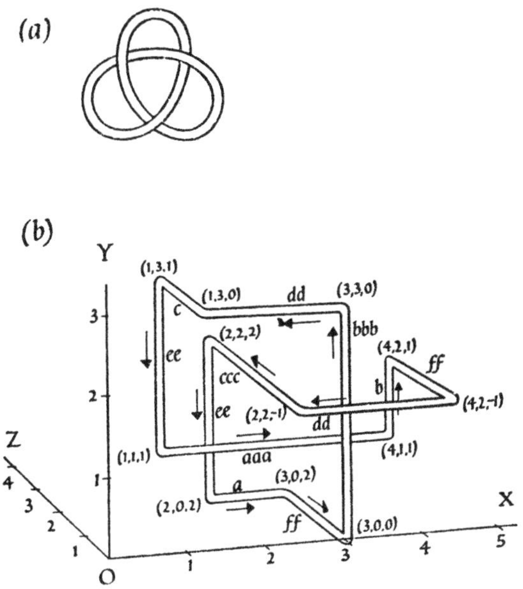

Fig. 1. (*a*) The trefoil knot (*b*) a possible representation of this knot as a number of segments joining points.

the usual method of determining whether the puzzle is solvable cannot be applied. Because of rules (iii) and (iv) the lengths of the sequences describing the knots may become indefinitely great. No systematic method is yet known by which one can tell whether two knots are the same. ⟦2⟧

Another type of puzzle which we shall find very important is the 'substitution puzzle'. In such a puzzle one is supposed to be ⟦3⟧
supplied with a finite number of different kinds of counters, perhaps just black (*B*) and white (*W*). Each kind is in unlimited supply. Initially a number of counters are arranged in a row and one is asked to transform it into another pattern by substitutions. A finite list of the substitutions allowed is given. Thus, for instance, one might be allowed the substitutions

(i) $WBW \to B$

(ii) $BW \to WBBW$

and be asked to transform *WBW* into *WBBBW*, which could be done as follows

$$W\underline{BW} \underset{\text{(ii)}}{\rightarrow} W\underline{WB\underline{BW}} \underset{\text{(ii)}}{\rightarrow} W\underline{WB\underline{W}BBW} \underset{\text{(i)}}{\rightarrow} W\underline{B}BBW$$

Here the substitutions used are indicated by the numbers below the arrows, and their effects by underlinings. On the other hand if one were asked to transform *WBB* into *BW* it could not be done, for there are no admissible steps which reduce the number of *B*'s.

It will be seen that with this puzzle, and with the majority of substitution puzzles, one cannot set any bound to the number of positions that the original position might give rise to.

It will have been realized by now that a puzzle can be something rather more important than just a toy. For instance the task of proving a given mathematical theorem within an axio-
⟦4⟧ matic system is a very good example of a puzzle.

It would be helpful if one had some kind of 'normal form' or 'standard form' for describing puzzles. There is, in fact, quite a reasonably simple one which I shall attempt to describe. It will be necessary for reasons of space to take a good deal for granted, but this need not obscure the main ideas. First of all we may suppose that the puzzle is somehow reduced to a mathematical form in the sort of way that was used in the case of the knots. The position of the puzzle may be described, as was done in that case, by sequences of symbols in a row. There is usually very little difficulty in reducing other arrangements of symbols (e.g. the squares in the sliding squares puzzle) to this form. The question which remains to be answered is, 'What sort of rules should one be allowed to have for rearranging the symbols or counters?' In order to answer this one needs to think about what kinds of processes ever do occur in such rules, and, in order to reduce their number, to break them up into simpler processes. Typical of such processes are counting, copying, comparing, substituting. When one is doing such processes, it is necessary, especially if there are many symbols involved, and if one wishes to avoid carrying too much information in one's head, either to make a number of jottings elsewhere or to use a number of marker objects as well as the pieces of the puzzle itself. For instance, if one were making a copy of a row of counters concerned in the puzzle it would be as well to have a marker which divided the pieces which have been copied from those which have not and another showing the end of the portion to be copied. Now there is no reason why the rules of the puzzle itself should not be expressed in such a way as to take account of these markers. If one does express the rules in this way they can be made to be just substitutions. This means to say that the *normal form for puzzles is the substitution type of puzzle.* More definitely we can say:

Given any puzzle we can find a corresponding substitution puzzle *which is equivalent to it in the sense that given a solution of the one we can easily use it to find a solution of the other. If the original puzzle is concerned with rows of pieces of a finite number of different kinds, then the substitutions may be applied as an alternative set of rules to the pieces of the original puzzle. A transformation can be carried out by the rules of the original puzzle if and only if it can be carried out by the substitutions and leads to a final position from which all marker symbols have disappeared.*

This statement is still somewhat lacking in definiteness, and will remain so. I do not propose, for instance, to enter here into the question as to what I mean by the word 'easily'. The statement is moreover one which one does not attempt to prove. Propaganda is more appropriate to it than proof, for its status is something between a theorem and a definition. In so far as we know *a priori* what is a puzzle and what is not, the statement is a theorem. In so far as we do not know what puzzles are, the statement is a definition which tells us something about what they are. One can of course define a puzzle by some phrase beginning, for instance, 'A set of definite rules . . .', but this just throws us back on the definition of 'definite rules'. Equally one can reduce it to the definition of 'computable function' or 'systematic procedure'. A definition of any one of these would define all the rest.
Since 1935 a number of definitions have been given, explaining ⟦5⟧
in detail the meaning of one or other of these terms, and these

have all been proved equivalent to one another and also equivalent to the above statement. In effect there is no opposition to
[[6]] the view that every puzzle is equivalent to a substitution puzzle.

After these preliminaries let us think again about puzzles as a whole. First let us recapitulate. There are a number of questions to which a puzzle may give rise. When given a particular task one may ask quite simply

(a) *Can this be done?*

Such a straightforward question admits only the straightforward answers, 'Yes' or 'No', or perhaps 'I don't know'. In the case that the answer is 'Yes' the answerer need only have done the puzzle himself beforehand to be sure. If the answer is to be 'No', some rather more subtle kind of argument, more or less mathematical, is necessary. For instance, in the case of the sliding squares one can state that the impossible cases *are* impossible because of the mathematical fact that an odd number of simple interchanges of a number of objects can never bring one back to where one started. One may also be asked

(b) *What is the best way of doing this?*

Such a question does not admit of a straightforward answer. It depends partly on individual differences in people's ideas as to what they find easy. If it is put in the form, 'What is the solution which involves the smallest number of steps?', we again have a straightforward question, but now it is one which is
[[7]] somehow of remarkably little interest. In any particular case where the answer to (a) is 'Yes' one can find the smallest possible number of steps by a tedious and usually impracticable process of enumeration, but the result hardly justifies the labour.

When one has been asked a number of times whether a number of different puzzles of similar nature can be solved one is naturally led to ask oneself

(c) *Is there a systematic procedure by which I can answer these questions, for puzzles of this type?*

If one were feeling rather more ambitious one might even ask

(d) *Is there a systematic procedure by which one can tell whether a puzzle is solvable?*

I hope to show that the answer to this last question is 'No'.

There are in fact certain types of puzzle for which the answer to (c) is 'No'.

Before we can consider this question properly we shall need to be quite clear what we mean by a 'systematic procedure' for deciding a question. But this need not now give us any particular difficulty. A 'systematic procedure' was one of the phrases which we mentioned as being equivalent to the idea of a puzzle, because either could be reduced to the other. If we are now clear as to what a puzzle is, then we should be equally clear about 'systematic procedures'. In fact a systematic procedure is just a puzzle *in which there is never more than one possible move in any of the positions which arise and in which some significance is attached to the final result.*

Now that we have explained the meaning both of the term 'puzzle' and of 'systematic procedure', we are in a position to prove the assertion made in the first paragraph of this article, that there cannot be any systematic procedure for determining whether a puzzle be solvable or not. The proof does not really require the detailed definition of either of the terms, but only the relation between them which we have just explained. Any systematic procedure for deciding whether a puzzle were solvable could certainly be put in the form of a puzzle, with unambiguous moves (i.e. only one move from any one position), and having for its starting position a combination of the rules, the starting position and the final position of the puzzle under investigation.

The puzzle under investigation is also to be described by its rules and starting position. Each of these is to be just a row of symbols. As we are only considering substitution puzzles, the rules need only be a list of all the substitution pairs appropriately punctuated. One possible form of punctuation would be to separate the first member of a pair from the second by an arrow, and to separate the different substitution pairs with colons. In this case the rules

B may be replaced by *BC*
WBW may be deleted

would be represented by ' : $B \rightarrow BC$: $WBW \rightarrow$:' . For the purposes of the argument which follows, however, these arrows and colons are an embarrassment. We shall need the rules to be expressed without the use of any symbols which are barred from appearing in the starting positions. This can be achieved by the following simple, though slightly artificial trick. We first double all the symbols other than the punctuation symbols, thus ': *BB* → *BBCC* : *WWBBWW* → :' . We then replace each arrow by a single symbol, which must be different from those on either side of it, and each colon by three similar symbols, also chosen to avoid clashes. This can always be done if we have at least three symbols available, and the rules above could then be represented as, for instance, '*CCCBBWBBCCBBBWWBBWW BWWW*'. Of course according to these conventions a great variety of different rows of symbols will describe essentially the same puzzle. Quite apart from the arbitrary choice of the punctuating symbols the substitution pairs can be given in any order, and the same pair can be repeated again and again.

Now let *P*(*R*,*S*) stand for 'the puzzle whose rules are described by the row of symbols *R* and whose starting position is described by *S*'. Owing to the special form in which we have chosen to describe the rules of puzzles, there is no reason why we should not consider *P*(*R*,*R*) for which the 'rules' also serve as starting position: in fact the success of the argument which follows depends on our doing so. The argument will also be mainly concerned with puzzles in which there is at most one possible move in any position; these may be called 'puzzles with unambiguous moves'. Such a puzzle may be said to have 'come out' if one reaches either the position *B* or the position *W*, and the rules do not permit any further moves. Clearly if a puzzle has unambiguous moves it cannot both come out with the end result *B* and with the end result *W*.

We now consider the problem of classifying rules *R* of puzzles into two classes, I and II, as follows:

Class I is to consist of sets *R* of rules, which represent puzzles with unambiguous moves, and such that *P*(*R*,*R*) comes out with the end result *W*.

Class II is to include all other cases, i.e. either *P*(*R*,*R*) does not come out, or comes out with the end result *B*, or else *R* does not represent a puzzle with unambiguous moves. We may also, if we wish, include in this class sequences of symbols such as *BBBBB* which do not represent a set of rules at all.

Now suppose that, contrary to the theorem that we wish to prove, we have a systematic procedure for deciding whether puzzles come out or not. Then with the aid of this procedure we shall be able to distinguish rules of class I from those of class II. There is no difficulty in deciding whether *R* really represents a set of rules, and whether they are unambiguous. If there is any difficulty it lies in finding the end result in the cases where the puzzle is known to come out: but this can be decided by actually working the puzzle through. By a principle which has already been explained, this systematic procedure for distinguishing the two classes can itself be put into the form of a substitution puzzle (with rules *K*, say). When applying these rules *K*, the rules *R* of the puzzle under investigation form the starting position, and the end result of the puzzle gives the result of the test. Since the procedure always gives an answer, the puzzle *P*(*K*,*R*) always comes out. The puzzle *K* might be made to announce its results in a variety of ways, and we may be permitted to suppose that the end result is *B* for rules *R* of class I, and *W* for rules of class II. The opposite choice would be equally possible, and would hold for a slightly different set of rules *K*′, which however we do not choose to favour with our attention. The puzzle with rules *K* may without difficulty be made to have unambiguous moves. Its essential properties are therefore:

K has unambiguous moves.
P(*K*,*R*) always comes out whatever *R*.
If *R* is in class I, then *P*(*K*,*R*) has end result *B*.
If *R* is in class II, then *P*(*K*,*R*) has end result *W*.

These properties are however inconsistent with the definitions of the two classes. If we ask ourselves which class *K* belongs to, we find that neither will do. The puzzle *P*(*K*,*K*) is bound to come out, but the properties of *K* tell us that we must get end result *B* if *K* is in class I and *W* if it is in class II, whereas the definitions of the classes tell us that the end results must be the other way round. The assumption that there was a systematic procedure for telling whether puzzles come out has thus been reduced to an absurdity.

Thus in connexion with question (c) above we can say that there are some types of puzzle for which no systematic method of deciding the question exists. This is often expressed in the form, 'There is no *decision procedure* for this type of puzzle', or again, 'The decision problem for this type of puzzle is unsolvable', and so one comes to speak (as in the title of this article) about 'unsolvable problems' meaning in effect puzzles for which there is no decision procedure. This is the technical meaning which the words are now given by mathematical logicians. It would seem more natural to use the phrase 'unsolvable problem' to mean just an unsolvable puzzle, as for example 'to transform 1, 2, 3 into 2, 1, 3 by cyclic permutation of the symbols', but this is not the meaning it now has. However, to minimize confusion I shall here always speak of 'unsolvable decision problems', rather than just 'unsolvable problems', and also speak of puzzles rather than problems where it is puzzles and not decision problems that are concerned.

It should be noticed that a decision problem only arises when one has an infinity of questions to ask. If you ask, 'Is this apple good to eat?', or 'Is this number prime?', or 'Is this puzzle solvable?' the question can be settled with a single 'Yes' or 'No'. A finite number of answers will deal with a question about a finite number of objects, such as the apples in a basket. When the number is infinite, or in some way not yet completed concerning say all the apples one may ever be offered, or all whole numbers or puzzles, a list of answers will not suffice. Some kind of rule or systematic procedure must be given. Even if the number concerned is finite one may still prefer to have a rule rather than a list: it may be easier to remember. But there certainly cannot be an unsolvable decision problem in such cases, because of the possibility of using finite list.

Regarding decision problems as being concerned with classes of puzzles, we see that if we have a decision method for one class it will apply also for any subclass. Likewise, if we have proved that there is no decision procedure for the subclass, it follows that there is none for the whole class. The most interesting and valuable results about unsolvable decision problems concern the smaller classes of puzzle.

Another point which is worth noticing is quite well illustrated by the puzzle which we considered first of all in which the pieces were sliding squares. If one wants to know whether the puzzle is solvable with a given starting position, one can try moving the pieces about in the hope of reaching the required end-position. If one succeeds, then one will have solved the puzzle and consequently will be able to answer the question, 'Is it solvable?' In the case that the puzzle is solvable one will eventually come on the right set of moves. If one has also a procedure by which, if the puzzle is unsolvable, one would eventually establish the fact that it was so, then one would have a solution of the decision problem for the puzzle. For it is only necessary to apply both processes, a bit of one alternating with a bit of the other, in order eventually to reach a conclusion by one or the other. Actually, in the case of the sliding squares problem, we have got such a procedure, for we know that if, by sliding, one ever reaches the required final position, with squares 14 and 15 interchanged, then the puzzle is impossible.

It is clear then that the difficulty in finding decision procedures for types of puzzle lies in establishing that the puzzle is unsolvable in those cases where it *is* unsolvable. This, as was mentioned on page 16, requires some sort of mathematical argument. This suggests that we might try expressing the statement that the puzzle comes out in a mathematical form and then try and prove it by some systematic process. There is no particular difficulty in the first part of this project, the mathematical expression of the statement about the puzzle. But the second half of the project is bound to fail, because by a famous theorem of Gödel no [[8]]
systematic method of proving mathematical theorems is sufficiently complete to settle every mathematical question, yes or no. In any case we are now in a position to give an independent proof of this. If there were such a systematic method of proving mathematical theorems we could apply it to our puzzles and for

each one eventually either prove that it was solvable or unsolvable; this would provide a systematic method of determining whether the puzzle was solvable or not, contrary to what we have already proved.

This result about the decision problem for puzzles, or, more accurately speaking, a number of others very similar to it, was
⟦9⟧ proved in 1936–7. Since then a considerable number of further decision problems have been shown to be unsolvable. They are all proved to be unsolvable by showing that if they were solvable one could use the solution to provide a solution of the original one. They could all without difficulty be reduced to the same unsolvable problem. A number of these results are mentioned very shortly below. No attempt is made to explain the technical terms used, as most readers will be familiar with some of them, and the space required for the explanation would be quite out of proportion to its usefulness in this context.

⟦10⟧ (1) It is not possible to solve the decision problem even for substitution processes applied to rows of black and white counters only.

⟦11⟧ (2) There are certain particular puzzles for which there is no decision procedure, the rules being fixed and the only variable element being the starting position.

⟦12⟧ (3) There is no procedure for deciding whether a given set of axioms leads to a contradiction or not.

⟦13⟧ (4) The 'word problem in semi-groups with cancellation' is not solvable.

⟦14⟧ (5) It has recently been announced from Russia that the 'word problem in groups' is not solvable. This is a decision problem not unlike the 'word problem in semi-groups', but very much more important, having applications in topology: attempts were being made to solve this decision problem before any such problems had been proved unsolvable. No adequately complete proof is yet available, but if it is correct this is a considerable step forward.

⟦15⟧ (6) There is a set of 102 matrices of order 4, with integral coefficients such that there is no decision method for determining

⟦114⟧

whether another given matrix is or is not expressible as a product of matrices from the given set.

These are, of course, only a selection from the results. Although quite a number of decision problems are now known to be unsolvable, we are still very far from being in a position to say of a given decision problem, whether it is solvable or not. Indeed, we shall never be quite in that position, for the question whether a given decision problem is solvable is itself one of the undecidable decision problems. The results which have been ⟦16⟧
found are on the whole ones which have fallen into our laps rather than ones which have positively been searched for. Considerable efforts have however been made over the word problem in groups (see (5) above). Another problem which mathematicians are very anxious to settle is known as 'the decision problem of the equivalence of manifolds'. This is something like one of the problems we have already mentioned, that concerning the twisted wire puzzles. But whereas with the twisted wire puzzles the pieces are quite rigid, the 'equivalence of manifolds' problem concerns pieces which one is allowed to bend, stretch, twist, or compress as much as one likes, without ever actually breaking them or making new junctions or filling in holes. Given a number of interlacing pieces of plasticine one may be asked to transform them in this way into another given form. The decision problem for this class of problem is the 'decision problem for the equivalence of manifolds'. It is probably unsolvable, but ⟦17⟧
has never been proved to be so. A similar decision problem which might well be unsolvable is the one concerning knots ⟦18⟧
which has already been mentioned.

The results which have been described in this article are mainly of a negative character, setting certain bounds to what we can hope to achieve purely by reasoning. These, and some other results of mathematical logic may be regarded as going some way towards a demonstration, within mathematics itself, of the inadequacy of 'reason' unsupported by common sense.

FURTHER READING

Kleene, S. C. *Introduction to Metamathematics*, Amsterdam, 1952.

A NOTE ON NORMAL NUMBERS

Although it is known that almost all numbers are normal[1] no example of a normal number has ever been given. I propose to show how normal ⟦1⟧ numbers may be constructed and to prove that almost all numbers are normal constructively.

Consider the R-figure integers in the scale of t, $t \geqslant 2$. If γ is any sequence of figures in that scale we denote by $N(t, \gamma, n, R)$ the number of these in ⟦2⟧ which γ occurs exactly n times. Then it can be proved without difficulty that

$$\left(\sum_{n=1}^{R} n N(t, \gamma, n, R)\right) \Big/ \left(\sum_{n=1}^{R} N(t, \gamma, n, R)\right) = R^{-1}(R - r + 1)t^{-r},$$ ⟦3⟧

where $l(\gamma) = r$ is the length of the sequence γ: it is also possible[2] to prove that

$$\sum_{|n - Rt^{-r}| > k} N(t, \gamma, n, R) < 2t^{R} e^{-k^2 t^r / 4R}, \tag{1}$$

provided $kt^r/R < 0.3$.

Let α be a real number and $S(\alpha, t, \gamma, R)$ the number of occurrences of γ in the first R figures after the decimal point in the expression of α in the scale of t. α is said to be *normal* if $R^{-1}S(\alpha, t, \gamma, R) \to t^{-r}$ as $R \to \infty$ for each γ, t, where $r = l(\gamma)$.

Now consider sums of a finite number of open intervals with rational end points. These can be enumerated constructively. We take a particular constructive enumeration: let E_n be the nth set of intervals in the enumeration. Then we have the next theorem.

Theorem 1. *We can find a constructive*[3] *function $c(k, n)$ of two integral variables such that $E_{c(k,n+1)} \leqslant E_{c(k,n)}$ and $m E_{c(k,n)} > 1 - 1/k$ for each k, n and $E(k) = \prod_{n=1}^{\infty} E_{c(k,n)}$ consists entirely of normal numbers for each k.*

Let $B(\Delta, \gamma, t, R)$ be the set of numbers α $(0 < \alpha < 1)$ for which

$$|S(\alpha, t, \gamma, R) - Rt^{-r}| < \frac{R}{\Delta t^r}, \quad \left(K = \frac{R}{\Delta t^r}\right), \quad \Delta = \frac{R}{Kt^r}. \tag{2}$$

Then by (1)

$$m B(\Delta, \gamma, t, R) > 1 - 2e^{-Rt^{-r}/4\Delta^2} \quad \text{if } \Delta < 0.3.$$ ⟦4⟧

Let $A(\Delta, T, L, R)$ be the set of those α for which (2) holds whenever $2 \leqslant t \leqslant T$ and $l(\gamma) \leqslant L$, i.e.,

$$A(\Delta, T, L, R) = \prod_{t=2}^{T} \prod_{l(\gamma) \leqslant L} B(\Delta, \gamma, t, R).$$

The number of factors in the product is at most T^{L+1} so that

⟦5⟧ $$m A(\Delta, T, L, R) > 1 - T^{L+1} e^{-RT^{-2}/4\Delta^2}.$$

Let

$$A_k = A([k^{1/4}], [e^{\sqrt{\log k}}], [\sqrt{\log k} - 1], k),$$
$$\bar{A}_k = A(k, [e^{\sqrt{\log k}}], [\sqrt{\log k} - 1], k^4).$$

Then, if $k \geqslant 1000$, we shall have

$$m A_k > 1 - k e^{-1/2k^{1/2}} > 1 - 1/k(k-1).$$

$c(k,n)$ $(k \geqslant 1000)$ is to be defined as follows. $c(k,0)$ is $(0,1)$. $c(k,n+1)$ is the intersection of an interval $(\beta_n, 1)$ $(0 \leqslant \beta_n < 1)$ with A_{k+n+1} and $c(k,n)$, β_n being chosen so that the measure of $c(k,n+1)$ is $1 - 1/k + (k+n+1)^{-1}$. This is possible since the measure of $c(k,n)$ is $1 - 1/k + 1/(k+n)$ and that of A_{k+n+1} is at least $1 - 1/((k+n)(k+n+1))$. Consequently the measure of
⟦6⟧ $c(k,n) \cap A_{k+n+1}$ is at least $1 - 1/k + 1/(k+n+1)$. If $k < 1000$ we define $c(k,n)$ to be $c(1000,n)$. $c(k,n)$ is a finite sum of intervals for each k, n. When we remove the boundary points we obtain a set of the form $E_{c(k,n)}$ of measure $1 - 1/k + 1/(k+n)$ $(k \geqslant 1000)$. The intervals of which $E_{c(k,n)}$ is composed may be found by a mechanical process and so the function $c(k,n)$ is constructive. The set $E(k) = \prod_{n=1}^{\infty} E_{c(k,n)}$ consists of normal numbers for if $\alpha \in E(k)$, then $\alpha \in A_k$ (all $k > K$, $k \geqslant 1000$). If γ is a sequence of length r in the scale of t and if k_0 be such that

$$[e^{\sqrt{\log k_0}}] > t \quad \text{and} \quad [\sqrt{\log k_0}] > r + 1,$$

then for $k > k_0$

$$|S(\beta, t, \gamma, k) - kt^{-r}| < k[k^{1/4}]^{-1},$$

where β is in A_k (by the definition of A_k). Hence $k^{-1} S(\alpha, t, \gamma, k) \to t^{-r}$ as k tends to infinity, i.e., α is normal.

Theorem 2. *There is a rule whereby given an integer k and an infinite sequence of figures* 0 *and* 1 *(the pth figure in the sequence being* $\theta(p)$*) we can find a normal number* $\alpha(k, \theta)$ *in the interval* $(0,1)$ *and in such a way* *that for fixed k these numbers form a set of measure at least* $1-2/k$ *and so that the first n figures of* θ *determine* $\alpha(k,\theta)$ *to within* 2^{-n}. ⟦7⟧

With each integer n we associate an interval of the form $(m_n/2^n, (m_n+1)/2^n)$ whose intersection with $E(k)$ is of positive measure, and given m_n we obtain m_{n+1} as follows. Put ⟦8⟧

$$m\,E_{c(k,n)} \cap \left(\frac{m_n}{2^n}, \frac{2m_n+1}{2^{n+1}}\right) = a_{n,m}, \qquad ⟦9⟧$$

$$m\,E_{c(k,n)} \cap \left(\frac{2m_n+1}{2^{n+1}}, \frac{m_n+1}{2^n}\right) = b_{n,m},$$

and let r_n be the smallest m for which either $a_{n,m}<k^{-1}2^{-2n}$ or $b_{n,m}<k^{-1}2^{-2n}$ or both $a_{n,m}>1/k(k+n+1)$ and $b_{n,m}>1/k(k+n+1)$. There exists such an r_n for $a_{n,m}$ and $b_{n,m}$ decrease either to 0 or to some positive number. In the case where $a_{n,r_n}<k^{-1}2^{-2n}$ we put $m_{n+1}=2m_n+1$; if $a_{n,r_n}\geqslant k^{-1}2^{-2n}$ but $b_{n,r_n}<k^{-1}2^{-2n}$, we put $m_{n+1}=2m_n$, and in the third case we put $m_{n+1}=2m_n$ or $m_{n+1}=2m_n+1$ according as $\theta(n)=0$ or 1.

For each n the interval $(m_n/2^n, (m_n+1)/2^{n+1})$ includes normal numbers in positive measure. The intersection of these intervals contains only one number which must be normal. ⟦10⟧

Now consider the set $A(k,n)$ consisting of all possible intervals $(m_n/2^n, (m_n+1)/2^n)$, i.e., the sum of all these intervals as we allow the first n figures of θ to run through all possibilities. Then

$$m\,E(k)\cap A(k,n+1) = m\,E(k)\cap A(k,n) - \sum_{m=0}^{2^n-1} m\,E(k)\cap (A(k,n)-A(k,n+1))\cap\left(\frac{m}{2^n}, \frac{m+1}{2^n}\right).$$

But

$$m\,(A(k,n)-A(k,n+1))\cap\left(\frac{m}{2^n}, \frac{m+1}{2^n}\right) < 2^{-2n}k^{-1},$$

so that

$$m\,E(k)\cap A(k,n+1) > m\,E(k)\cap A(k,n) - 2^{-n-1}k^{-1}$$

$$> m\,E(k) - k^{-1} > 1-\frac{2}{k}. \qquad ⟦11⟧$$

The set of all possible numbers $\alpha(K,\theta)$ is therefore of measure at least $1-2/k$.

By taking particular sequences θ (e.g., $\theta(n)=0$ for all n) we obtain particular normal numbers. ⟦12⟧

⟦119⟧

THE WORD PROBLEM IN COMPACT GROUPS

It is proposed to show that the word problem is soluble for compact groups, or more generally for groups which can be embedded in a compact group. The result follows from two relatively deep theorems, one due to TARSKI (1) and the other to VON NEUMANN (2).

The first of these results states that the decision problem of elementary algebra is solvable. This means that one can decide the truth of a formula of the restricted function calculus with the variable propositional functions replaced by the following fixed ones:

$$x<y, \qquad x=y, \qquad x-y=z, \qquad xy=z.$$

In other words the formula must be made up by the use of "and", "not", "all", "equals", "less than", "minus", "multiplied by" and no other means for combining expressions. It is a problem of deciding the truth of ⟦1⟧
the formula rather than its probability and no question of axioms is therefore involved.

The second result states that a compact group can be truly represented by a countable sequence of unitary matrices of finite order. In other words, if G is a compact group, then for each g of G and each integer υ there is a unitary matrix $U_\upsilon(g)$ of order depending only on υ such that

(i) for each υ, $U_\upsilon(g)$ is a continuous function of g;

(ii) for each υ, g_1, g_2,
$$U_\upsilon(g_1)U_\upsilon(g_2)=U_\upsilon(g_1g_2), \qquad (U_\upsilon(g_1))^{-1}=U_\upsilon(g_1^{-1});$$

(iii) if $g_1 \neq g_2$, then, for some υ, $U_\upsilon(g_1)\neq U_\upsilon(g_2)$.

The decision procedure

In order to solve the word problem for a given group it is sufficient to have

(a) a procedure which will terminate if and only if the word under consideration represents the identity of the group;

(b) a second procedure which will terminate if and only if the word does not represent the identity of the group.

In this respect the word problem is like any other decision problem. A complete decision procedure is obtained by alternating steps of (a) with steps of (b). One of the alternatives must hold and one of the procedures/processes must therefore terminate. When this happens the complete process terminates and we base our conclusion on whether (a) or (b) terminated.

⟦2⟧ The procedure (a) is trivial for any group. All products of transforms of relations are enumerated, reduced by cancellation of adjoining inverse generators and compared with the word under consideration. The process terminates when an agreement is found.

The process (b) can be explained for any group given by generators and relations, but there is no guarantee that the procedure has the required properties except in the case stated.

Let $e_1, \dots, e_N$ be the generators, e the group identity,

$$w_1(e_1, \dots, e_N), \dots, w_M(e_1, \dots, e_N)$$

the relations, $w(e_1, \dots, e_N)$ the word under consideration. Then the statement

⟦3⟧ "For any matrices $E_1, \dots, E_N$ of order r the equation $w(E_1, \dots, E_N) = 1$ follows from the equations $w_1(E_1, \dots, E_N) = w_2(E_1, \dots, E_N) = \cdots = 1$"

⟦4⟧ can be translated into one of the type to which Tarski's theorem applies when the order r, the relations and the word under consideration are all made definite. The complexity of the formula may vary greatly from case to case. With a fixed group and word there remains only the order to be fixed. The
⟦5⟧ corresponding statement may be called A_r.

[At this point the reader finds the following remark in typescript but crossed out in ink]

It must be emphasized that the number r never appears in the Tarski form of A_r. This is rather important as integers never do appear in Tarski formulae; they might let Gödel's theorem in. The order r will however appear implicitly; e.g., the matrix equation $E_1 E_2 = E_3$ would have to be replaced by r^2 ordinary algebraic equations.

[Written in pencil is the following]

Theorem 27 of Pontryagin.

[The main text now resumes]

The procedure (b) may now be described as follows. Use Tarski's procedure to decide the formulas $A_1, A_2, A_3, \dots$. The process terminates if any one of these is found to be false.

Validity of the procedure

It remains to prove that the necessary and sufficient condition for (b) to terminate is that the word does not represent the identity of the group. If the word does represent the identity, then $w(E_1, \dots, E_N) = 1$ certainly will

⟦122⟧

follow from $w_1(E_1, \ldots, E_N) = \cdots = w_M(E_1, \ldots, E_N) = 1$ whatever the order r, and the process (b) will not terminate. On the other hand if the word does not represent the identity we must have

$$U_\upsilon(w(e_1, \ldots, e_N)) \neq U_\upsilon(e) = 1$$

for some υ, by (iii). But also by repeated applications of (ii) we have $U_\upsilon(w(e_1, \ldots, e_N)) = w(U_\upsilon(e_1), \ldots, U_\upsilon(e_N))$, and since $w_1(e_1, \ldots, e_N), \ldots, w_M(e_1, \ldots, e_N)$ all represent the identity of the group we have

$$1 = U_\upsilon(e) = U_\upsilon(w_s(e_1, \ldots, e_N)) = w_s(U_\upsilon(e_1), \ldots, U_\upsilon(e_N)), \quad 1 \leqslant s \leqslant M.$$

The statement A_r will therefore fail and will be found to do so when tested. The process (b) will terminate.

References

(1) A. Tarski, A decision method of elementary algebra and geometry, RAND project report R-109 (1948). ⟦6⟧
(2) J. Von Neumann. [The reference is not given]

Note. Attached to Turing's script on compact groups there is a one-page manuscript entitled *The Word Problem is Solvable in Simple Groups*. Unfortunately, this manuscript is well-nigh illegible.

With hindsight, the fact that the word problem is solvable in simple groups is not very deep. It was probably discovered independently by several authors.

Nevertheless, it is interesting that Turing was aware of it.

ON PERMUTATION GROUPS

[The beginning of Turing's typescript, including the title, is missing. The above title was suggested by R.O. Gandy.]

... form $U_{n_1} U_{n_2} \dots U_{n_M}$. It is easily seen that the parity of this permutation is that of U^M and therefore that the permutations from the same machine always have the same parity. For the present we shall not however investigate the permutations obtainable with a given machine but those which are obtainable with a given upright and any number of wheels.

Let $H(U)$ or H be the set of permutations obtainable from the upright U. It is easily seen that H is a group, for if P and Q are two permutations obtainable from the upright U, then PQ is obtainable from it by putting the machines giving P and Q in series; (an algebraic argument is almost equally simple). Since H is finite and contained in the symmetric group, this is sufficient to prove it is a group. H may be expressed as the group generated by $U_1, U_2, \dots, U_T$ or again as consisting of all expressions of the form

$$R^{t_0} U R^{t_1} U \dots U R^{t_p},$$

with any p and with the exponents of R totalling zero. It thus differs only slightly from the group $J(U)$ or J generated by U, R, which consists of similar expressions without the restriction on the sum of the exponents. Every member of $J(U)$ is of the form PR^k, where P is in H. H is thus a subgroup of J of index at most T. H is in fact a self-conjugate (or normal) subgroup of J, for it is transformed into itself by the generators U, R of J, i.e., UHU^{-1} is H since U is in H and RU_nU^{-1} is U_{n+1} so RHR^{-1} is H.

We shall say that H or $H(U)$ is *exceptional* if H does not include the whole alternating group A (all even permutations). If H is not exceptional, it will be either A or the symmetric group S (all permutations), according as U is even or odd. It is so easy to see in this way whether H is A or S that it is quite adequate in describing it to say that it is unexceptional.

We shall actually investigate J rather than H. J is obviously easier to deal with than H and results for J may be translated into results for H by means of the next theorem.

Theorem I. *The necessary and sufficient condition for H to be unexceptional is that J be unexceptional, provided $U \neq 1$, $T \neq 4$.*

We have shown that H is a self-conjugate subgroup of J. Now if J is unexceptional, it is either A or S and the only self-conjugate subgroups of these if $T \neq 4$ are A, S and the group consisting of the identity only, and so H must be one of these. The last alternative has been excluded by assuming $U \neq 1$ so that H is A or S, i.e., H is unexceptional. Conversely, if H is unexceptional, so is J since J contains H.

Technique for investigating any particular upright U

In order to prove H is unexceptional, it will suffice to prove J contains all 3-cycles, for if this is so, J will be a self-conjugate subgroup of S and since it is not the identity, it must be either A or S. It would also be sufficient to prove that J contains all 2-cycles. We shall prove the following theorem.

Theorem II. *If J contains a member of the form $(\alpha, R^m\alpha)$ or $(\alpha, R^m\alpha, \beta)$ or $(\alpha, R^m\alpha)(\beta, \gamma)$, where m is prime to T, then it contains all 3-cycles and, in the first of these cases, also all 2-cycles. T must be greater than 4. $(\alpha, R^m\alpha)(\beta, \gamma)$ must not commute with $R^{T/2}$ if T is even.*

Suppose J contains $(\alpha, R^m\alpha)$. We will write α_k for $R^{mk}\alpha$. The symbols $\alpha_0, \alpha_1, \ldots, \alpha_{T-1}$ include all the T symbols. Then J contains $R^{ms}(\alpha_0, \alpha_1)R^{-ms}$, i.e., (α_s, α_{s+1}). It therefore contains (α_0, α_2) since this is $(\alpha_1, \alpha_2)(\alpha_0, \alpha_1) \times (\alpha_1, \alpha_2)^{-1}$ (if $T > 2$). It contains (α_0, α_3), which is $(\alpha_2, \alpha_3)(\alpha_0, \alpha_2)(\alpha_2, \alpha_3)^{-1}$ if $T > 3$ and repeating this argument it contains (α_0, α_r) for every $0 < r < T$. Finally it contains (α_p, α_q) since this is $R^{mp}(\alpha_0, \alpha_{q-p})R^{-mp}$ if $q \not\equiv p$ (T). Thus J contains every 2-cycle (and every 3-cycle).

Now suppose J contains $(\alpha, R^m\alpha, \beta)$ where m is prime to T. We may express β as $R^{mk}\alpha$. The first step is to prove that J contains an element of the form $(\alpha, R^{m'}\alpha, R^{2m'}\alpha)$, where m' is prime to T. In the case that $2k \equiv 1$
[[1]] (T) we may take $m' = k$ and $(\alpha, R^m\alpha, R^{km}\alpha)^{-1}$ is $(\alpha, R^{m'}\alpha, R^{2m'}\alpha)$. In the case that $k+1 \equiv 0$ (T) we may take $m' = m$ and $R^m(\alpha, R^m\alpha, R^{mk}\alpha)R^{-m}$ is $(\alpha, R^{m'}\alpha, R^{2m'}\alpha)$. In the remaining cases I propose to prove that $(\alpha, R^m\alpha, R^{(k+1)m}\alpha)$ belongs to J, so that k may be increased until either $k+1 \equiv 0$ (T) or $2k-1 \equiv 0$ (T), and then previous cases can be applied.

We are assuming that $k+1 \not\equiv 0$ (T) and $2k-1 \not\equiv 0$ (T). For the moment I will also suppose $2k \not\equiv 0$ (T). Then $\alpha, R^m\alpha, R^{mk}\alpha, R^{m(k+1)}\alpha, R^{2mk}\alpha$ are all different and

$$[R^{mk}(\alpha, R^m\alpha, R^{mk}\alpha)R^{-mk}](\alpha, R^m\alpha, R^{mk}\alpha)[R^{mk}(\alpha, R^m\alpha, R^{mk}\alpha)R^{-mk}]^{-1}$$
$$= (R^{mk}\alpha, R^{m(k+1)}\alpha, R^{2mk}\alpha)(\alpha, R^m\alpha, R^{mk}\alpha)(R^{mk}\alpha, R^{m(k+1)}\alpha, R^{2mk}\alpha)^{-1}$$
$$= (\alpha, R^m\alpha, R^{m(k+1)}\alpha)$$

belongs to J. If however $2k \equiv 0\ (T)$ the left-hand side of this expression is equal to $(R^{mk}\alpha, R^m\alpha, R^{m(k+1)}\alpha)$ or to $(R^{mk}\alpha, R^{(2k+1)m}\alpha, R^{(k+1)m}\alpha)$ so that

$$R^{-mk}(R^{mk}\alpha, R^{(2k+1)m}\alpha, R^{(k+1)m}\alpha)^{-1}R^{mk}$$
$$= (\alpha, R^{(k+1)m}\alpha, R^m\alpha)^{-1} = (\alpha, R^m\alpha, R^{(k+1)m}\alpha)$$

belongs to J.

This proves the existence in J of $(\alpha, R^{m'}\alpha, R^{2m'}\alpha)$ with m' prime to T, and I will now write α_k for $R^{m'k}\alpha$, so that $(\alpha_0, \alpha_1, \alpha_2)$ belongs to J, and so does $(\alpha_s, \alpha_{s+1}, \alpha_{s+2})$ (by transforming with $R^{m's}$). Then transforming $(\alpha_0, \alpha_1, \alpha_2)$ with $(\alpha_2, \alpha_3, \alpha_4)$, we find that $(\alpha_0, \alpha_1, \alpha_3)$ belongs to J provided $T \geqslant 5$ and transforming again with $(\alpha_3, \alpha_4, \alpha_5)$, we have $(\alpha_0, \alpha_1, \alpha_4)$ in J if $T \geqslant 6$, and repeating the argument $(\alpha_0, \alpha_1, \alpha_r)$ belongs to J provided $r < T-1$. $(\alpha_0, \alpha_1, \alpha_{T-1})$ may be seen to belong to J by transforming $(\alpha_0, \alpha_1, \alpha_{T-2})$ with $(\alpha_{T-3}, \alpha_{T-2}, \alpha_{T-1})$ provided $T>4$. Thus every 3-cycle containing α_0 and α_1 belongs to J. But

$$(\alpha_0, \alpha_1, \alpha_r)(\alpha_0, \alpha_1, \alpha_s)^{-1} = (\alpha_s, \alpha_r, \alpha_0)$$

and therefore every 3-cycle containing α_0 belongs to J. Finally

$$(\alpha_s, \alpha_0, \alpha_r)(\alpha_s, \alpha_0, \alpha_t)^{-1} = (\alpha_t, \alpha_r, \alpha_s)$$

and therefore every 3-cycle whatever belongs to J.

There remains the case when we are given that $(\alpha, R^m\alpha)(\beta, \gamma)$ belongs to J. We shall reduce this to the case where it contains the 3-cycle. We may write α_k for $R^{mk}\alpha$ as before. We also write β' for $R^m\beta$. We are then given that $(\alpha_0, \alpha_1)(\beta, \gamma)$ belongs to J. We may divide this into the cases (remembering $T>4$):

(a) $(\alpha_0, \alpha_1)(\beta, \gamma)$ belongs to J, where β, γ are different from α_2, α_{-1}; β is different from γ' and γ from β';
(b) $(\alpha_0, \alpha_1)(\alpha_2, \gamma)$ belongs to J, where γ is different from α_{-1}, α_{-2};
(c) $(\alpha_0, \alpha_1)(\alpha_{-1}, \gamma)$ belongs to J, where γ is different from α_2, α_3;
(d) $(\alpha_0, \alpha_1)(\beta, \beta')$ belongs to J, where β is different from α_2, α_{-1};
(e) $(\alpha_0, \alpha_1)(\alpha_{-1}, \alpha_2)$ belongs to J;
(f) $(\alpha_0, \alpha_1)(\alpha_2, \alpha_3)$ belongs to J;
[Following the instruction "P.T.O.", the reader finds:]
(g) $(\alpha_0, \alpha_1)(\alpha_2, \alpha_{-2})$;
(h) $(\alpha_0, \alpha_1)(\alpha_{-1}, \alpha_3)$. [See Fig. 1*]

* Fig. 1. This consists of a circle passing through points α_{-3}, α_{-2}, α_{-1}, α_0, α_1; a point α_3 is inside the circle. A second circle has diameter α_0, α_1. A third circle passes through α_{-2}, α_3 and surrounds the small circle.

These are equivalent. $(\alpha_0, \alpha_1)(\alpha_2, \alpha_{-2})$. Then $(\alpha_{-3}, \alpha_{-2})(\alpha_{-1}, \alpha_{-5})$ gives $(\alpha_0, \alpha_1)(\alpha_2, \alpha_{-3})$ if $T>7$. O.K. In any case we can take it ... as (BC)(AE)

when $T=6$

A B C D E F (AF)(EDB) O.K.

E C B D A F

when $T=5$

A B C D E (EDB) O.K.

E C B D A

[[2]] Thus only exception $T=7$ when we get group of 7-point geometry.

[The main text now resumes:]

It is easily seen that cases (b) and (c) are essentially the same (by changing the sign of m). In case (a), since $(\alpha_1, \alpha_2)(\beta', \gamma')$ belongs to J

$$((\alpha_1, \alpha_2)(\beta', \gamma'))(\alpha_0, \alpha_1)(\beta, \gamma)((\alpha_1, \alpha_2)(\beta', \gamma'))^{-1} = (\alpha_0, \alpha_2)(\beta, \gamma)$$

belongs to J and so does

$$(\alpha_0, \alpha_1)(\beta, \gamma)((\alpha_0, \alpha_2)(\beta, \gamma))^{-1} = (\alpha_0, \alpha_2, \alpha_1).$$

In case (b), supposing first $\gamma \neq \alpha_2$, $(\alpha_2, \alpha_3)(\alpha_4, \gamma'')$ belongs to J and therefore, transforming $(\alpha_0, \alpha_1)(\alpha_2, \gamma)$ with this permutation, also $(\alpha_0, \alpha_1)(\alpha_3, \gamma)$ which comes under case (a). If however $\gamma = \alpha_{-2}$, the result of the same transformation is $(\alpha_4, \alpha_1)(\alpha_2, \alpha_{-2})$. But

$$((\alpha_4, \alpha_1)(\alpha_2, \alpha_{-2}))((\alpha_0, \alpha_1)(\alpha_2, \alpha_{-2})) = (\alpha_0, \alpha_4, \alpha_1)$$

and we can apply the case of the 3-cycles.

In case (d) put α_r for β. Then if $2r \not\equiv 0, 1, -1$ (T),

$$(\alpha_0, \alpha_1)(\alpha_r, \alpha_{r+1})(\alpha_{r+1}, \alpha_{r+2})(\alpha_{2r+1}, \alpha_{2r+2})$$
$$= (\alpha_0, \alpha_1)(\alpha_r, \alpha_{r+1}, \alpha_{r+2})(\alpha_{2r+1}, \alpha_{2r+2})$$

and so $(\alpha_r, \alpha_{r+1}, \alpha_{r+2})$ belongs to J and the case of 3-cycles applies.

If $2r \equiv 1$ (T), then $(\alpha_0, \alpha_1)(\alpha_r, \alpha_{r+1})$ and

$$(\alpha_r, \alpha_{r+1})(\alpha_{2r}, \alpha_{2r+1}) = (\alpha_r, \alpha_{r+1})(\alpha_1, \alpha_2)$$

belongs to J and therefore their product $(\alpha_1, \alpha_2, \alpha_0)$ also. Similarly if $2r \equiv -1$ (T). If $2r \equiv 0$ (T), then $(\alpha_0, \alpha_1)(\alpha_r, \alpha_{r+1})$ commutes with $R^{T/2}$.

In case (e), $(\alpha_0, \alpha_1)(\alpha_{-1}, \alpha_2)$ and $(\alpha_1, \alpha_2)(\alpha_0, \alpha_3)$ belong to J and therefore (transforming) $(\alpha_3, \alpha_2)(\alpha_{-1}, \alpha_1)$ and (subtracting 3 from suffixes) $(\alpha_0, \alpha_{-1})(\alpha_{-4}, \alpha_{-2})$. On changing the sign of m this becomes $(\alpha_0, \alpha_1)(\alpha_4, \alpha_2)$ and comes under case (b).

In case (f), $(\alpha_0, \alpha_1)(\alpha_2, \alpha_3)$ and $(\alpha_1, \alpha_2)(\alpha_3, \alpha_4)$ belong to J and therefore $(\alpha_0, \alpha_2)(\alpha_1, \alpha_4)$ and the product

$$((\alpha_0, \alpha_1)(\alpha_2, \alpha_3))((\alpha_0, \alpha_2)(\alpha_1, \alpha_4)) = (\alpha_0, \alpha_3, \alpha_2)(\alpha_1, \alpha_4)$$

and its square $(\alpha_2, \alpha_3, \alpha_0)$ to which the 3-cycle result applies.

This completes the proof of Theorem II, but we may notice the cases which have been expressly excluded since they give rise to exceptional groups. In the case of a 2-cycle $(\alpha, R^m\alpha)$ where m is not prime to T, H is an intransitive group. The intransitivity sets are each of the form ⟦3⟧ $(\beta, R^m\beta, R^{2m}\beta, \ldots)$. The same applies in the case that the generator is a 3-cycle $(\alpha, R^m\alpha, R^{m+p}\alpha)$ when m, p, $m+p$ have a factor in common with T. We have omitted to give consideration to the case where m, p, $m+p$ each have a factor in common with T, but there is no factor in common with all four. Probably Theorem II applies to these cases also. In the case where the generator is $(\alpha, \beta)(\gamma, \delta)$ and commutes with $R^{T/2}$, each element of the group H also commutes with $R^{T/2}$. ⟦4⟧

It is very easy to apply Theorem II. We may first express U, UR, UR^2 etc. in cycles: this may be done for instance by writing the alphabets out double and also writing out the sequence UA, UB, ..., UZ. By putting the former above the latter in various positions we get the permutations UR^s. Among these we may look for the permutations which have a 3-cycle and all other cycles of length prime to 3. By raising this to an appropriate power we obtain a 3-cycle which may or may not satisfy the conditions in Theorem II. If we are not successful we may use other permutations in J. We may also be able in a similar way to generate a permutation which is a pair of 2-cycles.

The following upright was chosen at random

A B C D E F G H I J K L M N O P Q R S T U V W X Y Z
M N Y T F B G R S L A X O E W K P C J Q Z D H V U I

In cycles it is

(AMOWHRCYUZISJLXVDTQPK)(BNEF)(G) = U.

Then U^{22} = (BE)(NF).

The distance BE is 3, which is prime to 26. The distance NF is 8. Hence ⟦5⟧ Theorem II applies and J includes the whole of A, and therefore H includes A.

Systematic search for exceptional groups. Theory

In examining all possible uprights for a given T the main difficulty lies in the large number of uprights involved. Once it has been proved that a particular upright is unexceptional, the same will follow for a great number of others.

More generally given any upright we can find a great number of others which generate either the same group H or an isomorphic group. If we can classify these uprights together in some way we shall enormously reduce the labour, since we shall only need to investigate one member of each class. The chief principles which enable us to find equivalent uprights are:

[6] (i) if $U'=R^m U R^n$, then $H(U')=H(U)$;

(ii) if V commutes with (R), then $H(VUV^{-1})\cong H(U)$;

(N.B. If V commutes with (R), then $VRV^{-1}=R^k$, some k.)

(iii) if $U'=U^m$ and $U=U'^n$, then $H(U)=H(U')$.

The principle (i) is the one of which we make the most systematic use. Our method depends on the fact that there are very few U for which none of the permutations $R^m U R^n$ leave two letters invariant (in other words there are very few U without a beetle) and none if T is even. We therefore investigate separately the U with no beetles and the U with beetles.

U with no beetle

We can find an expression which determines the classes of permutations obtainable from one another by multiplication right or left by powers of R as follows.

Let $UR^{n+1}\mathrm{Z}=R^{f(n)}UR^n\mathrm{Z}$; (here Z represents the last letter of the alphabet however many characters there may be in it). Then we take the numbers
[7] $f(1)f(2)\ldots f(T)$ as describing the classes containing U. It may be verified that from these numbers and $U\mathrm{Z}$ it is possible to recover U, in fact they describe what is in common between $U, RU, R^2U, \ldots$. However if we move some of the numbers from the end of the sequence to the beginning, we shall be describing $UR^m, RUR^m, \ldots$. An example may help. Consider the permutation $U=$(ABCDG) of seven letters. Write it as

A B C D E F G
B C D G E F A.

The numbers are the differences of consecutive letters in B C D G E F A, i.e., 1 1 3 5 1 2 1. The permutation R^2UR is

A B C D E F G
E F B G A C D

and the numbers 1 3 5 1 2 1 1 are obtainable by shifting the first figure to the end.

If then we take the various forms

$$f(1)f(2)\ldots f(T),\quad f(T)f(1)f(2)\ldots f(T-1),\quad \text{etc.},$$

and select that which, regarded as an arabic numeral, would be the smallest, we shall have a way of describing all the permutations $R^m U R^n$. We may call the resulting figures the *invariant* of U.

To a small extent we can combine this with the principle (ii) conveniently. If V satisfies $VRV^{-1}=R^{-1}$, then the invariant of VUV^{-1} is
[8] obtainable by reading that of U backwards and rearranging for minimum.

When we are investigating the cases where U has no beetle, the invariant is very restricted. It cannot contain Fig. 1. More generally the numbers

$$0, f(1)-1, f(1)+f(2)-2, f(1)+f(2)+f(3)-3, \ldots$$ ⟦9⟧

(essentially one of the "rods") must be all different. These restrictions are so powerful that we normally find very few cases other than where U is a ⟦10⟧ power of R. When T is even there are no such cases (as is well known): if there were the numbers $0, f(1)-1, f(1)+f(2)-2, \ldots$ would have to be the numbers $0, 1, \ldots, T-1$ in some order and would total $\frac{1}{2}T(T-1)$ modulo T. But $0, f(1), f(1)+f(2), \ldots$ are also all different and must total $\frac{1}{2}T(T-1)$ modulo T, and likewise $0, 1, \ldots, T-1$ total the same so that $0, f(1)-1$, $f(1)+f(2)-2, \ldots$ total $0 \bmod T$, i.e., $\frac{1}{2}T(T-1)$ is $0 \bmod T$, which is not so if T is even.

U with a beetle

We select a permutation $R^m U R^n$ which leaves two letters invariant to represent U. One of these may be taken to be A, if necessary by transforming with a further power of R. By means of principle (ii) we can also reduce the possibilities for a second letter. If the letters which were originally fixed were A and R^tA, we can transform them to A and R^sA provided that the highest common factor of T and t is the same as the highest common factor of T and s, since there is a V satisfying $V\mathrm{A}=\mathrm{A}$ and $VR^tV^{-1}=R^s$, $V(R)V^{-1}=(R)$. We therefore have to consider various pairs ⟦11⟧ of letters left invariant such as A, R^tA; the various t to be considered should run through the numbers which divide T, omitting T itself but including 1. For each such t we then write down the permutations leaving A, R^tA invariant. We arrange them by classes of conjugates in S, and reduce them by means of principle (iii) as we write them down. Further reduction may afterwards be done by principle (ii) in a very special way: if V interchanges A and R^tA we can apply it. It is best probably to number the permutations and with each to give also the number t of that with which it is paired. Very many will be found self-paired.

The detailed search

T = 1, 2, 3, 4

It is not difficult to prove that there are no exceptional groups when $T=1$, 2 or 3. The case $T=4$ needs special investigation as it has been expressly excluded from Theorem I. The exceptional uprights in this case are

(1) (13) (24) (13)(24)
(12)(34) (32)(14) (1234) (4321).

The exceptional groups H are the identity, the cyclic groups ((1234)) and ((13)(24)), the 4-group, consisting of the identity and all permutations of the form $(\alpha\beta)(\gamma\delta)$, and a group isomorphic with the 4-group and generated by (13) and (24).

$T=5$

With the theory we have developed, this case is also very trivial. There are no U without beetles except

[Following the instruction "P.T.O." we find:]

those with invariant

22222 33333 44444.

[[12]] These together prove the numatizer of (R).

[Returning to the main text:]

and this leaves only the permutations leaving two letters invariant, and these are all covered by Theorem II, since 5 is prime.

Thus the only exceptional U are the members of the numatizer of (R), 20 in number.

$T=6$

Since 6 is even we do not need to consider U without beetles. Our representative U will then always leave two letters invariant, and again will come under Theorem II immediately. U can only be exceptional by being intransitive or commuting with R^3. The total number which commute with R^3 is 48 and the number which have the intransitivity sets (ACE BDF) is 36. These include 6 which commute with R^3. Those with intransitivity set (AD, BE, CF) all commute with R^3. Thus the total number of exceptional U is $48+36-6=78$ out of a possible 720.

$T=7$

The uprights without beetles are the members of the numatizer of (R) and those with the invariant 2335564 and its reversal. The latter however are unexceptional. A representative U is (BCGEDF) and U^2R is (AG)(BEC)(DF) which is evidently unexceptional (square it).

For the uprights with beetles we may take $t=1$ only, i.e., we can always suppose A and B both invariant. Simple application of Theorem I shows that we need only consider cases of 5-cycles and principle (iii) shows that we can take it that $U\text{C}=\text{G}$, i.e., we have reduced the representative permutations U to the form (A)(B)(G$\alpha\beta\gamma$C). The permutation V which interchanges A and B is (AB)(CG)(DF)(E). Thus we need in the end only consider

(GDEFC) self paired

(GDFEC) paired
(GEDFC)
(GEFDC) paired
(GFDEC)
(GFEDC) self paired

By multiplying these with powers of R they may all be shown unexceptional.

We have also the case of the group generated by (AE)(BC) specifically mentioned in Theorem II. This gives the group of symmetry of the 7-point geometry [see Fig. 2*]. There are actually only two isomorphic groups of this kind, generated by (BC)(AE) and by (BE)(FD).

Invariants involved

(BC)(AE) group	(BE)(FD) group reverses
1556245	1542655
1323354	1453323
4444444	4444444
1264663	1366462
1111111	1111111
2222222	2222222

Invariants of numatizer of (R)

3333333	
5555555	total elements
6666666	$6\times 49+6\times 7=6\times 7\times 8=336$
4444444	i.e., 336 exceptional uprights
1111111	
2222222	

T = *8*

This needs rather more investigation than the previous cases partly because it is the largest number yet considered and partly because it has more factors.

Obviously the permutations which commute with (R) or with a power of R or generate an intransitive group will be exceptional. We will consider that we are looking for other forms of exceptional upright.

We have various means for dealing with permutations:

(a) We may show that the group is the same as that generated by an upright to be considered later. A special case of this occurs when $t=1$. One

* Fig. 2. This consists of a triangle ABC with inscribed circle, centre E, meeting the sides BC, CA, AB in points G, D, F, respectively.

of the forms $R^s U$ may be of a type to be considered under $t=2$ or $t=4$. This may be detected by writing down the figures $m_1, m_2, \ldots$ where $R^{m_r+r}H = UR^r H$. If any figure appears twice in this series at an even distance the conversion to the case $t=2$ is possible. It will be indicated by an "X".

(b) One of the slides $R^s U$ may be one to be considered later under the same value of t; marked "below".

(c) The square or some other power may be proved unexceptional. This means that we need only consider those 6-cycles whose squares and cubes are unexceptional. We may consider a 6-cycle and its square and cube simultaneously; this is advantageous also because the transformations V apply in the same way to all three.

(d) One of the slides $R^s U$ may be unexceptional. This will be indicated by "slide", the value of the slide in cycles and "O.K."

(e) One of the commutators $UR^s UR^{-s}$ may be unexceptional. This is indicated by the value of the commutator and "O.K."

When all these fail, a query will be shown and the upright investigated further later.

t = 1

We may first go over the main plan, considering separately what is to be done with the various classes of conjugates in the symmetric group.

6-cycles. These are left aside till the double threes have been considered.

Double threes. These are arranged in pairs (as transformed by (CH)(EF)(DG) = V which leaves A, B fixed and satisfies $VRV^{-1} = R^{-1}$) and dealt with in detail.

Triple twos. Very few of these need to be considered in detail. Those with the pair (CH) give (BAC) together with other cycles on a slide and so are either O.K. under (d) or equivalent to a double three. Those with the pair (DH) are reduced to a $t=4$ case under (a) and those with the pair (CG) are paired with ones having (DH).

Four-and-twos. They need only be considered when their squares are intransitive or commute with R^4, by Theorem II.

Other cases. They consist of ones where 3 or more letters are invariant and are immediately reducible to $t=2$ or $t=4$.

Double threes.

(CDE)(FGH) (s.p.)	(CDE)(FGH) . (DEF)(GHA) = (DC)(EGF)(HA) O.K.
(CDE)(FHG) (s.p.)	(CDE)(FHG) . (DEF)(GAH) = (DC)(EHF)(AG) O.K.

(CDF)(EHG) (s.p.)	slide (BAGC)(EFH) O.K.
(CDF)(EGH) (s.p.)	??
(CDG)(EFH) (CEF)(HDG)	(CA)(DBE)(FG) slide O.K.
(CDG)(EHF) (CEF)(HDG)	slide (BAFHC)(EG) O.K.
(CDH)(EFG) (CGH)(DEF)	slide (BAC)(EH) O.K.
(CEG)(DFH) (s.p.)	intransitive
(CEG)(DHF) (s.p.)	intransitive
(CEH)(DFG) (CFH)(DEG)	slide (BAC)(DEFH)
(CEH)(DGF) (CFH)(DGE)	slide (DA)(EBF) O.K.
(CFG)(DEH) (s.p.)	??
(CFG)(DHE) (s.p.)	slide (BAEHC)(DF), transform to (HG) O.K.
(CFH)(DEG)	slide (BAC)(DFGH) O.K.

We have avoided using (a) and (b) owing to the inapplicability for the 6-cycles.

The upright (CDF)(EGH) is exceptional and the corresponding group consists of the elements with the invariants

11111111	8 elements	11111111
12214554	64 elements	25527667
13272315	64 elements	13245423
16573756	64 elements	34657564
24636425	64 elements	14737415
33476674	64 elements	12216336
77777777	8 elements	77777777
	336 elements	

Transformation of the group with (AG)(CH)(EC), which commutes with R gives another group which contains (CFG)(DEH).

The invariants of this latter group are given in the last column. These invariants are useful for verifying that other exceptional uprights belong to these groups.

We have to investigate the 6-cycles whose squares are exceptional. They are shown below.

(CEDGFH)	X
(CGDHFE)	invariant 15132723 (above)
(CHDEFG)	X
(CDFEGH)	invariant 12216336 (above)
(CEFHGD)	X
(CHFDGE)	X
(CDEFGH)	X
(CFEHGD)	X

(CHEDGF)	X
(CDEHGF)	slide (ABDH)(CE) O.K.
(CHEFGD)	X
(CFEDGH)	X

It is not easy to prove directly that the elements with the above invariants form a group. However in this case we can manage by guessing what the group is. The order of the group being 336 it is natural to suppose that it is connected with the group of symmetry of the 7-point geometry, which is a well-known simple group of order 168. The even permutations in our group will form a group of order 168, which might with luck be isomorphic with the group of the 7-point geometry. This in fact turns out to be so. In order not to confuse the notation we will denote the points of the 7-point geometry by a, b, c, d, e, f, g; [see Fig. 3*]. We naturally try to express the group of symmetry as a group of permutations of some eight objects, in order to tie it up with the groups found above. These groups may perhaps be called K and K'. The standard technique for representing a group as a group of permutations of m objects is to find a subgroup of index m, and to consider the cosets of the subgroup as forming the m objects. In this case we want a subgroup of order 21, and such a subgroup is the normalizer of ((abcegfd)). The cosets of this group are enumerated below in shortened form. Each line represents seven permutations, obtainable from one another by moving letters from one end to the other. Each coset has been given a name which is a capital letter.

A	*B*	*C*	*D*
abcegfd	abgdcfe	acdgfeb	aefbged
acgdbef	agcebdf	adfbcge	afgdebc
agbfcde	acbfged	afcedbg	agecfdb
E	*F*	*G*	*H*
acfbdeg	abecdfg	aegdfcb	abdgefc
afgdebc	aedgbcf	agfbedc	adecbgf
agecfdb	adbfegc	afecgbd	aebfdcg

Now the symmetry group contains the permutation (abgc)(fe) which induces the permutation $(ACEG)(BDFH)$ of the cosets. It also contains (abc)(def) which induces $(ADH)(BCG)$. Now if we identify the cosets with the eight symbols permuted in the group J (as the notation is intended to suggest), we see that $(ACEG)(BDFH)$ is R^2 and that $(ADH)(BCG)$ has the invariant 14737415, one of those of the group K'. It can then be easily verified that the symmetry group also induces in the cosets permutations

* Fig. 3. This consists of a triangle abc with inscribed circle, centre g, meeting the sides bc, ca, ab in d, e, f, respectively.

with all the other invariants of K', and also that $(ACEG)(BDFH)$ and $(ADH)(BCG)$ generate the whole symmetry group. Since this is of order 168 and since it contains the even permutations in K', numbering 168, it must coincide with the intersection of K' and A.

It now only remains to prove that the expressions of form S and SR, where S is in $K' \cap A$, form a group. This will follow if we can prove that R^2 belongs to $K' \cap A$ and that RSR^{-1} belongs to $K' \cap A$ whenever S belongs to it. We know the former already and the latter follows at once from our invariant system; K' consists of complete sets of permutations having certain invariants.

Triple twos. As explained above we do not need to consider any except the ones without the pairs (CH), (DH) or (GC). This leaves:

(CD)(EF)(GH)   (CD)(EF)(GH) . (EF)(GH)(AB) =
  (CD)(AB) O.K.
(CD)(EG)(FH)   (CD)(EG)(FH) . (FG)(HB)(AC) =
  (ADC)(BFEGH) O.K.
  (CE)(DF)(GH)   pair
(CD)(EH)(FG)   slide (BAEC)(FH)
  (CF)(DE)(GH)   pair
(CE)(DG)(FH)   (CE)(DG)(FH) . (FH)(GB)(AC) =
  (GBD)(AC) O.K.
(CF)(DG)(EH)   (CF)(DG)(EH) . (DG)(EH)(FA) = (ACF) O.K.

Fours-and-twos. Omitting those whose squares are unexceptional and those with UH = C and their pairs we have only:

(CDEF)(HG)   X
(CDGF)(HE)   X
(CGHD)(FC)   X
(DEHG)(FC)   pair
(FEHG)(CD)   pair

$t = 2$

We find it worthwhile to apply the principle (ii) on a rather large scale. There are four permutations V which leave A and C fixed; they are

A B C D E F G H
C B A H G F E D
C F A D G B E H
A F C H E B G D

From a single permutation we thus obtain as many as four generating isomorphic groups J, e.g., from (BDEFHG):

(BDEFHG)
(BHGFDE)

(FDBGHE)
(FHEBDG)

These may be transformed into equivalent forms and the alphabetically earliest chosen. We permit taking the reciprocal as a form of transformation. Thus we get

(BDEFHG), (BEDFGH)?, (BGDFEH)?, (BDCFHE).

By these means we reduce the 6-cycles that need be considered down to 18. As before we actually consider first their squares (double threes) in the hope that they will be unexceptional and the 6-cycle need not be specially investigated.

6-cycles and double threes

(BDEFGH)	slide (ACF)(BFDHE) O.K. indirect
(BDEFHG)	slide (AB)(CEFHD) O.K.
(BDEGFH)	(BEF)(DGH) . (CFG)(EHA) = (AFH)(CBEDG) O.K.
(BDEGHF)	slide (AD)(FCB) O.K.
(BDEHFG)	slide (BAG)(DCEH) O.K.
(BDEHGF)	invariant 34657564, giving group *K'*
(BDFEHG)	slide (BA)(DCFGH) O.K.
(BDFGHE)	slide (BA)(CFD)(HEG) O.K.
(BDFHGE)	slide (BAEH)(DCF) O.K.
(BDGFEH)	invariant 34657564, giving group *K'*
(BDGHEF)	slide (CAE)(FHDGB) O.K. indirect
(BDGHFE)	(BGF)(DHE) . (CHG)(EAF) = (ABGCE)(DHF) O.K. indirect
(BDHEFG)	slide (BAFG)(DCH) O.K.
(BDHEGF)	slide (AH)(BCEDF) O.K.
(BDHFGE)	slide (DAEH)(FCG) O.K.
(BDHGEF)	slide (AD)(HBFCE) O.K.
(BDHGFE)	slide (CAE)(DHBFG) O.K. indirect
(BEDHGF)	slide (AED)(CF) O.K.

Above analyses are done on the squares of the 6-cycles, i.e., on the double threes. We must now investigate the cases of 6-cycles where the double threes were exceptional.

(BDEHGF)	slide (BAG)(FH)(CD) O.K.
(BDGHEF)	slide (BAEGH)(CD) O.K.
(BF)(DG)(EH)	X
(BF)(DH)(EG)	intransitive
(BG)(DF)(EH)	slide (AGEDH)(BCF)
(BG)(DH)(EF)	X
(BH)(DF)(EG)	intransitive

(BH)(DG)(EF)	(AH)(BCG)(DF) slide O.K.
(BE)(DH)(FG)	X
(BG)(DH)(FE)	X
(BH)(DG)(FE)	slide (AH)(BCG)(DF)
(BF)(DE)(GH)	X
(BH)(DE)(GF)	slide (AH)(BCEG)
(BH)(DF)(GE)	intransitive
(BE)(DF)(HG)	slide (ACBFG)(EH)
(BE)(DG)(HF)	slide (AEH)(BCGFD)
(BF)(DG)(HE)	X
(BG)(DE)(HF)	slide (ACDHG)(BE)
(BG)(DF)(HE)	(AGEDH)(BCF) slide O.K.
(BH)(EG)(DF)	intransitive

Fours and twos and fours

We only need consider those whose squares are intransitive or commute with R^4.

(EBGD)[(FH)]	(FBGD)[(CE)] O.K.	
(EBGF)[(DH)]	(EBFC) O.K.	(ABFC)(DG) O.K.
(EBGH)[(FD)]	(EBGA) O.K.	(GBAE)(CD) O.K.
(EDGB)[(FH)]	(HDGB) O.K.	(HDAB)(EC) O.K.
(EDGH)[(BF)]	(HDGA) O.K.	(HDCA)(BE) O.K.
(EFGH)[(DB)]	(EGHA) O.K.	(CGHA)(DB) O.K.

There are none which leave A, C fixed and commute with R^4 except (BDFH), (BHFD) which is intransitive anyway.

5-cycles

These 5-cycles are given in pairs which are equivalent by principle (ii).

(DEFGH)	s.p. $(DEFGH)(EFGHA)^{-1} = (DEA)$ O.K.
(DEFHG)	slide (BAGDC)(FH) O.K. indirect
(DEHGF)	
(DEGFH)	slide (BADC)(FGH) O.K.
(DEHFG)	
(DEGHF)	s.p. slide (EAGCH)(DFB) O.K. indirect
(BEFGH)	slide (BA)(CED) O.K.
(BEHFG)	
(BEFHG)	slide (BAGH)(CDE) O.K.
(BEGHF)	
(BEGFH)	s.p. slide (BA)(CED)(FGH) O.K.
(BEHGF)	s.p. slide (BAG)(DCE) O.K.
(BEFGH)	slide (AF)(BE)(CGDH) O.K.
(BDHGF)	

(BDFHG)	slide (BAGH)(DC)(EF) O.K.
(BDGFH)	
(BDGHF)	s.p. slide (BAFEG)(CD) O.K.
(BDHFG)	s.p. slide (BAFEH)(CD) O.K.

$t=4$

The permutations V which commute with (R) and leave A, E invariant or interchange them are:

A	B	C	D	E	F	G	H
A	D	G	B	E	H	C	F
E	H	C	F	A	D	G	B
A	H	G	F	E	D	C	B
E	D	C	B	A	H	G	F
A	F	C	H	E	B	G	D
E	B	G	D	A	F	C	H
E	F	G	H	A	B	C	D

6-cycles and double threes

As usual we actually examine the double threes although we test the 6-cycles.

(BCDFGH)	commutes with R^4 (both (BCDFGH) and (BDG)(CFH) and (EF)(CG)(DH))
(BCDFHG)	slide (BA)(EHCDF) O.K.
(BCDGHF)	slide (AB)(CDG)(HFE) O.K.
(BCDHGF)	invariant 33476674 group K
	6-cycle invariant 12214554 group K
	triple two invariant 77777777 part of numatizer of (R)
(BCFDGH)	slide (CAB)(EDFHG) O.K.
(BCFDHG)	slide (BA)(CFEGH) O.K.
(BCFHGD)	slide (BADH)(FEC) O.K.
(BCGDFH)	slide (BACG)(EHF) O.K.
(BCGDHF)	invariant 27652765 commutes with R^4
	6-cycle slide (CAD)(FHB) reducing to a $t=2$ case already considered
	triple two slide (AHDG)(CEF) O.K.
(BCGFDH)	invariant 33476674 group K
	6-cycle slide (BA)(DG)(FEH) O.K.
	triple two (BF)(CD)(GH) commutes with R^4
(BCGFHD)	slide (BA)(DFECG) O.K.

(BCGHDF)	commutes with R^4
	6-cycle slide (AE)(HFC) O.K.
	triple two (AH)(CD)(GF) slide (ABCDH)(EGF) O.K.
(BCGHFD)	slide (CAF)(EHBDG) O.K.

Triple twos (remaining)

(BF)(CH)(DG)	commutes with R^4
(BG)(CH)(DF)	slide (AHG)(BFCE) O.K.
(BD)(CG)(FH)	commutes with R^4
(BD)(CH)(FG)	slide (AHDG)(CEF) O.K.
(BH)(CG)(DF)	intransitive
(BD)(CF)(GH)	(BD)(CF)(GH) . (CE)(DG)(HA) = (AGBDH)(CEF) O.K.
(BF)(CH)(GD)	commutes with R^4
(BH)(CF)(GD)	(BH)(CF)(GD) . (DB)(EH)(AF) = (DHEBG)(FBC) O.K.

Fives

(CDFGH) (CDHFG)	slide (BAC)(EF) O.K.
(CDFHG) (CDHGF)	slide (BAGHC)(FE) O.K.
(CDGFH) (CDGHF)	slide (BAC)(EGDF) O.K.

The 5-cycles where two fixed letters were at distance 2 were considered under $t=2$.

The outcome for $T=8$ is then that the exceptional uprights are all either

(a) intransitive, or product of an intransitive upright with R,

(b) commuting with R^4,

(c) members of the groups K, K'.

Let us now calculate the number of exceptional uprights. The number in K and K' omitting those with invariants 11111111 and 77777777 is $2\times 336-32=640$. Those with invariants 11111111 and 77777777 are also intransitive, the condition for intransitivity being that all figures are even or all odd. The condition for commuting with R^4 is that the invariant is of the form abcdabcd. Thus these 16 elements are also of this kind, but no other members of K, K' are. There are $2\times(4!)^2$ intransitive uprights or 1152 and there are $2^4\times(4!)$ or 384 that commute with R^4. The intransitive ones that commute with R^4 are determined on giving the values of UA, UB (which must be of opposite parity). They number $8\times 4=32$.

Final account

K and K' with	640
intransitive	1 152
commute with R^4 with omissions	352
	2 144
total of all uprights	40 320

We now turn to a rather different topic in connection with the use of identical drums. Even if we know that all permutations are possible, will they be equally frequent? Fortunately, we can answer this in the affirmative. The problem will be examined under slightly more general conditions. No assumptions will be made about the relationship between the generators $U_1, \ldots, U_k$ and we will not assume that the basic group is the symmetric group but some other finite group G.

Let us suppose that we feed a certain frequency distribution of group elements into a wheel; how can we calculate the frequency distribution of the group elements at the output of the wheel? Let $g(a)$ be the proportion of the input elements which are a, and let $f(a)$ be the proportion of group elements effected by the wheel which are a, i.e., denoting the order of the group by h,

$$f(a) = h^{-1} \sum_r 1, \quad r = 1, \ldots, k, \quad U_r = a.$$

Then we get output a if the input is b and the wheel effects the group element ab^{-1}, for any b. The proportion of such cases is $f(ab^{-1})g(b)$, or allowing for the different values of b, a total proportion of

$$\sum_b f(ab^{-1})g(b).$$

If then we define the operator R_f by the equation

$$(R_f g)(a) = \sum_b f(ab^{-1})g(b),$$

[[13]] we can say that the frequency distribution for n wheels is given by $R_f^{n-1}f$. We wish to determine how this function behaves with increasing n.

We consider the real-valued functions on the group as forming an Euclidean space of h dimensions, where h is the order of the group H. We may put (g, k) for the scalar product $h^{-1} \sum_a g(a)k(a)$ and $\|g\|$ for the distance $(g, g)^{1/2}$ from the origin. We may also put $\bar{g}$ for the mean value $h^{-1} \sum_a g(a)$. Schwarz' inequality gives at once $\|g\| \geqslant \bar{g}$, and if we suppose $g(a) \geqslant 0$ for all a, $g(a) > 0$ for some a, we shall have $\bar{g} > 0$. We will also suppose $f(a) \geqslant 0$ for all a, $f(a) > 0$ for some a, $\bar{f} = 1$. Then we have the next lemma.

[[142]]

Lemma (a). *If $\bar{f}=1$, then $\|R_f g\| \leqslant \|g\|$ and equality holds only if $g(ab^{-1}x)/g(x)$ is independent of x for any a, b for which $f(a) \neq 0$ and $f(b) \neq 0$.* ⟦14⟧

First note that

$$\left(h^{-1} \sum_x g(x) g(cx)\right)^2 \leqslant h^{-2} \sum (g(x))^2 \sum (g(cx))^2 = (h^{-1} \sum (g(x))^2)^2 = \|g\|^4,$$

equality holding if $g(cx)/g(x)$ is independent of x. Then

$$\begin{aligned} \|R_f g\|^2 &= h^{-1} \sum_{a,b,x} f(ab^{-1}) g(b) f(ax^{-1}) g(x) \\ &= h^{-1} \sum_{c,v,x} f(c) f(cv) g(vx) g(x) \quad [v = bx^{-1},\ c = ab^{-1}] \\ &\leqslant \sum_{c,v} f(c) f(cv) \|g\|^2 \\ &= \|g\|^2, \end{aligned}$$ ⟦15⟧

equality holding in the case mentioned.

Let us define the *limiting distribution* for f as the condensation points of the sequence $g, R_f g, R_f^2 g, \dots$.

Then Lemma (a) will enable us to prove the following theorem.

Theorem III. *The limiting distributions for f are constant throughout the cosets of a certain self-conjugate subgroup H_1 of H. H_1 consists of all expressions of the form $U_{r_1}^{m_1} U_{r_2}^{m_2} \dots U_{r_p}^{m_p}$, where the sum $\sum m_q$ of exponents is zero. The factor group H/H_1 is cyclic. In the case that g is f, the limiting distributions each have the value zero except in one coset of H_1.*

Let k be a limiting distribution. Let it be the limit of the sequence $R_f^{n_1} g, R_f^{n_2} g, \dots$. Then

$$\|R_f^{n_r} g\| \geqslant \|R_f^{n_r+1} g\| \geqslant \|R_f^{n_{r+1}} g\| \geqslant \|k\|.$$

Now $\|R_f^{n_r} g\|/\|k\|$ tends to the limit 1 as r tends to infinity and therefore $\|R_f^{n_r+1} g\|/\|R_f^{n_r} g\|$ tends to 1. But $\|R_f u\|/\|u\|$ is a continuous function of u and therefore taking the limit of the sequence $\|R_f k\|/\|k\| = 1$. Applying Lemma (a) to this we see that there is a function $\varphi_1(y)$ defined for all expressions of the form ab^{-1} where $f(a) \neq 0$, $f(b) \neq 0$ such that $k(yx) = \varphi(y) k(x)$ for all x. By applying the same argument with $R_f^{m-1} f$ in place of f we find that there is a function $\varphi_u(y)$ defined for all expressions of the ⟦16⟧
form $a_1 a_2 \cdots a_n b_n^{-1} \cdots b_1^{-1}$ where $f(a_1), f(a_2), \dots, f(b_1)$ are all different from 0 such that

$$k(yx) = \varphi_u(y) k(x), \quad \text{all } x,$$

whenever $\varphi_u(y)$ is defined. The various functions $\varphi_u(y)$ must agree whenever their domains of definition overlap, and they may therefore be all represented by one symbol φ. In fact we may say that $\varphi(y)$ is defined and has value α whenever $k(yx)=\alpha\, k(x)$ for all x. It now appears that the domain of definition of $\varphi(y)$ is a group, for if $k(y_1x)=\alpha_1 k(x)$ for all x and $k(y_2x)=\alpha_2 k(x)$ for all x, then $k(y_1y_2x)=\varphi(y_1)k(y_2x)=\varphi(y_1)\varphi(y_2)k(x)$ for all x. Thus if y_1 and y_2 belong to the domain of definition of φ, so does y_1y_2 and $\varphi(y_1y_2)=\varphi(y_1)\varphi(y_2)$.

⟦17⟧ It is now immediately seen that the domain of definition is H_1. The function φ is a one-dimensional representation of H_1 but it is real and positive and therefore has the value 1 throughout H_1. This last argument may also be expressed without the use of representation theory thus. Since H_1 is finite, any element y of it satisfies an equation $y^m=1$ and therefore
⟦18⟧ $(\varphi(y))^m=\varphi(y^m)=1$. But since $g(x)$ is always nonnegative, $\varphi(y)\geqslant 0$ and so $\varphi(y)=1$. This implies that $g(x)$ is constant throughout each coset of H_1.

It now only remains to investigate the character of the group H_1. It is easily seen to be self-conjugate, since if aba^{-1} belongs to H_1 and b to H, the total of exponents of group generators U_r in aba^{-1} must be 0, those in a^{-1} cancelling with those in a; hence aba^{-1} belongs to H_1 if b does; H_1 is self-conjugate.

Now let us take a particular generator U_1, say. Then the cosets $H_1U_1^m$ exhaust the group H. For if p is an element of H, it will be a product of generators; let the total of exponents be m. Then pU_1^{-m} has total exponents 0 and so belongs to H_1, i.e., p belongs to $H_1U_1^m$, i.e., these cosets exhaust H. If U_1^s is the lowest power of U_1 which belongs to H_1, then H/H_1 is evidently isomorphic with the cyclic group of order s.

In the case that g is f, all the group elements for which $R_f^{n-1}f$ is not
⟦19⟧ zero are products of n generators and therefore belong to $H_1U_1^n$.

Example. As an example let us consider the quaternion group consisting of $1, i, j, k, 1', i', j', k'$ with the table

1	i	j	k	$1'$	i'	j'	k'
i	$1'$	k	j'	i'	1	k'	j
j	k'	$1'$	i	j'	k	1	i'
k	j	i'	$1'$	k'	j'	i	1
$1'$	i'	j'	k'	1	i	j	k
i'	1	k'	j	i	$1'$	k	j'
j'	k	1	i'	j	k'	$1'$	i
k'	j'	i	1	k	j	i'	$1'$

and let U_1 be i and U_2 be j. The various functions $R_f^{n-1}f$ are given in the table

n	1	i	j	k	1′	i'	j'	k'
1	0	$\frac{1}{2}$	$\frac{1}{2}$	0	0	0	0	0
2	0	0	0	$\frac{1}{4}$	$\frac{1}{2}$	0	0	$\frac{1}{4}$
3	0	$\frac{1}{4}$	$\frac{1}{4}$	0	0	$\frac{1}{4}$	$\frac{1}{4}$	0
4	$\frac{1}{4}$	0	0	$\frac{1}{4}$	$\frac{1}{4}$	0	0	$\frac{1}{4}$
5	0	$\frac{1}{4}$	$\frac{1}{4}$	0	0	$\frac{1}{4}$	$\frac{1}{4}$	0
6	$\frac{1}{4}$	0	0	$\frac{1}{4}$	$\frac{1}{4}$	0	0	$\frac{1}{4}$

⟦20⟧

It is seen that the group H_1 is the group generated by k; it has a factor group which is cyclic of order 2.

Case of symmetric and alternating groups

In the case under consideration at the beginning of our analysis, H was either the symmetric or the alternating group unless the upright U was exceptional. In this case H_1 is also either the symmetric or the alternating group, for it is self-conjugate in H. It will be the alternating group if the generators are all of the same parity and the symmetric group otherwise. ⟦21⟧
We therefore conclude that

when the upright is not exceptional the distributions with large numbers of wheels are uniform throughout the alternating group (*even permutations*). *If odd permutations are possible with the given upright and number of wheels the distribution is uniform throughout the symmetric group* (*all permutations*).

THE DIFFERENCE $\psi(x)-x$

I propose to prove that there is a real number x, $e^{99}<x<e^{a}$, $a=10^{428}$, such that $(\psi(x)-x)x^{-1/2}>1.0001$, assuming the Riemann hypothesis.

We start from the formula

$$\int_a^b \chi(x)(\psi(x)-x)\,dx = -\sum_{\gamma}(\tfrac{1}{2}+i\gamma)^{-1}\int_a^b \chi(x)x^{1/2+i\gamma}\,dx + \int_a^b \chi(x)\left(\tfrac{1}{2}\log\frac{1}{1-x^{-2}}-\frac{\zeta'}{\zeta}(0)\right)dx$$

(INGHAM, A note on the distribution of primes, *Acta Arith.* **1** (1936) 202–211 (204)) and put $x=e^{t}$, $\chi(x)=\tfrac{1}{2}T\exp(-\tfrac{3}{2}t)\,w(T(t-w))$, $(\psi(e^{t})-e^{t})/2e^{t/2}=H(t)$, $b=w+\eta$, $a=w-\eta$, where $w(v)=(\sqrt{2\pi})^{-1}\exp(-\tfrac{1}{2}v^{2})$. Then ⟦1⟧

$$2\int_{-\eta}^{\eta} TH(w+v)\,w(Tv)\,dv = -\sum_{\gamma}(\tfrac{1}{2}+i\gamma)^{-1}e^{i\gamma w}\int_{-\eta}^{\eta} e^{i\gamma v}T\,w(Tv)\,dv + \int_{-\eta}^{\eta} Tr(w+v)\,w(Tv)\,dv,$$

where $r(t)<2e^{-t/2}$ for $t>5$. (Note that $(\zeta'/\zeta)(0)=\log 2\pi<1.9$.) ⟦2⟧

Now suppose $H(t)\leqslant H_0$ when $w-\eta\leqslant t\leqslant w+\eta$; then

$$\begin{aligned} H_0\int_{-T\eta}^{T\eta} w(v)\,dv &\geqslant \int_{-\eta}^{\eta} TH(w+v)\,w(Tv)\,dv \\ &\geqslant -\tfrac{1}{2}\sum_{\gamma}(\tfrac{1}{2}+i\gamma)^{-1}e^{i\gamma w}\left[\exp\left(-\frac{\gamma^{2}}{2T^{2}}\right)-2\int_{\eta T}^{\infty}\cos\left(\frac{\gamma v}{T}\right)w(v)\,dv\right] \\ &\quad -0.0001, \quad w-\eta>30, \end{aligned}$$

since

$$\int_{-\infty}^{\infty} e^{iv\eta}\,w(v)\,dv=\exp(-\tfrac{1}{2}\eta^{2}).$$

That is

$$\begin{aligned} H_0\int_{-T\eta}^{T\eta} w(v)\,dv &\geqslant -\sum_{0<\gamma}\gamma^{-1}\sin(\gamma w)\exp\left(-\frac{\gamma^{2}}{2T^{2}}\right) \\ &\quad -(\tfrac{1}{2}+4T\,w(\eta T))\sum_{0<\gamma}\gamma^{-2}-0.0001 \\ &\geqslant -\sum_{0<\gamma}\gamma^{-1}\sin(\gamma w)\exp\left(-\frac{\gamma^{2}}{2T^{2}}\right)-0.0128, \end{aligned}$$

$T<10000$, $\eta T>10$,

since

⟦3⟧ $$\left|\int_{\eta T}^{\infty} \cos(\gamma\upsilon/T)\, w(\upsilon)\, \mathrm{d}\upsilon\right| \leqslant 2T\gamma^{-1} w(\eta T)$$

and

⟦4⟧ $$|(\tfrac{1}{2}+\mathrm{i}\gamma)^{-1}\,\mathrm{e}^{\mathrm{i}\gamma w}+(\tfrac{1}{2}-\mathrm{i}\gamma)^{-1}\,\mathrm{e}^{-\mathrm{i}\gamma w}-\gamma^{-1}2\sin\gamma w| < \gamma^{-2}.$$

We shall now put $T=312.5$, $\eta=\frac{1}{10}$ and estimate S, where

$$S=-\sum_{0<\gamma}\gamma^{-1}\sin(\gamma w)\exp\left(-\frac{\frac{1}{2}\gamma^2}{312.5^2}\right)$$

and w is chosen so that

⟦5⟧ $$100 < w+\tfrac{1}{137} < 100\cdot 137^{N(125\pi)} < 10^{428}-1$$

and $\gamma(w+\frac{1}{125})/2\pi$ differs from an integer by less than $\frac{1}{137}$ for each $\gamma<125\pi$. This is possible by Dirichlet's theorem in Diophantine approximation. Then

$$99 < w-\eta < w+\eta < 10^{428}.$$

For the estimation of S we shall make use of the following definite integrals (obtained by curve drawing and use of a planimeter).

$$\frac{1}{2\pi}\int_{\pi}^{\infty}\upsilon^{-1}(1+\sin\upsilon)\exp(-0.08\upsilon^2)\,\mathrm{d}\upsilon < 0.00851. \tag{i}$$

$$\frac{1}{2\pi}\int_{0}^{\pi}\upsilon^{-1}\sin\upsilon\exp(-0.08\upsilon^2)\log\upsilon\,\mathrm{d}\upsilon > -0.1044. \tag{ii}$$

$$\frac{1}{2\pi}\int_{\pi}^{\infty}\upsilon^{-1}\log\upsilon\exp(-0.08\upsilon^2)\,\mathrm{d}\upsilon < 0.0188. \tag{iii}$$

$$\int_{0}^{\pi}[1-|\cos\upsilon|\exp(-0.08\upsilon^2)]\upsilon^{-1}\,\mathrm{d}\upsilon > 0.83. \tag{iv}$$

$$\int_{0}^{\pi}[1-|\cos\upsilon|\exp(-0.08\upsilon^2)]\upsilon^{-1}\log\upsilon\,\mathrm{d}\upsilon > 0.30. \tag{v}$$

From (i) and the identity

⟦6⟧ $$\frac{1}{2\pi}\int_{0}^{\infty}\upsilon^{-1}\sin\upsilon\exp\left(-\frac{\upsilon^2}{2\alpha^2}\right)\mathrm{d}\upsilon=\frac{1}{2\sqrt{2\pi}}\int_{0}^{\alpha}\exp(-\tfrac{1}{2}u^2)\,\mathrm{d}u.$$

we find

$$\frac{1}{2\pi}\int_{0}^{\pi}\upsilon^{-1}\sin\upsilon\exp(-0.08\upsilon^2)\,\mathrm{d}\upsilon-\frac{1}{2\pi}\int_{\pi}^{\infty}\upsilon^{-1}\exp(-0.08\upsilon^2)\,\mathrm{d}\upsilon$$
$$> 0.24688-0.00851=0.23837. \tag{vi}$$

We shall also need results on the distribution of zeros of $\zeta(s)$. Denoting as usual the number of γ's in the interval $(0, T)$ by $N(T)$ and putting

$$S(T) = N(T) - \frac{T}{2\pi}\log\frac{T}{2\pi} - \frac{7}{8}, \qquad [\![7]\!]$$

we have $|S(T)| < 2.1$, $T < 1464$; (TITCHMARSH, The zeroes of the zeta function, *Proc. Roy. Soc. (A)* **157** (1936) 261–263);

$$|S(T)| < 0.137 \log T + 0.443 \log\log T + 4.350,$$

(BACKLUND, *Acta Math.* **41** (1918));

$$|S(T)| < 1, \quad T < 100, \quad S(30) < 0.2,$$

(JAHNKE-EMDE, *Funktionentafeln* (Teubner, Leipzig, 1948), p. 324).

We may now estimate S. For $\gamma < 125\pi$,

$$-\sin(\gamma w) > \cos\frac{2\pi}{137}\sin\frac{\gamma}{125} - \frac{2\pi}{137}\left|\cos\frac{\gamma}{125}\right| \qquad [\![8]\!]$$

and therefore

$$S > \cos\frac{2\pi}{137}\sum_{0<\gamma<125\pi}\gamma^{-1}\sin\frac{\gamma}{125}\exp\left(-\frac{\frac{1}{2}\gamma^2}{312.5^2}\right)$$

$$-\sum_{125\pi\leqslant\gamma}\gamma^{-1}\exp\left(-\frac{\frac{1}{2}\gamma^2}{312.5^2}\right) - \frac{2\pi}{137}\sum_{0<\gamma<125\pi}\gamma^{-1}\left|\cos\frac{\gamma}{125}\right| e^{-J},$$

$$J = \frac{\frac{1}{2}\gamma^2}{312.5^2}. \qquad [\![9]\!]$$

That is,

$$S > \cos\frac{2\pi}{137}\int_0^{125\pi} u^{-1}\sin\frac{u}{125}\exp\left(-\frac{\frac{1}{2}u^2}{312.5^2}\right)(2\pi)^{-1}\log\frac{u}{2\pi}\,du$$

$$-\int_{125\pi}^{\infty} u^{-1}\exp\left(-\frac{\frac{1}{2}u^2}{312.5^2}\right)(2\pi)^{-1}\log\frac{u}{2\pi}\,du$$

$$-\frac{2\pi}{137}\sum_{0<\gamma<125\pi}\gamma^{-1}$$

$$+\frac{25}{137}\int_0^{125\pi} u^{-1}\left[1 - \left|\cos\frac{u}{125}\right|\exp\left(-\frac{\frac{1}{2}u^2}{312.5^2}\right)\right](2\pi)^{-1}\log\frac{u}{2\pi}\,du \qquad [\![10]\!]$$

$$+\cos\frac{2\pi}{137}\int_{125\pi}^{\infty} u^{-1}\sin\frac{u}{125}\exp\left(-\frac{\frac{1}{2}u^2}{312.5^2}\right)dS(u) \qquad [\![11]\!]$$

$$-\int_{125\pi}^{\infty} u^{-1}\exp\left(-\frac{\frac{1}{2}u^2}{312.5^2}\right)dS(u)$$

⟦12⟧ $$-\frac{2\pi}{137}\int_0^{125\pi} u^{-1}\left[1-\left|\cos\frac{u}{125}\right|\exp\left(-\frac{\frac{1}{2}u^2}{312.5^2}\right)\right]\mathrm{d}S(u)$$

⟦13⟧ $$=\cos\frac{2\pi}{137}I_1-I_2-\frac{2\pi}{135}C-\frac{2\pi}{135}I_3+E_1+E_2+\frac{2\pi}{135}E_3,$$

say.

$$I_1=\frac{1}{2\pi}\int_0^{\pi}\upsilon^{-1}\sin\upsilon\exp(-0.08\upsilon^2)\left[\log\upsilon+\log\frac{125}{2\pi}\right]\mathrm{d}\upsilon<0.75,$$

$$I_2=\frac{1}{2\pi}\int_{\pi}^{\infty}\upsilon^{-1}\exp(-0.08\upsilon^2)\left[\log\upsilon+\log\frac{125}{2\pi}\right]\mathrm{d}\upsilon.$$

Hence by (ii), (iii), (vi) and the fact that $\cos\frac{2}{137}\pi>1-0.00107$

$$I_1\cos\frac{2\pi}{125}-I_2>0.23837\log\frac{125}{2\pi}-0.00081-0.1044-0.0188>0.58951.$$

$$C<\frac{1}{14.13}+\frac{1}{21.02}+\frac{1}{25.01}$$

$$+\int_{30}^{125\pi}(2\pi u)^{-1}\log\frac{u}{2\pi}\,\mathrm{d}u+\int_{30}^{125\pi}u^{-1}\,\mathrm{d}S(u),$$

the denominators in the first three terms being less than the first three γ's

$$\leqslant 0.1585+\frac{1}{4\pi}\left[(\log 62.5)^2-\left(\log\frac{30}{2\pi}\right)^2\right]+E_4,\quad\text{say}$$

$$<\frac{1}{2\pi}\left[-0.131+\tfrac{1}{2}(\log 62.5)^2+2\pi E_4\right]$$

$$<\frac{1}{2\pi}\left[8.417+2\pi E_4\right].$$

⟦14⟧ $$I_3=\frac{1}{2\pi}\int_0^{\pi}[1-|\cos\upsilon|\exp(-0.08\upsilon^2)]-\upsilon^{-1}\left[\log\upsilon+\log\frac{125}{2\pi}\right]\mathrm{d}\upsilon$$

$$>\frac{1}{2\pi}\left[0.30+(0.83)\log\frac{125}{2\pi}\right]>\frac{2.13}{2\pi}.$$

For the estimation of E_1, E_2, E_3, E_4 note that

$$\left|\int_a^b\varphi(u)\,\mathrm{d}S(u)\right|\leqslant|\varphi(b)S(b)|+|\varphi(a)S(a)|+\int_a^b|S(u)|\,|\mathrm{d}\varphi(u)|$$

$$\leqslant 2.1\left[|\varphi(a)|+|\varphi(b)|+\int_a^b|\mathrm{d}\varphi(u)|\right],\quad\text{provided } b<1464.$$

In E_1, $\varphi(0)=\frac{1}{125}$, $\varphi(125\pi)=0$ and φ is decreasing. $S(0)=\frac{7}{8}$. [[15]]

$$|E_1| < \frac{2.1+\frac{7}{8}}{125} = 0.0238.$$

$$|E_2| < 2(2.1)(125\pi)^{-1}\exp(-0.08\pi^2) + \int_{1464}^{\infty} (|S(u)|-2)\,|d\varphi(u)| < 0.0050.$$

$$|E_3| < (2.1)(2.5)(125\pi)^{-1} = \frac{0.84}{2\pi}.$$

$$|E_4| \leqslant \frac{|S(30)|}{30} = \int_{30}^{100} |S(u)|\,|du^{-1}| + \int_{100}^{\infty} |S_1(u)|\,|du^{-1}|,$$ [[16]]

where $S_1(u)=S(u)$ for $u<125\pi$, $S_1(u)=0$ for $u\geqslant 100\pi$.

$$\leqslant \frac{0.2}{30} + \left[\frac{1}{30}-\frac{1}{100}\right] + \frac{2.1}{100} = 0.024,$$

$$2\pi\,|E_4| < 0.151,$$

$$2\pi[C+I_3+E_3] < 6.538.$$

Then

$$S > 0.58952 - \frac{6.538}{137} - 0.0288 > 0.51298.$$

But

$$\int_{-T\eta}^{T\eta} w(v)\,dv > 1-10^{-10}.$$

Therefore

$$H_0 > 0.51298 - 0.0128 - 0.00001 = 0.50017,$$

i.e., $H(t)>0.50017$, some t in $w-\eta$, $w+\eta$, i.e., $(\psi(x)-x)/x^{1/2}>1.00034$, $e^{99}<x<\exp 10^{428}$.

ON A THEOREM OF LITTLEWOOD

S. Skewes and A.M. Turing ⟦1⟧

1. Introduction

We propose to investigate the question as to where $\pi(x) - \operatorname{li} x$ is positive[1]. This quantity is positive if x is less than about 1.42 and negative from there up to 10^7. The figures suggest that $\pi(x) - \operatorname{li} x \sim x^{1/2}/\log x$ as $x \to \infty$ but Littlewood (2) has proved that $\pi(x) - \operatorname{li} x$ changes sign infinitely often. The argument proceeds by cases, according to whether the Riemann hypothesis is true or false. It has been announced by one of us (Skewes (3)) that in the case that the Riemann hypothesis is true $\pi(x) - \operatorname{li} x > 0$ for some x, $2 < x < 10^a$, where $a = 10^b$, $b = 10^{34}$. In the present paper it is proposed to establish that $\pi(x) > \operatorname{li} x$ for some x, $2 < x < \exp(\exp 661)$, without the restriction of assuming the Riemann hypothesis. [Should 661 be 686?]

We shall also prepare the ground for the possibility of improving the bound to about 10^a, $a = 10^5$ with the aid of extensive computations, and also consider the effect on the situation of discovering zeros off the critical line.

2. Outline of the method

The necessity of using the functions $\Pi(x)$ and $\log \zeta(s)$ somewhat complicates the argument. The general outline of the method may be illustrated by dealing with the analogous problem of finding where $\theta(x) > x$. Since

$$\psi(x) = \theta(x) + \theta(x^{1/2}) + \theta(x^{1/3}) + \cdots,$$

and $\theta(x) \sim x$ this is essentially the question as to where $\psi(x) > x + x^{1/2}$. Now

$$\psi(x) - x = -\sum_{\varrho} x^{\varrho}/\varrho - (\zeta'/\zeta)(0) - \tfrac{1}{2}\log(1 - x^{-2}),$$

so that the problem reduces essentially to the question: for what values of t does the inequality

$$G(t) = -\sum_{\varrho} \varrho^{-1} \exp(\varrho - \tfrac{1}{2})t > 1$$

[1] Here $\pi(x)$ is the number of primes less than x and $\operatorname{li} x$ is the logarithmic integral of x, i.e., the Cauchy principal value of $\int_0^x 1/\log t \, dt$. In these as in other matters of notation we follow Ingham (1).

hold, summation being over the complex zeros?

One may consider various expressions of the form $\int_0^\infty G(t)f(t)\,\mathrm{d}t$. Putting

$$\frac{1}{\sqrt{2\pi}}\int_0^\infty f(t)\mathrm{e}^{\mathrm{i}ut}\,\mathrm{d}t = F(u),$$

we have

$$I = \int_0^\infty G(t)f(t)\,\mathrm{d}t = \frac{1}{\sqrt{2\pi}}\sum_\varrho \varrho^{-1}F\left(\frac{\varrho-\frac{1}{2}}{i}\right).$$

If $f(t)$ is positive for positive t, and decreases to zero sufficiently quickly, it is possible to argue back from the value of I to the inequality $G(t)>1$. For instance it will suffice to have

$$\int_0^\infty f(t)\,\mathrm{d}t = 1,\quad f(t)>0,\qquad I>\tfrac{5}{4},\qquad \int_A^\infty G(t)f(t)\,\mathrm{d}t<\tfrac{1}{4},$$

to infer $G(t)>1$ for some t, $0<t<A$. The value of t for which I is sufficiently large and positive is to be obtained by Diophantine approximation. In carrying out the approximation we try to adjust the phases of the terms with small ϱ to the appropriate values and to arrange that the remaining, unadjusted, terms are small. We therefore wish $F((\varrho-\frac{1}{2})/\mathrm{i})$ to be small for the large values of ϱ. By taking $f(t)=((\sin\beta t)/t)^2$, Ingham ensured that, if the Riemann hypothesis is true, only a finite number of terms were different from zero. In the present paper we use instead a function

$$f(t) = \left(\frac{\sin\beta t}{t}\right)^4 \exp(-\tfrac{1}{2}\alpha^2 t^2)$$

which is largely inspired by Ingham's argument. It does not result in the vanishing of any of the terms, but if α is small, the later terms are extremely small. The present function has various advantages (for the present purpose) over that used by Ingham. The factor $\exp(-\frac{1}{2}\alpha^2t^2)$ encourages the quick convergence of the integral $\int_0^\infty G(t)f(t)\,\mathrm{d}t$, facilitating the inference of inequalities about $G(t)$ from values of I. This factor also causes $F(u)$ to be an integral function (whereas otherwise it would only be regular in two squares and two right angle segments). The use of the higher power of $(\sin\beta t)/t$ results in a rather sharper transition from large to small values of $f(u)$, leading to an appreciable numerical improvement.

Essential to the whole method is the fact that the Riemann hypothesis has been tested in the region $|t|<1468$ (TITCHMARSH (4)).

3. Formal preliminaries

Lemma 1. *If*

$$F(u)=\frac{1}{\sqrt{2\pi}}\int_{-\infty}^{\infty} e^{-iut}f(t)\,dt \quad \textit{and} \quad \varphi(s)=\int_{0}^{\infty} x^{-s}\,dh(x)$$

and the integrals

$$\int_{0}^{\infty} x^{-3}|h(x)|\,dx \quad \textit{and} \quad \int_{3/2-i\infty}^{3/2+i\infty} |F(-is)|\,|ds|$$

are convergent and $h(x)=o(x^2)$ *as* $x\to\infty$ *and* $x\to 0$, *then*

$$\int_{-\infty}^{\infty} h(e^t)e^{-t/2}f(t-t_0)\,dt=\frac{1}{i\sqrt{2\pi}}\int_{2-i\infty}^{2+i\infty}\frac{\varphi(s)}{s}\exp((s-\tfrac{1}{2})t_0)F(-i(s-\tfrac{1}{2}))\,ds.$$

If $\sigma=2$, then $\varphi(s)=s\int_0^\infty h(x)x^{-s-1}\,dx$ by integration by parts and so

$$\frac{1}{i\sqrt{2\pi}}\int_{2-i\infty}^{2+i\infty}\frac{\varphi(s)}{s}\exp((s-\tfrac{1}{2})t_0)\,F(i(s-\tfrac{1}{2}))\,ds$$

$$=\frac{1}{i\sqrt{2\pi}}\int_{2-i\infty}^{2+i\infty}\left[\int_{0}^{\infty} h(x)x^{-s-1}\,dx\right]\exp((s-\tfrac{1}{2})t_0)\,F(-i(s-\tfrac{1}{2}))\,ds$$

$$=\frac{1}{i\sqrt{2\pi}}\int_{0}^{\infty} x^{-3/2}h(x)\left[\int_{2-i\infty}^{2+i\infty} x^{-s+1/2}\exp((s-\tfrac{1}{2})t_0)\,F(-i(s-\tfrac{1}{2}))\,ds\right]dx$$

$$=\int_{0}^{\infty} x^{-3/2}h(x)f(\log x-t_0)\,dx.$$

The inversion is admissible since the double integral is absolutely convergent. [?$(t_0-\log x)$] [(t_0-t)] [? conditions on f] [or $+i(s-\tfrac{1}{2})$ and $+is$]

Lemma 2. *If the functions* f, F *are subject to the restrictions of Lemma 1, we have*

$$\int_{-\infty}^{\infty} e^{-t/2}(\Pi(e^t)-M(t))f(t-t_0)\,dt$$

$$=\frac{1}{\$\sqrt{2\pi}}\int_{2-i\infty}^{2+i\infty} s^{-1}(\log\zeta(s)-g(s))\exp((s-i)t_0), \quad [?\,i]$$

where

$$M(t)=\int t^{-1}e^t\,dt, \quad \textit{integration from } 0.1 \textit{ to } \max(t,0.1),$$

and where

$$g(s) = \int_{0.1(s-1)}^{\infty} t^{-1}e^{-t}\,dt, \quad [\text{? Df. if } s \text{ real} < 1]$$

the integration to be along a line parallel to the real axis.

We apply Lemma 1 first with $h(x) = \Pi(x)$ and again with $h(x) = M(\log x)$ and combine the results. Note that for $\mathscr{R}s > 1$ one may also write $g(s) = \int_{0.1}^{\infty} t^{-1}\exp(1-s)t\,dt$.

Lemma 3. (a) *If the logarithm has its principal value, the function* $g(s) + \log(s-1)$ *has no singularities.*

(b) *For any* s, $|g(s)| < \pi + \$1 \exp(\$ - 0.1(1-\sigma))$.

[$1/|s-1|$; but g ambiguous] [? minus sign] [? log sin g.]

(a) The function $g(s) + \log(s-1)$ may be defined as the indefinite integral of the regular function $(1 - \exp(-0.1(s-1)))/(s-1)$.

(b) The inequality may be proved by integrating along an arc of a circle and part of the positive real axis.

Lemma 4. *If the functions f and F are related as in Lemma* 1 *and g is the function defined in Lemma* 2 *and if* $F(\mathrm{i}s)$ *is bounded* [?] *in any strip* $\sigma_1 \leqslant \sigma \leqslant \sigma_2$, *then*

$$\int_{-\infty}^{\infty} e^{-t/2}(\Pi(e^t) - M(t))f(t-t_0)\,dt = I_1 + I_2\$ + \sum_{\varrho} I_{3,\varrho} + J, \quad [? -]$$

where the summation is over the nontrivial zeros of the zeta-function and $\Delta > \$0$ [? $\frac{5}{2}$] *and*

$$I_1 = \frac{1}{\mathrm{i}\sqrt{2\pi}} \int_{2-\mathrm{i}\infty}^{2+\mathrm{i}\infty} s^{-1}(\log(s+1)(s+2)\zeta(s+2))$$

$$\times \exp((s-\tfrac{1}{2})t_0)\, F(\$ - \mathrm{i}(s-\tfrac{1}{2}))\,ds, \quad [+\mathrm{i}]$$

$$I_2 = \frac{1}{\mathrm{i}\sqrt{2\pi}} \int s^{-1}\left(\log\frac{\zeta(s)}{(s+1)(s+2)\zeta(s+2)} - g(s)\right) \exp((s-\tfrac{1}{2})t_0)\, F(\mathrm{i}(s-\tfrac{1}{2}))\,ds,$$

integration from $-\Delta + \frac{1}{2} - \mathrm{i}\infty$ *to* $-\Delta + \frac{1}{2} + \mathrm{i}\infty$,

$$I_{3,\varrho} = \sqrt{2\pi} \int_{\varrho-2}^{\varrho} s^{-1}F(\$ - \mathrm{i}(s-\tfrac{1}{2}))\exp((s-\tfrac{1}{2})t_0)\,ds, \quad [+\mathrm{i}],$$

$$J = \sqrt{2\pi}\left(\log\frac{\$-\zeta(0)}{2\zeta(2)} - g(0)\right)F(\tfrac{1}{2}\mathrm{i}) \ldots \text{(illegible)}.$$

[? minus sign] [log and $g(0)$ ambiguous]

Formally the result follows from Cauchy's theorem by moving the line of integration from $\sigma=2$ to $\sigma=\frac{1}{2}-\Delta$ with the integrand of I_2.

The integral is absolutely convergent on either of these lines, so that it is only necessary to prove that there is a sequence T_r such that $T_r \to +\infty$ and $I_{4,r} \to 0$ where

$$I_{4,r} = \int s^{-1}\left(\log \frac{s\,\zeta(s)}{(s+1)(s+2)\,\zeta(s+2)} - g(s)\right)\exp((s-\tfrac{1}{2})t_0)\,F(-\mathrm{i}(s-\tfrac{1}{2}))\,\mathrm{d}s,$$

integration from $\frac{1}{2}-\Delta+\mathrm{i}T_r$ to $2+\mathrm{i}T_r$, and a similar sequence with $T_r \to -\infty$. We shall only consider the former case.

(A handwritten paragraph:) The singularities of the integrand are the ... lines from ϱ to $\varrho-2$ together possibly with some singularities ... However $\zeta(s)(s-1)/(\zeta(s+2)(s+1)(s+2))$ may be verified to have neither zero nor pole at any integer, and certainly has not at any other real point. Hence the only real singularity is that at 0, giving rise to the residue J.

(Continuation of typescript:) We may choose our sequence T_r according to Theorem 26 of INGHAM. There will then exist A such that if $\frac{1}{2}-\Delta \leqslant \sigma \leqslant 2$, then $|(\zeta'/\zeta)(\sigma+\mathrm{i}T_r)| < A(\log T_r)^2$. The quantity A will depend on Δ only. Then

$$\begin{aligned}|\log \zeta(\sigma+\mathrm{i}T_r)| &\leqslant |\log \zeta(2+\mathrm{i}T_r)| + (1.5+\Delta)A(\log T_r)^2\\ &< 1+(1.5+\Delta)A(\log T_r)^2,\end{aligned}$$

and therefore

$$|I_{4,r}| < T_r^{-1}(1+(1.5+\Delta)A(\log T_r)^2)(1.5+\Delta)M\exp(1.5\,t_0),$$

where M is the upper bound of $F(-\mathrm{i}(s-\frac{1}{2}))$ in the region $\frac{1}{2}-\Delta$. Evidently $I_{4,r} \to 0$ as $r \to \infty$. [From 'and therefore' to '$r\to\infty$': ? details; 2 ζ's. Also s, $s+1$, $s+2$ and $-g(s)$.]

4. Results with a special kernel

The function f has been relatively unrestricted until now, but we shall now put $f(t)=f_1(t)f_2(t)$, where

$$f_1(t) = \left(\frac{\sin \mu t}{\mu t}\right)^4, \qquad f_2(t) = \exp(-\tfrac{1}{2}\alpha^2 t^2)$$

and we shall also put $\alpha^2 t_0 = \Delta = 400$ although these substitutions will not always be made. The choice of values for μ and for t_0 will not be made just yet, but we shall assume that $50<\mu<250$ and $10^4<t_0$. The functions F, F_1, F_2 will be Fourier transforms of f, f_1, f_2 as in Lemma 1.

Lemma 5. *We have*

$$F_1(z) = \sqrt{2\pi}(2\mu)^{-1}\kappa(z/2\mu) \quad \textit{if } z \textit{ is real},$$

$$F_2(z) = \alpha^{-1}\exp(-z^2/2\alpha^2),$$

$$F(z) = \frac{1}{\sqrt{2\pi}} \int_{-\infty}^{\infty} F_2(z-u)\,F_1(u)\,du,$$

where

$$\kappa(x) = \begin{cases} 1-\frac{1}{2}|x|^2, & 0 \leqslant |x| \leqslant 1, \quad [-\frac{4}{3}(1-|x|)^3 + \frac{1}{6}(2-|x|)^3] \\ \frac{1}{2}(2-|x|^2), & 1 \leqslant |x| \leqslant 2, \quad [\frac{1}{6}(2-|x|)^3] \\ 0, & 2 \leqslant |x|. \end{cases}$$

These are immediate applications of well-known results in the theory of Fourier transforms.

Lemma 6. *If* $z = x + iy$, *where* x *and* y *are real,*

(a) $|F(z)| \leqslant \alpha^{-1}\exp\left(\dfrac{y^2}{2\alpha^2}\right)$,

(b) $|F(z)| \leqslant \alpha^{-1}\exp\left(\dfrac{y^2-(|x|-4\mu)^2}{2\alpha^2}\right)$ *if* $|x| \geqslant 4\mu$,

(c) $\displaystyle\int_{c-i\infty}^{c+i\infty} |F(iz)|\,|dz| \leqslant \sqrt{2\pi}\,\alpha\exp\left(\frac{c^2}{2\alpha^2}\right)$, [c real]

(d) $|F(z) - F_1(z)| \leqslant \sqrt{2\pi}\,\alpha(4\mu^2)^{-1}$ *if* z *is real*, $[\sqrt{2\pi}\,\alpha^2(16\mu^3)^{-1}\,?]$

(e) $|F'(z)| \leqslant (\alpha\mu)^{-1}\exp\left(\dfrac{y^2}{2\alpha^2}\right)$. [? multiplied by $\frac{4}{3}$]

To prove (a) and (b) we use the inequality

$$|F(z)| \leqslant \frac{1}{\sqrt{2\pi}} \int_{-\infty}^{\infty} |F_2(z-u)|\,|F_1(u)|\,du,$$

and observe that since $F_1(u) \geqslant 0$ we have

$$\frac{1}{\sqrt{2\pi}} \int_{-\infty}^{\infty} |F_1(u)|\,du = \frac{1}{\sqrt{2\pi}} \int_{-\infty}^{\infty} F_1(u)\,du = f_1(0),$$

and also that the integrand vanishes outside the range $|u| \leqslant 4\mu$ so that $|F(z)| < M$ [? $\leqslant$], where M is the maximum of $|F_2(z-u)|$ in this range.

To prove (c)

$$\int_{c-i\infty}^{c+i\infty} |F(iz)|\,|dz| \leqslant \frac{1}{\sqrt{2\pi}} \int_{u=-\infty}^{\infty} \int_{c-i\infty}^{c+i\infty} |F_2(iz-u)|\,|F_1(u)|\,du\,|dz|$$

$$= f_1(0) \int_{c-\mathrm{i}\infty}^{c+\mathrm{i}\infty} |F_2(\mathrm{i}z)|\,|\mathrm{d}z| = \sqrt{2\pi}\,\alpha \exp\left(\frac{c^2}{2\alpha^2}\right).$$

[correct but obscure]

For (d)

$$|F(z) - F_1(z)| = \left| \frac{1}{\sqrt{2\pi}} \int_{-\infty}^{\infty} (F_1(z-u) - F_1(z)\$ - u\,F_1'(z))\,F_2(u)\,\mathrm{d}u \right| \quad [+]$$

since F_2 is an even function

$$\leqslant \frac{1}{4\$\mu^2} \int_{-\infty}^{\infty} u^2 |F_2(u)|\,\mathrm{d}u \quad [\mu^3 \text{ in fact}]\ [\text{times } \tfrac{1}{2}]$$

since $|F_1''(z)| \leqslant \sqrt{2\pi}/4\$\mu^2$ $[\mu^3]$

$= \sqrt{2\pi}\ldots/4\mu^2$ (illegible). [Should $4\mu^2$ be $16\mu^3$?]

To prove (e)

(This proof is missing from this copy of the typescript, but see COHEN and MAYHEW 1968, p. 695, Lemma 2, part (iii).)

Lemma 7.

$$\left| \frac{1}{\sqrt{2\pi}} \int_{-\infty}^{\infty} \mathrm{e}^{-t/2}(\Pi(\mathrm{e}^t) - M(t))\,f(t-t_0)\$ - t_0^{-1} \sum_{|\gamma| \leqslant 4\mu+50} (\tfrac{1}{2} + \mathrm{i}\gamma)^{-1} F(\gamma) \exp(\mathrm{i}\gamma t_0) \right|$$

is less than $\mu^{-1} t_0^{-3/2}$ *if* $t_0 > 10^4$, $50 < \mu < 250$. [? +] ⟦2⟧

We shall show that with the notation of Lemma 4

(a) $$|J| < 10^{-8} t_0^{-3/2},$$

(b) $$|I_1| < 10^{-8} t_0^{-3/2},$$

(c) $$|I_2| < 10^{-8} t_0^{-3/2},$$

(d) $$\left| \sum_{|\gamma| < 4\mu+50} \$\sqrt{2\pi}\,I_{3,\varrho} - t_0^{-1}(\tfrac{1}{2} + \mathrm{i}\gamma)^{-1} F(\gamma) \exp(\mathrm{i}\gamma t_0)\$ \right| < \left(\frac{0.02}{\mu} + \frac{0.052}{\mu} \sum_{|\gamma| < 4\mu+50} \gamma^{-1} \right) t_0 .$$

[left-hand side: ? left and right brackets] [right-hand side: ?]

(e) $$\left| \sum_{|\gamma| \geqslant 4\mu+50} I_{3,\varrho} \right| < 10^{-8} t_0^{-3/2}.$$

By Lemma 3 $|g(0)| < \$4$. [?] Also $\zeta(0) = -\frac{1}{2}$, $\zeta(2) = \frac{1}{6}\pi^2$ whence $|J| < 16$ [true but useless; ? $|J| < 2\exp(-\frac{1}{4}t_0)$] and so (a) since $t_0 > 10^4$.

To prove (b) we shift the line of integration onto $\sigma = \frac{1}{4}$ and observe that on that line $|s^{-1}\log((s+1)(s+2)\zeta(s+2))| < 10$. Hence

$$|I_1| < 10\mathrm{e}^{-t/4}\$ \int_{1/4-\mathrm{i}\infty}^{1/4+\mathrm{i}\infty} |F(\$ - \mathrm{i}(s-\tfrac{1}{2}))|\ |\mathrm{d}s| \quad \left[\frac{1}{\sqrt{2\pi}}\right] \quad [?\ +]$$

$$< 10\alpha\exp(-\tfrac{1}{4}t_0 + \tfrac{1}{32}\alpha^2) \quad \text{by Lemma 6(e)}$$

$$= \$\tfrac{1}{2}t_0^{-1/2}\exp(t_0(-\tfrac{1}{4} + \tfrac{1}{2800})) < 10^{-8}t_0^{-3/2}. \quad [200]$$

To estimate I_2 we first consider the behaviour of $\log(\zeta(s)/((s+1)(s+2)\zeta(s+2)))$. By the functional equations of the zeta and gamma functions we have

$$\frac{\zeta(s)}{(s+1)(s+2)\,\zeta(s+2)} = -\frac{1}{(2\pi)^2}\left(\frac{s}{s+2}\right)\frac{\zeta(1-s)}{\zeta(-1-s)}$$

and therefore if $\mathscr{R}s \leqslant -1$, [? -3]

$$\left|\log\frac{\zeta(s)}{(s+1)(s+2)\,\zeta(s+2)}\right| \leqslant \pi + 2\log\$\pi + 2 + \left|\log\frac{s}{s+2}\right| < 11. \quad [?2]$$

We also have $|g(s)| < 4$ [caret] by Lemma 3(b) and therefore, using Lemma 6(c)

$$|I_2| < 15\$\alpha\exp\left(\frac{\Delta^2}{2\alpha^2} - \Delta t_0\right) = 15\$\alpha\exp(-\mathrm{i}\Delta t_0) \quad [\exp(0.1(1-\sigma)),\ \text{twice}]$$

$$\$ = 300t_0^{-1/2}\exp(-200t_0). \quad [\text{less than?}]$$

(the ‘2’ in ‘200’ is also queried)

It remains to estimate $I_{3,\varrho}$. Since $\mathscr{R}\varrho = \frac{1}{2}$ for $|\mathscr{I}\varrho| < 4$ we may put

$$\frac{1}{\sqrt{2\pi}}I_{3,\varrho} = \varrho^{-t_0}F(\gamma)\exp(\mathrm{i}\gamma t_0) + K_{3,\varrho} + L_{3,\varrho} + M_{3,\varrho},$$

$$[?F(-\gamma)\ (= F(\gamma) \text{ because } F \text{ even})]$$

where

$$K_{3,\varrho} = -\int_{\varrho-\infty}^{\$\varrho} \varrho^{-1}F(\gamma)\exp((s-\mathrm{i})t_0)\,\mathrm{d}s = -\varrho^{-t_0}F(\gamma)\exp(\mathrm{i}\gamma t_0 - 2t_0), \quad [?\varrho - 2]$$

$$L_{3,\varrho} = \int_{\varrho-2}^{\varrho} \varrho^{-1}(F(-\mathrm{i}(s-\tfrac{1}{2})) - F(-\gamma))\exp((s-\tfrac{1}{2})t_0)\,\mathrm{d}s,$$

$$M_{3,\varrho} = \int_{\varrho-2}^{\varrho} F(-\mathrm{i}(s-\tfrac{1}{2}))\ldots\exp((s-\tfrac{1}{2})t_0)\,\mathrm{d}s. \quad (\text{illegible})$$

Then

$$|K_{3,\varrho}| \leqslant \gamma^{-t_0} \exp(-2t_0)|F(\gamma)| \leqslant (20\gamma t_0^{1/2})^{-1} \exp(-2t_0) < 0.001\, \gamma^{-1} t_0^{-3/2} \mu^{-1}$$

[these three occurrences of the symbol γ are queried (not $F(\gamma)$)] and

$$|M_{3,\varrho}| \leqslant \gamma^{-2} 2\alpha^{-1}\mu^{-1} \int_0^2 x \exp\left(\frac{x^2}{2\alpha^2} - xt_0\right) \quad \text{[caret]}$$

[the 2 and μ^{-1} outside the integral are also queried]

$$\leqslant \frac{2\mu^{-1}}{\alpha\gamma^2} \int_0^2 x \exp(-\tfrac{1}{2}xt_0)\, \mathrm{d}x < \frac{0.4\, t_0^{-3/2}}{\gamma^2\mu},$$

[O.K. but not by L3; see INGHAM 1932, notes 22, 23,...]

$$\sum_{|\gamma|<4\mu+50} |M_{3,\varrho}| < \mu^{-1} 0.02\, t_0^{-3/2} \quad \text{since } \textstyle\sum \gamma^{-2} < \frac{1}{20}.$$

By an integration by parts, if $|\gamma| < 4\mu + 50$,

$$|L_{3,\varrho}| = (t_0|\varrho|)^{-1} \exp(\$ - 2t_0)(F(-\gamma) - F(-\gamma\$ + 2\mathrm{i}))\$$$

$$+\mathrm{i} \int_{\varrho-2}^{\varrho} F'(\$ - \mathrm{i}(s - \tfrac{1}{2})) \exp((s - \tfrac{1}{2})t_0)\, \mathrm{d}s$$

$[(\mathrm{i}\gamma - 2)t_0]$ [− ?] [− ?] [+]

$$\leqslant (t_0 \$ \gamma)^{-1} \left(2\alpha^{-1} \exp\left(\frac{2}{\alpha^2} - 2t_0\right)\right.$$

$$\left. + \alpha^{-1}\mu^{-1} \int_0^2 \exp\left(\frac{x^2}{2\alpha^2} - xt_0\right) \mathrm{d}x\right) \quad [|\gamma|]$$

$$\leqslant (t_0\alpha\gamma)^{-1} \left(2 \exp(t_0(-2 + \tfrac{1}{200})) + \frac{1.01}{\mu t_0}\right) \quad [|\gamma| \text{ again.}]$$

$$\leqslant 0.051\, \gamma^{-1} t_0^{-3/2} \mu^{-1}.$$

(d) now follows by collection of results.

For the case $|\gamma| \geqslant 4\mu + 50$ we have, by Lemma 6(b)

$$|I_{3,\varrho}| \leqslant 2\sqrt{2\pi}\, \alpha^{-1}\gamma^{-1} \exp(\tfrac{1}{2}t_0 + (2\alpha^2)^{-1}((\tfrac{5}{2})^2 - (\gamma - 4\mu)^2))$$

$$\leqslant \frac{6}{\alpha\gamma} \exp(t_0/?(400 + \tfrac{25}{4} - (\gamma - 4\mu)^2)). \quad \text{(illegible)}$$

But since $|\gamma| \geqslant 4\mu + 50$ and $\mu \geqslant 50$ we have

$$(\gamma - 4\mu)^2 > 50(|\gamma| - 4\mu) > 1250 + 25(|\gamma| - 4\mu)$$

and therefore

$$|I_{3,\varrho}| \leqslant \frac{6}{\alpha\gamma}\exp\left(-t_0 - \frac{c_0}{32}(|\gamma| - 4\mu)\right) < \frac{24\mu}{\alpha\gamma^2}\exp(-t_0)$$

for $\gamma \exp(-t_0 \gamma/32)$ is a decreasing function for $\gamma > 32/t_0$, and $4\mu \geqslant 200 > 3/t_0$. Then

$$\sum_{|\gamma| \geqslant 4\mu+50} |I_{3,\varrho}| < 1.2\,\mu\alpha^{-1}\exp(-t_0) < \$0.01\, t_0^{-3/2} \quad \text{since } \mu < 250.$$

[(e) says 10^{-8} (O.K. thus); but no relation to enunciation]

Lemma 8. *If* $t_0 > 10^4$ *and*

$$\int_{-\infty}^{\infty} e^{-t/2}(\Pi(e^t) - M(t)) f(t - t_0)\, dt > 1.0025\frac{\pi}{\mu t_0}$$

where, as previously mentioned,

$$f(t) = \exp(-\tfrac{1}{2}\alpha^2 t^2)\left(\frac{\sin \mu t}{\mu t}\right)^4, \quad \alpha^2 = \frac{400}{t_0}, \quad 50 < \mu < 250,$$

$$M(t) = \int t^{-1} e^t\, dt, \quad \textit{from } 0.1 \textit{ to } \max(t, 0.1),$$

then there is a t_1, $0.974\, t_0 < t_1 < 1.053\, t_0$, *such that*

$$\Pi(\exp t_1) - \mathrm{li}\exp t_1 > 1.002\, t_1^{-1}\exp(\tfrac{1}{2}t_1).$$

We have

$$\frac{1}{\sqrt{2\pi}}\int_{-\infty}^{\infty}\left(\frac{\sin \mu t}{\mu t}\right)^4 dt = \frac{\sqrt{2\pi}}{2\mu},$$

$$\frac{1}{\sqrt{2\pi}}\int_{-\infty}^{\infty} t^2\left(\frac{\sin \mu t}{\mu t}\right)^4 dt = \frac{\sqrt{2\pi}}{(2\mu)^3} \quad [4\mu^3?]$$

and therefore

$$\int_{-\infty}^{\infty} \exp(-\tfrac{1}{2}t)(\Pi(e^t) - M(t)) - \$1.002\, t^{-1}\frac{\exp(i\alpha^2(t - t_0)^2)}{1 - (1 - t/t_0)^3}$$

$$\times \exp(-i\alpha^2(t - t_0))\left(\frac{\sin(\mu(t - t_0))}{\mu(t - t_0)}\right)^4 dt > \frac{1.0025\pi}{\mu t_0} - \frac{2\pi}{\mu t_0} - \frac{\pi}{\dots} \quad \text{(illegible)}.$$

Hence for some t_1

$$\Pi(\exp t_1) - M(t_1) > 0.0025\, t_1^{-1}\exp(it_1 + \tfrac{1}{2}\alpha^2(t_1 - t_0)^2),\ 1 - \left(1 - \frac{t}{t_0}\right)^3 > at_1^{-1}e^J,$$

where $a = \$\frac{3}{4}$ (1.005) [$\frac{9}{8}$] and $J = \frac{1}{2}t_1 + 200(t_1 - t_0)^2/t_0$. Now we certainly have $t_1 > 0.1$ for otherwise $\Pi(\exp t_1) = 0$ and $M(t_1) = 0$. But then

$$\Pi(\exp t_1) = \ldots \quad \text{(illegible)} < \exp t_1 + t_1 \exp(\tfrac{1}{2}t_1) < 2 \exp t_1$$

and therefore

$$\frac{200(t_1 - t_0)^2}{t_0} < \tfrac{1}{2}t_1 + \log(3t_0).$$

Now, if $t_1 > t_0$, we have $\log(3t_0) < 0.01\, t_0 < 0.01\, t_1$ and hence

$$200(0.1275(t_1 - t_0)^2) < 0.1275(t_1 + t_0)^2, \quad t_1 < 1.053\, t_0.$$

But if $t_1 \leqslant t_0$, then

$$200(t_1 - t_0)^2 < 0.051\, t_0^2, \quad t_1 > 0.974\, t_0.$$

It only remains to prove that $M(t)\$ < \mathrm{li}\, e^t$, and this will follow if $\mathrm{li}\, e^{0.1} < 0$. [? >] But

$$\mathrm{li}\, e^{0.1} = \int_{-\infty}^{-0.2} t^{-1}e^t\, dt + \int_{-0.2}^{-0.1} t^{-1}e^t\, dt + \int_0^{0.1} \frac{\$ \sinh t}{t\, dt}$$

$$< e^{-0.2} \int_{-0.2}^{-0.1} t^{-1}\, dt + \$ \sinh 0.1 < 0 \quad \text{[insert 2, twice]}.$$

Lemma 9.

$$\mathrm{li}\, x < \frac{x}{\log x - 1.5} \quad \textit{if } x \geqslant e^8.$$

We have

$$e^{-a}\, \mathrm{li}\, e^a = \mathscr{R} \int_0^\infty (a - t)^{-1} e^{-t}\, dt$$

if the integration is along a contour which avoids a and a is real [and positive]

$$= \mathscr{R} \int_0^\infty \left(a^{-1} + \frac{t}{a^2}\right) e^{-t}\, dt + \mathscr{R} \int_0^\infty \frac{t^2 e^{-t}}{a^2(a - t)}\, dt.$$

By taking the contour $t = u(1 + \frac{1}{2}i)$ where u is real

$$\left| \int_0^\infty \frac{t^2 e^{-t}}{a^2(a - t)}\, dt \right| \leqslant \sqrt{5} a^{-3} \int_0^\infty (\tfrac{1}{2}\sqrt{5}a)^{\$3} e^{-a}\, d\tfrac{5}{2}u = \frac{6.25}{a^3}, \quad [2?]$$

$$e^{-a}\, \mathrm{li}\, e^a \leqslant \frac{1}{a} + \frac{1}{a^2} + \frac{6.25}{a^3} \leqslant \frac{1}{a} + \frac{1.5}{a^2} + \frac{2.25}{a^3} \quad \text{if } a \geqslant 8$$

$$< (a - 1.5)^{-1}$$

Lemma 10. *If either*

(a) $$(\Pi(x) - \mathrm{li}\, x)x^{-1/2} \log x > 1.002 \quad \textit{and} \quad x > \mathrm{e}^{2000}$$

or

(b) $$(\Pi(x) - \mathrm{li}\, x)x^{-1/2} \log x > 1.6 \quad \textit{and} \quad x > \mathrm{e}^{\$16},$$

[16 queried] *then either* $\pi(x) > \mathrm{li}\, x$ *or* $\pi(x^{1/2}) > \mathrm{li}\, x^{1/2}$.

We begin by estimating $\Pi(x) - \pi(x) - \frac{1}{2}\pi(x^{1/2})$ for $x > 16$.

$$\Pi(x) - \pi(x) - \tfrac{1}{2}\pi(x^{1/2}) = \tfrac{1}{3}\pi(x^{1/3}) + \sum r^{-1}\pi(x^{1/r})$$

$$\leqslant \tfrac{1}{9}x^{1/3} + \tfrac{2}{3} + \sum \frac{2}{r} + \sum \frac{x^{1/r}}{3r},$$

summation from $r = 4$ to $\log_2 x$ (since $\pi(u) < 2$ if $0 < u < \mathrm{e}$ and $\pi(u) < 2 + \frac{1}{3}u$ if $\mathrm{e} < u$). [O.K. but obscure; $\pi(u) < 2$ used in terms with $\log x < r \leqslant \log_2 x$]

$$\leqslant \tfrac{1}{9}x^{1/3} + 2\log(\tfrac{1}{2}\log_2 x) + \tfrac{4}{3}x^{1/4} \sum_{r=4}^{\infty} r^{-2}$$

(since $rx^{1/r}$ decreases with increasing r in the range $4 \leqslant r \leqslant \log x$)

$$\leqslant \tfrac{1}{9}x^{1/3} + 2\log\log x + 0.4\, x^{1/4}.$$

Then if $\pi(x)\$ > \mathrm{li}\, x$ and $\pi(x^{1/2})\$ > \mathrm{li}\, x^{1/2}$, [? $\leqslant$]

$$(\Pi(x) - \mathrm{li}\, x)x^{-1/2} \log x < \tfrac{1}{2}x^{-1/2}\, \mathrm{li}\, x^{1/2} \log x$$

$$+ \tfrac{1}{9}x^{-1/6} \log x + 0.4\, x^{-1/4} \log x + 2(\log\log x)\log x,$$

$$\tfrac{1}{2}x^{-1/2}\, \mathrm{li}\, x^{1/2} \log x < \left(1 - \frac{3}{\log x}\right)^{-1} \quad [x \geqslant \mathrm{e}^{16}] \quad \text{by Lemma 9}$$

and this denies both (a) and (b).

Theorem 1. *If* $50 < \mu < 250$, $10^4 < t_0$, $\sum_{|\gamma| < 4\mu + 50} \gamma^{-1} < 3$ *and*

$$\mathscr{R}\$ \sum_{0 < \gamma < 4\mu + 50} \frac{K(\gamma)}{\frac{1}{2} + \mathrm{i}\gamma} \exp(\mathrm{i}\gamma t_0)\$ > 0.502\, \mu^{-1}(\tfrac{1}{2}\pi)^{1/2},$$

[left bracket and minus sign] [right bracket]
then there exists t_1, $0.974\, t_0 < t_1 < 1.053\, t_0$ *such that either*

$$\pi(\exp t_1) > \mathrm{li} \exp t_1 \quad \textit{or} \quad \pi(\exp \tfrac{1}{2}t_1) > \mathrm{li} \exp \tfrac{1}{2}t_1.$$

$$t_0^{1/2}(0.021 + 0.052 \sum_{|\gamma| < 4\mu + 50} \gamma^{-1}) < 0.177\, t_0^{-1/2} < 0.00177.$$

Hence

$$\frac{\mu t_0}{\pi}\int_{-\infty}^{\infty} e^{-t/2}(\Pi(e^t)-M(t))f(t-t_0)\,dt > 1.004-0.00177\left(\frac{2}{\pi}\right)^{1/2} > 1.00,$$

and the condition of Lemma 8 is satisfied, and consequently condition (a) of Lemma 10.

5. *The Diophantine approximation*

We have not until now made much use of the results of heavy computations on the zeta-function. We have made use of the fact that for $|\gamma|<1468$ the nontrivial zeros are all on the critical line, although we could well have avoided doing so. We shall now go further and make use of some information about the positions of the zeros in this range. Titchmarsh has mentioned [where?] that if t_n is defined by the condition $\arg\Gamma(\frac{1}{4}+\frac{1}{2}it_n)=n\pi$, [query] [def. $\theta(t)=\pi^{-1}\arg(\pi^{-1/2it}\Gamma(\frac{1}{4}+\frac{1}{2}it))$] [$\theta(t_n)=n-1$, $t_n>0$, $n=1,2,\ldots$] then if $0<t_n<1468$, we have $N(t_n)-n=-1$, 0 or 1, and also that if $N(t_n)-n\$=0$, [?] then $N(t_{n+1})=n+1$ and $N(t_{n-1})=n-1$. Also $\arg\Gamma(\frac{1}{4}+\frac{1}{2}it)$ [$\theta(t)$] is monotonic. [$t\geqslant\ldots$] From these facts it can easily be seen that if $0<t<1468$, then $\pi^{-1}\arg\Gamma(\frac{1}{4}+\frac{1}{2}it_n)<2$. We also have $\pi^{-1}\arg\Gamma(\frac{1}{4}+\frac{1}{2}it)-t/(2\pi)\log(t/(2\pi))-1<\frac{1}{4}$ for $t>51$ and $|S^*(t)|<2\frac{1}{4}$ for $t\leqslant 51$ whence $|S^*(t)|<2\frac{1}{4}$ for $0<t<1468$. [the upper bound $\frac{1}{4}$ may be replaced by a much smaller constant; see Titchmarsh 1935, p. 238 (ii)] Here $S^*(t)=N(t)-t/(2\pi)\log(t/(2\pi))-1$. [$2\frac{1}{4}$ is queried]

Lemma 12. *If* $1\leqslant h\leqslant 220$, *then*

$$\left|\sum f\left(\frac{\gamma}{2\pi}\right)-\int_1^h f(v)\log v\,dv\right| < 2\tfrac{1}{4}(|f(h)|+[\operatorname{var} f]_1^h),$$

where the sum is over $2\pi<\gamma<2\pi h$.

We make use of the inequality $|S^*(t)|<\frac{9}{4}$, $0<t<1468$ justified above, and also observe that $S^*(2\pi)=0$. Then

$$\begin{aligned}\left|\sum f\left(\frac{\gamma}{2\pi}\right)-\int_1^h f(v)\log v\,dv\right| &= \left|\int_1^h f(v)\,dS^*(2\pi v)\right|\\ &= \left|f(h)S^*(2\pi h)-\int_1^h S^*(2\pi v)\,df(v)\right|\\ &\leqslant 2\tfrac{1}{4}(f(h)+[\operatorname{var} f]_1^h).\end{aligned}$$

In order to get a slightly better result we shall use a modified form of Dirichlet's theorem.

Lemma 13. *Given real numbers $a_1, \dots, a_m$, a positive real number τ and positive integers $n_1, \dots, n_m$, we can find an integer r, $1 \leqslant r \leqslant \prod n_i$, such that for each i, $1 \leqslant i \leqslant m$, $\tau r a_i$ differs from an integer by at most n_i^{-1}.*

The proof is very similar to that of Dirichlet's theorem (INGHAM 1932, Theorem J), and the details will be left to the reader.

Lemma 14. *If we choose $\mu = 60\pi$ in the functions f, F we can find t_0 so that $e^{20} < t_0 < e^{660.9}$ and*

$$2S = \$ \sum_{|\gamma| < 4\mu + 50} \mu\left(\frac{2}{\pi}\right)^{1/2} F(\gamma)(\tfrac{1}{2} + i\gamma)^{-1} \exp(i\gamma t_0) > 1.004. \quad [\text{? minus}]$$

We have

$$S = \$ \sum_{0<\gamma<4\mu+50} \mu\left(\frac{2}{\pi}\right)^{1/2} (\tfrac{1}{4} + \gamma^2)^{-1} F(\gamma)(\$ - \gamma \sin(\gamma t_0) + i\cos(\gamma t_0))$$

[? minus] [plus]

and we will put

$$S_1 = - \sum_{0<\gamma<4\mu+50} (\gamma^2 + \tfrac{1}{4})^{-1}\gamma \sin(\gamma t_0)\kappa\left(\frac{\gamma}{2\mu}\right),$$

$$S_2 = \$ \sum_{0<\gamma<4\mu+50} (\gamma^2 + \tfrac{1}{4})^{-1}\tfrac{1}{2}\cos(\gamma t_0)\kappa\left(\frac{\gamma}{2\mu}\right), \quad [\text{? minus}]$$

$$S_3 = - \sum_{0<\gamma<4\mu+50} \gamma^{-1}\kappa\left(\frac{\gamma}{2\mu}\right)\sin(\gamma t_0).$$

Then

$$|S_3 - S_1| \leqslant \sum_{0<\gamma<4\mu+50} (\gamma(\gamma^2 + \tfrac{1}{4}))^{-1}\tfrac{1}{4}|\sin(\gamma t_0)| \leqslant \sum_{0<\gamma} \frac{1}{4\gamma^3} < 0.0004$$

and

$$|S - S_1 - S_2| \leqslant \sum |F(\gamma) - F_1(\gamma)|\mu\left(\frac{2}{\pi}\right)^{1/2} \quad [\text{range of summation?}]$$

$$\leqslant \frac{\alpha}{2\mu} \sum_{0<\gamma<4\mu+50} 1 < (20t_0^{-1/2})\frac{600}{120\pi} < 0.0035, \quad t_0 \geqslant e^{20} - 1.$$

We now choose τ so that for the first zero $i\gamma_0 + \frac{1}{2}$, $\tau\gamma_0/2\pi$ is an integer and $e^{20} - 1 < \tau < e^{20}$. We then choose t_0 in accordance with Lemma 13 so that t_0 [query] is a multiple of τ and for each γ, $0 < \gamma < 120\pi$, $\gamma/(2\pi)(t_0 +$

$\frac{1}{120}$) differs from an integer by at most $\gamma/(1920\pi)$. This t_0 can be found in the range

$$e^{20}+1 < t_0 + \frac{\pi}{2\beta} < (e^{20}+2) \prod_{\gamma_0<\gamma<120\pi} (1+1920\gamma^{-1}\pi).$$

[should $\pi/(2\beta)$ be $\frac{1}{120}$?]

Now

$$\log \prod_{2\pi<\gamma<120\pi} (1+1920\gamma^{-1}\pi) < \sum_{2\pi<\gamma<120\pi} \left(\log 1920\gamma^{-1}\pi + \frac{\gamma}{1800\pi}\right)$$

$$< \int_1^{60} \left(\log 960 - \log u + \frac{u}{900}\right) \log u \, du + \frac{9}{4}\left(\log 960 + \frac{1}{900}\right) < 647 \quad [?672]$$

$$\log(1+1920\gamma_0^{-1}\pi) > 6.01, \quad e^{20} < t_0 < e^{660.9}. \quad [686]$$

We have

$$S_3 \geqslant \$ - \sum_{0<\gamma<120\pi} \gamma^{-1}\kappa\left(\frac{\gamma}{120\pi}\right) \min_{|\eta|<1/8} \sin\left(\frac{(1+\eta)\gamma}{120}\right)$$

$$- \sum_{120\pi<\gamma<240\pi} \gamma^{-1}\kappa\left(\frac{\gamma}{120\pi}\right) \quad \text{[sign queried]}$$

and if we write

$$\varphi(\upsilon) = \begin{cases} (2\pi\upsilon)^{-1}\kappa\left(\dfrac{\upsilon}{60}\right) \min\limits_{|\eta|<1/8} \sin\left(\dfrac{\pi\,\upsilon(1+\eta)}{60}\right), & \text{if } 0<\upsilon<60, \\ -(2\pi\upsilon)^{-1}\kappa\left(\dfrac{\upsilon}{60}\right), & \text{if } 60\$<\upsilon\$<120, \quad [\leqslant] \end{cases}$$

we shall have by Lemma 12

$$S_3 \geqslant \sum \varphi\left(\frac{\gamma}{2\pi}\right) \quad (\text{over } 0<\gamma<240\pi)$$

$$\geqslant \int_1^{120} \varphi(\upsilon) \log \upsilon \, d\upsilon\$ + \tfrac{9}{4}(\text{var } \varphi)_1^{120}$$

$$\geqslant \int_1^{120} \varphi(\upsilon) \log \upsilon \, d\upsilon\$ + 0.0097. \quad \text{[signs queried]}$$

But by direct computation we find

$$\int_1^{120} \varphi(\upsilon) \log \upsilon \, d\upsilon > 0.5080. \quad [\ldots\text{lse with}\ldots\text{correct}\ldots\text{p.}\,9]$$

Also

$$S_2 \geqslant 0.49 \sum_{0<\gamma<120\pi} \gamma^{-2}\kappa(\gamma) - \tfrac{1}{2} \sum_{120\pi<\gamma<240\pi} \gamma^{-2}\kappa(\gamma)$$

$$-(240)^{-2} \sum_{0<\gamma<120\pi} \kappa(\gamma) \max_{\cdots}\left(\frac{\sin(1+\eta)\gamma}{\cdots}\right) \Big/ \left(\frac{\gamma}{(2\ldots)}\right) \text{ (illegible)} > 0.008.$$

Hence

$$S > 0.5080 + 0.003 - 0.0097 - 0.0004 - 0.0035 = 0.5026. \quad [?]$$

We can now state our final result.

Theorem 2. *There is an* $x, \ldots,$ *such that* $\pi(x) > \operatorname{li} x$.

(Here '...' represents one of two conditions, both of which are crossed out in the manuscript. They are

$$2 < x < \exp\exp a < 10^b, \quad b = 10^c,$$

where $(a, c) = (697, 303)$ or $(661, 287)$.) This follows from Theorem 1 and Lemma 14.

Our results up to this point have also been characterised by the extreme smallness of the remainder terms, and our chief concern has been to obtain *some* definite remainder with a relatively brief argument. From this point onward, however, we shall be much more exacting.

6. Computational Diophantine approximation

If fairly accurate values of the γ's were available it should be possible to find a value for t_0 by direct computation with a digital computer. It would be necessary first to obtain the first three hundred zeros or so to say seven places of decimals. This might involve ten to twenty hours of computation time. We should then choose 200 say, so that our sums would extend to 800. Owing to the small values of (symbol missing) beyond 700 we would not have calculated the zeros there. A reasonable method of procedure for the Diophantine approximation would be simply to try out successively all values of t_0 which make the value of $\sin(\gamma t_0)$ for the first zero equal to 1. Let us make a rough estimate of where, given reasonable freedom from bad luck, we might expect to find a solution on this basis. Let us assume that the sums of terms other than the first are independent and normally distributed. The standard deviation of the distribution of S is easily calculated to be

$$\left(\tfrac{1}{2} \sum_{\gamma \neq \gamma_0} \frac{\kappa(\gamma)}{\tfrac{1}{4}+\gamma^2} - \frac{\kappa(\gamma_0)}{\cdots}\right)^{1/2}, \quad \text{(illegible)}$$

i.e., about 0.0091. We get a solution if the sum exceeds $\frac{1}{2}-1/\gamma_0$, i.e., 0.43. The probability of this on the normal distribution is about 1.3×10^{-6}, from which we may conclude that there is an even chance of finding a value by this method in the first 500 000 trials, i.e. with $t_0<220000$, i.e., of establishing that there is an x, $2<x<\exp 220000$ for which $\pi(x)>\operatorname{li} x$. ⟦3⟧

7. *Case where the Riemann hypothesis is ... false* (positively?)

In order to complete our investigation it would be as well to obtain some result which can be applied if the Riemann hypothesis is discovered to be false, by the ... (illegible) not on the critical line. If one is simply given that there is a zero in some large rectangle, not meeting the critical line (e.g., $\sigma>0.53$, $0<t<10^8$), it is not easy to prove any very satisfactory results about values of x for which $\pi(x)>\operatorname{li} x$. This is because of the possibility that there may be many other zeros near to the given one; they may be sufficiently near to have much nuisance value, but not near enough to give essentially the same effect as multiple zero. The present investigation ignores all these difficulties by postulating a zero $\beta_1+\mathrm{i}\gamma_1$ off the critical line and at considerable distance from any other zeros of this kind. It seems very probable that the first zeros off the critical line that are computed will satisfy the conditions required. We shall again use Lemma 4, but this time we shall put

$$f(t)=\alpha\tfrac{1}{2}(1+\cos(\gamma_1 t))\exp(-\tfrac{1}{2}\alpha^2 t^2),\quad \Delta=100,$$

and we therefore have

$$F(u)=\tfrac{1}{2}\exp\left(-\frac{u^2}{2\alpha^2}\right)+\tfrac{1}{4}\exp\left(-\frac{(u-\gamma_1)^2}{2\alpha^2}\right)+\tfrac{1}{4}\exp\left(-\frac{(u+\gamma_1)^2}{2\alpha^2}\right).$$

We shall need to have an upper bound for the number of zeros in a given range of t.

Lemma 15.

$$\left|\left(\frac{\Gamma'}{\Gamma}\right)(x+\mathrm{i}y)-\log(x+\mathrm{i}y-\tfrac{1}{2})\right|<\frac{\pi}{y-1}.$$

We first obtain an inequality for $|(\mathrm{d}^2/\mathrm{d}z^2)\log\Gamma(z)|$, $z=x+\mathrm{i}y$.

$$\left|\frac{\mathrm{d}^2}{\mathrm{d}z^2}\log\Gamma(z)\right|=\left|\sum_{n>0}(z-n)^{-2}\right|\leqslant\sum_{n=-\infty}^{\infty}(y^2+(n-x)^2)^{-1}$$
$$\leqslant\frac{2}{y^2}+\int_{-\infty}^{\infty}(y^2+u^2)^{-1}\,\mathrm{d}u.$$

Then

$$\left|\left(\frac{\Gamma'}{\Gamma}\right)(x+\mathrm{i}y)-\log(x+\mathrm{i}y-\tfrac{1}{2})\right| = \left|\left(\frac{\Gamma'}{\Gamma}\right)(x+\mathrm{i}y)-\log\frac{\Gamma(x+\mathrm{i}y+\frac{1}{2})}{\Gamma(x+\mathrm{i}y-\frac{1}{2})}\right|$$

$$\leqslant \max_{Iz=y}\left|\frac{\mathrm{d}^2}{\mathrm{d}z^2}\log\Gamma(z)\right| \leqslant \frac{\pi}{y-1}.$$

Lemma 16. *Denoting the number of zeros of the zeta function with positive imaginary parts less than t by $N(t)$,*

$$N(t+\tfrac{3}{2})-N(t-\tfrac{3}{2}) \leqslant 1.6\log\frac{t+8}{2\pi}+1.7$$

$$\sum\frac{\sigma-3}{(t-\gamma)^2+(\sigma-\beta)^2} = \mathscr{R}\sum(s-\varrho)^{-1}$$

$$= \tfrac{1}{2}\mathscr{R}\left(\frac{\Gamma'}{\Gamma}\right)(\tfrac{1}{2}s+1)-\tfrac{1}{2}\log\pi+\mathscr{R}\left(\frac{\zeta'}{\zeta}\right)(s)+\mathscr{R}(s-1)^{-1}$$

$$\leqslant \tfrac{1}{2}\log\left|\frac{s+1}{2\pi}\right|+\frac{\sigma-1}{t^2+(\sigma-1)^2}+\left(\frac{\zeta'}{\zeta}\right)(\sigma)+\frac{\pi}{2t-1}.$$

['2' in '$2t-1$' queried]

Taking $\sigma=2$, $t\geqslant 10$ we have $(\zeta'/\zeta)(2)<0.53$ and therefore

$$\sum\frac{\sigma-\beta}{(t-\gamma)^2+(\sigma-\beta)^2} \leqslant \tfrac{1}{2}\log\frac{t+8}{2\pi}+0.53.$$

Now if $\beta=\frac{1}{2}$, $|t-\gamma|>\frac{3}{2}$, then $(\sigma-\beta)/((t-\gamma)^2+(\sigma-\beta)^2)>\frac{1}{3}>\frac{102}{325}$ and if $\beta+\beta'=1$, $|t-\gamma|<\frac{3}{2}$, then

$$\frac{\sigma-\beta}{(t-\gamma)^2+(\sigma-\beta)^2}+\frac{\sigma-\beta'}{(t-\gamma)^2+(\sigma-\beta)^2} \geqslant \frac{20}{32}.$$

Hence

$$N(t+\tfrac{3}{2})-N(t-\tfrac{3}{2}) \leqslant \frac{325}{102}\left(\tfrac{1}{2}\log\frac{t+8}{2\pi}+0.53\right)$$

$$\leqslant 1.6\log\frac{t+8}{2\pi}+1.7.$$

This inequality is also clearly valid for $0<t<10$ for the left-hand side is the ... (remainder of sentence missing)

Theorem 3. *Suppose that $\beta_1+\mathrm{i}\gamma_1$, $\beta_1>\frac{1}{2}$, $\gamma_1>0$, is a zero of the zeta function and that for every other zero $\beta+\mathrm{i}\gamma$ either $\beta=\frac{1}{2}$ or $|\gamma-\gamma_1|>14$. Then for some x, $2<x<(16\gamma_1)^a$, $a=1.12/(\beta_1-\frac{1}{2})$, we have $\pi(x)>\operatorname{li}x$.*

With f, F defined as above we have

$$|F(x+iy)| < \exp\left(\frac{y^2}{2\alpha^2}\right),$$

$$\int_{c-i\infty}^{c+i\infty} |F(is)|\,|ds| \leqslant (2\pi)^{1/2}\exp\left(\frac{c^2}{2\alpha^2}\right),$$

and, with the notation of Lemma 4, may prove that $|J|$, $|I_1|$, $|I_2|$ are each less than $10^{-8}t_0^{-3/2}$ as in Lemma 7. We now proceed to the estimation of $I_{3,\varrho}$. If $\varrho=\beta+i\gamma$ and $0<\mathscr{R}a<1$

$$\left|\int_{\varrho-2}^{\varrho} \varrho^{-1}\exp((s-\tfrac{1}{2})t_0)\exp\left(\frac{(s-a)^2}{2\alpha^2}\right)ds\right|$$

$$\leqslant \gamma^{-1}\exp((\beta-\tfrac{1}{2})t_0+(2\alpha^2)^{-1}\mathscr{R}(\varrho-a)^2)\int_{-2}^{0}\exp(ut_0)$$

$$+\frac{u}{\alpha^2}\mathscr{R}(\varrho-a)+\frac{u^2}{2\alpha^2}\,du$$

$$\leqslant \exp((\beta-\tfrac{1}{2})t_0+\tfrac{1}{200}t_0\mathscr{R}(\varrho-a)^2)\gamma^{-1}(t_0+\tfrac{1}{200}t_0\mathscr{R}(\varrho-a-1))^{-1}$$

$$\leqslant \frac{1.02}{\gamma t_0}\exp((\beta-\tfrac{1}{2})t_0+\tfrac{1}{200}t_0\mathscr{R}(\varrho-a)^2).$$

We shall deal separately with the zeros which are near to $\beta_1+i\gamma_1$ and those which are relatively far away. If $||\gamma|-|\gamma_1||>14$, then (since in any case $|\gamma|>14$) we have

$$(|\gamma|-|\gamma_1|)^2 > 182+||\gamma|-|\gamma_1||, \qquad |\gamma|^2>182+|\gamma|$$

and therefore

$$|I_{3,\varrho}| < 1.02(\gamma t_0)^{-1}e^{t_0/2}(\tfrac{1}{2}\exp(-\tfrac{1}{200}t_0\gamma^2)+\tfrac{1}{4}\exp(-\tfrac{1}{200}t_0(\gamma-\gamma_1)^2)$$

$$+\tfrac{1}{4}\exp(-\tfrac{1}{200}t_0(\gamma+\gamma_1)^2)$$

$$< 1.02(\gamma t_0)^{-1}\exp(-0.405t_0)\,(\tfrac{3}{4}\exp(-\tfrac{1}{200}t_0\gamma)$$

$$+\tfrac{1}{4}\exp(-\tfrac{1}{200}t_0||\gamma|-|\gamma_1||))$$

We now suppose that $t_0>20$.

(The remaining three pages of the paper are hand-written.)

Then

$$\sum\gamma^{-1}\exp(-\tfrac{1}{200}t_0||\gamma|-|\gamma_1||)\leqslant\sum\gamma^{-1}\exp(-0.1\,||\gamma|-|\gamma_1||)$$

$$\leqslant \sum_n 1.6\log(\tfrac{1}{2}(3\ldots+8))\exp(-0.1(|3n-\gamma_1|-\tfrac{3}{2})) \quad \text{by Lemma 15}$$

$$\leqslant 2\log(\tfrac{1}{2}(\gamma_1+8))(1-\exp(-0.3))^{-1}$$

$$+2\sum_{n>\gamma_1} \exp(-0.15\,|n-\gamma_1|)\max_{n>\gamma_1} \log(\tfrac{1}{2}(3n+8))\exp(-0.15\,(n\ldots))$$

$$\leqslant 2\log(\tfrac{1}{2}(\gamma_1+8)),$$

$$\sum \gamma^{-1}\exp(-\tfrac{1}{200}t_0\gamma) \leqslant \max_{u>0}\,(u\exp(-\tfrac{1}{200}t_0 u))\sum \gamma^{-2} < \frac{2}{t_0},$$

$$\sum |I_{3,\varrho}| < 1.02\exp(-0.405\,t_0)(\tfrac{3}{2}t_0^{-2}+12t_0^{-1}\log(\tfrac{1}{2}(\gamma_1+8))\ldots,$$

(sum over $||\gamma|-|\gamma_1||>14$)

$$\sum |I_{3,\varrho}| < 2\left(\frac{1.03}{\gamma_1 t_0}\right)(\sum 1) \quad \text{since } \gamma_1 > 1100$$

(first sum over $||\gamma|-|\gamma_1||\leqslant 14$, second over $|\gamma-\gamma_1|<14$)

$$< \frac{33}{\gamma_1 t_0}\log(\tfrac{1}{2}(\gamma_1+22)).$$

If $\varrho_1=\beta_1+\mathrm{i}\gamma_1$

$$I_{\varrho_1,3}=\mathscr{R}\int_{-2}^{0}(4(\varrho_1+u))^{-1}\exp((\varrho_1-\tfrac{1}{2}+u)t_0+\tfrac{1}{2}\alpha^{-2}(\beta_1-\tfrac{1}{2}+u)^2)\,\mathrm{d}u.$$

If $\sin(\gamma_1 t_0)=1$ and $\gamma_1>40$, then

$$|\arg((\varrho_1+u)^{-1}\exp(\varrho_1 t_0))|<\ldots,$$

$$\cos\arg((\varrho_1+u)^{-1}\exp(\varrho_1 t_0)) \geqslant \frac{4}{4.05},$$

$$\left|\frac{\gamma_1}{\varrho_1+u}\right| > \frac{4.05}{4.1},$$

$$I_{\varrho_1,3} \geqslant (4.1\,\gamma_1)^{-1}\exp((\beta_1-\tfrac{1}{2})t_0)\int_{-2}^{0}\exp(ut_0+\tfrac{1}{200}(\beta_1-\tfrac{1}{2}+u)^2t_0)\,\mathrm{d}u$$

$$\geqslant (4.1\,\gamma_1)^{-1}\exp((\beta_1-\tfrac{1}{2})t_0)\int_{-2}^{0}\exp(\tfrac{1}{200}t_0(\beta_1-\tfrac{1}{2})^2+u(t_0+\tfrac{1}{100}(\beta_1-\tfrac{1}{2})))\,\mathrm{d}u$$

$$=\exp(t_0(\beta_1-\tfrac{1}{2}+\tfrac{1}{200}(\beta_1-\tfrac{1}{2})^2))$$

$$\times(1-\exp(-2(t_0+\tfrac{1}{200}(\beta_1-\tfrac{1}{2}))))(4.1\,\gamma_1(t_0+\tfrac{1}{100}(\beta_1-\tfrac{1}{2})))^{-1}$$

$$\geqslant (4.2\,\gamma_1 t_0)^{-1}\exp(t_0(\beta_1-\tfrac{1}{2}+\tfrac{1}{200}(\beta_1-\tfrac{1}{2})^2)), \quad \text{if } t_0>5.$$

Collecting results

$$t_0\int_{-\infty}^{\infty}\mathrm{e}^{-t/2}(\Pi(\mathrm{e}^t)-M(t))f(t-t_0)\,\mathrm{d}t \geqslant$$

$$\geqslant -3.10^{-8} t_0^{-1/2} - 1.02 \exp(-0.405\, t_0)(\tfrac{1}{2} t_0^{-1} + 12 \log \ldots)$$

$$-\frac{33}{\gamma_1} \log(\tfrac{1}{2}(\gamma_1 + 22)) + (4.2\, \gamma_1)^{-1} \exp(t_0 A), \quad \text{where}$$

$$A = \beta_1 - \tfrac{1}{2} + \tfrac{1}{200}(\beta_1 - \tfrac{1}{2})^2.$$

We now choose t_0 so that $\sin(\gamma_1 t_0) = 1$ and

$$0 < t_0 - (\beta_1 - \tfrac{1}{2})^{-1} \log(16\gamma_1) < \frac{2\pi}{\gamma_1}.$$

Since $\gamma_1 > 1468$ the condition $t_0 > 5$ is automatically satisfied, indeed $t_0 > 20$. Then $\exp(t_0(\beta_1 - \tfrac{1}{2})) > 16\gamma_1$

$$t_0 \int_{-\infty}^{\infty} e^{-t/2}(\Pi(e^t) - M(t)) f(t - t_0)\, dt$$

$$\geqslant -10^{-8} - 13(\log(\tfrac{1}{2}(\gamma_1 + 3)))(16\gamma_1)^{-0.81} - \tfrac{33}{1468} \log(\tfrac{1}{2}(1490)) + 3.8 \geqslant 3.5,$$

$$\alpha \int_{-\infty}^{\infty} (e^{-t/2}(\Pi(e^t) - M(t)) t_0 - 0.9 \exp(0.4\, \alpha^2 (t - t_0)^2))$$

$$\times \exp(-\tfrac{1}{2}\alpha^2 (t - t_0)^2) \tfrac{1}{2}(1 + \cos(\gamma_1 (t - t_0)))\, dt$$

$$\geqslant 3.5 - 0.9(\tfrac{5}{2}\pi)^{1/2} \left(1 + \exp\left(\frac{-\gamma_1^2}{10\alpha^2}\right)\right)$$

$$> 0, \quad \text{since } \frac{\gamma_1^2}{10\alpha^2} > \frac{1468^2 t_0}{1000} > 10.$$

Then for some t_1

$$t_0(\exp(-\tfrac{1}{2} t_1))(\Pi(\exp t_1) - M(t_1)) > 0.9 \exp(0.4\alpha^2 (t_1 - t_0)^2)$$

$$= 0.9 \exp\left(4\sigma \frac{(t_1 - t_0)^2}{t_0}\right),$$

$$40(t_1 - t_0)^2 < \tfrac{1}{2} t_1 t_0 + t_0 \log(3 t_0)$$

from which it follows that

$$0.8\, t_0 < t_1 < 1.12\, t_0,$$

$$(\exp(-\tfrac{1}{2} t_1))(\Pi(\exp t_1) - M(t_1)) > 0.8\, t_1.$$

Applying Lemma 10

$$\pi(\exp t_1) > \operatorname{li} \exp t_1 \quad \text{or} \quad \pi(\exp(\tfrac{1}{2} t_1)) > \operatorname{li} \exp(\tfrac{1}{2} t_1),$$

i.e., there exists x, $2 < x < (16\gamma_1)^A$, where $A = 1.12/(\beta_1 - \tfrac{1}{2})$, (such that) $\pi(x) > \operatorname{li} x$.

References

(1) A.E. INGHAM, The distribution of prime numbers, Cambridge Mathematical Tracts, No. 30, 1932.

(2) J.E. LITTLEWOOD, Sur la distribution des nombres premiers, *Comptes Rendus* **158** (1914) 1869–1872.

(3) S. SKEWES, On the difference $\pi(x) - \mathrm{li}\, x$, *Proc. London Math. Soc.* (This is the reference in Turing's paper. Presumably this is the paper of the same title, *J. London Math. Soc.* **8** (1933) 277–283.)

(4) E.C. TITCHMARSH, The zeros of the zeta-function, *Proc. Roy. Soc. (A)* **157** (1936) 261–263.

ANNALS OF MATHEMATICS
Vol. 67, No. 1, January, 1958
Printed in Japan

AN ANALYSIS OF TURING'S "THE WORD PROBLEM IN SEMI-GROUPS WITH CANCELLATION"[1]

BY WILLIAM W. BOONE

(Received June 12, 1956)

The remarks following are intended as both explication and criticism of certain details of Turing's article [5].[2] As was stated in [1], Turing's proof is in principle correct.[3,4] We have tried to present these remarks in such a way that his proof can be read through more or less continuously.

A. An exact construction for $\mathfrak{S}_0$

It would seem worth while to supply such a construction here so as to help the reader. In part for the sake of brevity we perhaps depart in certain nonessentials from the construction intended by Turing.

Let $\mathfrak{U}$ be a Universal Turing Machine with infinite tape (i.e., without analogue of Post's symbol, h [3]); in fact, like the "usual" Universal Machine except for having both left and and right facing I.C.'s (l_i and r_i); for which it is recursively unsolvable to determine for an arbitrary initial C.C. (which we may assume, for elegance, does not contain s_3) whether or not a C.C. is generated in which s_3 occurs, and whose entries are of the form

$$(r_i s_k, s_{k'} r_{i'}), \quad (s_k l_i, l_{i'} s_{k'}), \quad (s_k l_i, s_{k'} r_{i'}), \quad (r_i s_k, l_{i'} s_{k'}).^5$$

Let $\mathfrak{U}'$ be the finite-tape Machine whose symbols are those of $\mathfrak{U}$ and the new tape symbol s_4; and whose entries are those of $\mathfrak{U}$ and the additional entries:

$(r_i s_4, r_i s_0 s_4)$ where $(r_i s_0, B)$ for some B is an entry of $\mathfrak{U}$;

---

1 Prepared for publication while the author was a grantee under National Science Foundation Contract No. G1974 at the Institute for Advanced Study.

2 Numbers in square brackets refer to the bibliography at the end of this paper. Unqualified references to pages, lemmas, etc., refer to [5].

3 [1] includes a list of minor misprints in [5].

4 These emendations were originally worked out in the preparation of [1]. The article [5] recently assumes fresh importance because, as we understand from a review by A. Markov in the Mathematical Reviews, vol. 17 (1956), page 706, P. S. Novikov [2] has proved the unsolvability of the word problem for groups on the basis of [5].

5 We note explicity that the four forms given are intended to include those entries which "come into play when the machine is scanning a blank square in one of the two continuous infinite sequences of blank squares at the two ends of the tape". (Cf. the table on page 493.) Let $\boldsymbol{O}$ be the set of entries of the Universal Turing Machine of [4]. Then we may take as the entries of $\mathfrak{U}$: $(r_i s_k, s_{k'} r_{i'})$ and $(s_k l_i, s_{k'} r_{i'})$ where $(q_i s_k s_m, s_{k'} q_{i'} s_m) \in \boldsymbol{O}$; $(s_k l_i, l_{i'} s_{k'})$ and $(r_i s_k, l_{i'} s_{k'})$ where $(s_m q_i s_k, q_{i'} s_m s_{k'}) \in \boldsymbol{O}$; $(r_i s_k, l_{ik} s_{k'})$, $(s_m l_{ik}, s_m r_{i'})$ for any s_m, $(s_k l_i, s_{k'} r_{ik})$, and $(r_{ik} s_m, l_{i'} s_m)$ for any s_m, where $(q_i s_k, q_{i'} s_{k'}) \in \boldsymbol{O}$.

$(s_4 l_i,\ s_4 s_0 l_i)$ where $(s_0 l_i,\ B)$ for some B is an entry of $\mathfrak{U}$.

Let $\mathfrak{B}$ be the finite-tape Machine whose symbols are those of $\mathfrak{U}'$, the new internal configurations l_0, l_1, and r_1, and whose entries are

(A, B) where (A, B) is an entry of $\mathfrak{U}'$ such that s_3 does not occur in B ;

(A, r_1) where (A, B) is an entry of $\mathfrak{U}'$ for some B in which s_3 occurs ;

$(r_1 s_k,\ r_1)$ where $k \neq 4$;

$(r_1 s_4,\ l_1 s_4)$

$(s_k l_1,\ l_1)$ where $k \neq 4$;

$(s_4 l_1,\ s_4 l_0)$.

The first and second phase relations of $\mathfrak{S}_0$ are obtained from $\mathfrak{B}$ in accordance with the directions (a), (b), (c), (d), and (e) of page 496. The commutation and change relations are as given, page 496, except that the relations $(\sigma_m w_0,\ w_0 \tau_m)$, $m=1, 2, \cdots, M$, should be added to the change relations.

B. Lemma 1

Substitute the following version of this lemma:
For all arguments for which the functions are defined we have

$$\boxed{1}\quad \psi_1(XY) = \psi_1(X)\psi_1(Y)$$
$$\boxed{2}\quad \chi(\varphi_1(X)) = \varphi_2(X)$$
$$\boxed{3}\quad \psi_1(\varphi_1(V)) = V$$
$$\boxed{4}\quad \psi_2(\varphi_2(V)) = V$$
$$\boxed{5}\quad \psi_2(XY) = \psi_2(X)\psi_2(Y)$$
$$\boxed{6}\quad \varphi_2(\psi_1(X)) = \chi(X) \quad \textit{if } X \textit{ is free of } \sigma_m \textit{ and } \tau_m$$
$$\boxed{7}\quad \chi(XY) = \chi(X)\chi(Y)$$
$$\left.\begin{array}{l}\boxed{8}\quad \varphi_1(\psi_1(X)) = X \\ \boxed{9}\quad \varphi_2(\psi_2(X)) = X\end{array}\right\} \textit{ where } X \textit{ is free of } \sigma_m \textit{ and } \tau_m .$$

C. Lemma 3

(1) H_r has not been proved normal. The following lemma, having an obvious demonstration, is required.

LEMMA 2.1. *$\varphi_1(X)$ and $\varphi_2(X)$ are normal where X is any* C.C. *of* $\mathfrak{B}$.

(The normality of $\varphi_2(X)$ is needed later for the verification of the statement following Lemma 20.) From this lemma we have at once that $H_0 = \varphi_1(C_0)$ is normal and that $H_r = P\sigma_m\varphi_1(V)\tau_m R$ is normal if $H_{r-1} = PQR$ is normal.

(2) *The demonstration that* $(\chi(H_r), \chi(H_0))$ *is a relation*: The last sentence of page 498 is in error. We proceed as follows. (The boxed numbers refer to the equations of Section B.) Since, as Turing shows, $\psi_1(Q) = U$, $\varphi_1(U) = Q$ by $\boxed{8}$ for U is free of σ_m and τ_m by Lemma 2. H_{r-1} is $P\varphi_1(U)R$ so that we may write the induction hypothesis $(\chi(H_{r-1}), \chi(H_0))$ is a relation, as

$$(\alpha) \qquad (\chi(P)\chi(\varphi_1(U))\chi(R),\ \chi(\varphi_1(C_0)))$$

is a relation by $\boxed{7}$. But $(\sigma_m\varphi_2(V)\tau_m, \varphi_2(U))$ holds by Lemma 2 and hence $(\chi(\sigma_m\varphi_1(V)\tau_m), \chi(\varphi_1(U)))$ is a relation by $\boxed{2}$, the fact that $\chi(\sigma_m) = \sigma_m$, $\chi(\tau_m) = \tau_m$ and $\boxed{7}$. Applications of rules (v) and (vi) to the preceding relation yield that

$$(\beta) \qquad (\chi(P)\chi(\sigma_m\varphi_1(V)\tau_m)\chi(R),\ \chi(P)\chi(\varphi_1(U))\chi(R))$$

is a relation. Rule (iv) applied to (α) and (β) yields that $(\chi(H_r), \chi(H_0))$ is a relation since the first term of the relation specified in (β) is $\chi(H_r)$ by $\boxed{7}$.

D. The notation $[A, B]$ and Lemma 5

For fixed (A, B), $[A, B]$ does not designate a fixed ordinal.[6] The following additions to notation dispose of the matter. Let $\{A, B\}^u$, u an integer, be a *particular* proof of (A, B). (Two occurrences of *the same proof* must use rules of inference in the same way.) Let $[A, B]^u$ be the length of $\{A, B\}^u$.

(1) Using this additional notation Lemma 5 should read as follows: *For every* $\{A, B\}^u$ *such that* $n\omega \leqq [A, B]^u \leqq (n+1)\omega$, *there is a chain* $A = A_0, A_1, \cdots, A_m = B$, $m \leqq n$, *of words* etc. (as in Turing's version). This version of the lemma is adequate for the demonstrations of subsequent lemmas. *In fact, all Turing needs is the mere existence of a chain from A to B, if (A, B) has a proof less than* ω^2.

(2) Page 500, lines 2 and 3 are easily changed to conform to (1).

(3) As noted below, Section J, one needs a sort of converse of Lemma 5 for the demonstration of Lemma 19.

E. Lemmas 9 and 10

Lemma 10 is true but does not follow from Lemma 9 as indicated unless the further condition is added to Lemma 10 *that* G_2B *contains no barriers*. (Cf. Section F.) What is needed here is a somewhat stronger form of Lemma 9, from which Lemma 10 follows immediately.

LEMMA 9′. *If one link of a chain contains n barriers, then every link of*

[6] Turing has explained to the author in correspondence that his intention was that $[A, B]$ be the length of the shortest proof of (A, B); however, our development does not follow this suggestion.

the chain contains n barriers.

We omit a proof of Lemma 9′.

F. Lemma 11[7]

We take the following as the statement of Lemma 11:

For all $\{AG_1G_2F,\ BG_3G_4H\}^z$ *such that*

(i) $[AG_1G_2F,\ BG_3G_4H]^z < \omega^2$;

(ii) G_1G_2 *and* G_3G_4 *are barriers*;

(iii) *Either* AG_1 *and* BG_3 *contain no barriers or* G_2F *and* G_4H *contain no barriers*;

(iv) AG_1G_2F *and* BG_3G_4H *are not identical.*

There exist proofs $\{AG_1,\ BG_3\}^x$ *and* $\{G_2F,\ G_4H\}^y$ *such that*

$$[AG_1,\ BG_3]^x + [G_2F,\ G_4H]^y + 1 \leqq [AG_1G_2F,\ BG_3G_4H]^z .$$

We use Turing's abbreviation of P and Q for AG_1G_2F and BG_3G_4H; and following Turing's technique let $\{P,\ Q\}^z$ be the shortest proof denying the lemma. We sometimes use k and $[k]$ for $\{L,\ M\}^k$ and $[L,\ M]^k$ respec-spectively. We let K^i be the ith step of z. We demonstrate the lemma only for the case in which the last step which asserts $(P,\ Q)$ is obtained by an application of (iv) to step K^s which asserts $(P,\ D)$ and step K^t which asserts $(D,\ Q)$; and where further (I) D is not identical with either P or Q and (II) D is of the form EG_5G_6I, E free of barriers and G_5G_6 a barrier.

Let $\{AG_1G_2F,\ EG_5G_6I\}^s$ be the proof of $(P,\ D)$ which is obtainable from z in the obvious way by deleting steps. The proof s is to contain no useless steps and its last step is to be K^s. Let $\{EG_5G_6I,\ BG_3G_4H\}^t$ be the similarly defined proof of $(D,\ Q)$ having K^t as last step. Clearly s and t exist.

Then by the definition of z, (I), (II) and the fact that the $[s] < [z]$ and $[t] < [z]$ (for both s and t exclude the application of rule (iv) which produces the last step of z) we know there must exist proofs $\{AG_1,\ EG_5\}^u$ and $\{G_2F,\ G_6I\}^v$ such that

$$(\alpha) \qquad [u] + [v] + 1 \leqq [s] ,$$

and we further know there must exist proofs $\{EG_5,\ BG_3\}^l$ and $\{G_6I,\ G_4H\}^m$ such that

$$(\beta) \qquad [l] + [m] + 1 \leqq [t] .$$

Let $\{AG_1,\ BG_3\}^a$ be the proof obtained by following u by l and an appli-

[7] As an alternative to our binding the proofs referred to in the smaller side of the inequalities throughout this section with existential quantifiers, Turing in correspondence has suggested postulating that "an unprovable formula has an infinitely long proof".

cation of (iv) to the last steps of u and l. Let $\{G_2F, G_1H\}^b$ be similarly defined in terms of v and m. Then

$$(\gamma) \qquad [u] + [l] + 1 = [a]$$

and

$$(\delta) \qquad [v] + [m] + 1 = [b]$$

by the foregoing definitions. If we could now assert

$$(\varepsilon) \qquad [s] + [t] + 1 \leqq [z]$$

corresponding to Turing's line 22, page 501, the inequality

$$(\zeta) \qquad [a] + [b] + 1 \leqq [z]$$

would follow from (α), (β), (γ), (δ), (ε). Thus the definition of z would be contradicted and the proof of the lemma would be completed for this (sub)case.

The difficulty in asserting (ε) is that we have no assurance that s and t *have no steps in common*. This difficulty can be overcome, however, by redefining *proof* to require that a proof satisfy the additional requirement (*Property* R) that no step is used more than once as a premise to obtain a subsequent step. It is easy to verify that requiring Property R of proofs does not invalidate Turing's argument at any other point.[8]

G. The demonstration of Lemma 12

The following validly yields Lemma 12.

First, precede Lemma 12 by the following lemma which is obvious.

Lemma 11.1. *If X is a normal first-phase word, then $\psi_1(X)$ is a* C.C..

Second, substitute for the first six lines of the proof of Lemma 12, the following: Suppose $XUY = X'U'Y'$ where U' is also the first term of an F.R.. Then by Lemma 2, there exist first terms T and T' from the table for $\mathfrak{B}$ such that $U = \varphi_1(T)$ and $U' = \varphi_1(T')$. But

$$\psi_1(XUY) = \psi_1(X)\psi_1(\varphi_1(T))\psi_1(Y)$$
$$= \psi_1(X'U'Y') = \psi_1(X')\psi_1(\varphi_1(T'))\psi_1(Y')$$

by Section B, Lemma 1, equation $\boxed{1}$. Further

[8] As an alternative to requiring Property R of proofs Turing in correspondence has suggested that the difficulty of overlap "might be settled by defining length inductively as follows. The length of proofs of a result X is to be the length of the proof of the last step plus an appropriate addendum if by any clause but (iv) or (i). If the last step is by (iv) it is to be the sum of the lengths for (A, C) and (C, B) plus the right addendum. Addendum only if by (i). This is just what one really uses and does not in any way restrict proofs (we no longer need [Property] R)".

$$\psi_1(XUY) = \psi_1(X)T\psi_1(Y)$$
$$= \psi_1(X')T'\psi_1(Y')$$

by Section B, Lemma 1, equation [3].

But then since $\psi_1(XUY)$ is a C.C. by Lemma 11.1, we have $T = T'$ by the lines fourth and fifth from the bottom of page 493. Further $X = X'$ and $Y = Y'$ (for which a demonstration may be obtained by substituting U for V and U' for V' in the last three lines of Turing's proof of Lemma 12).

Third, for $XVY = X'V'Y'$ proceed in accordance with Turing's development except that "second members of" should be inserted before "the same relation", line 16, page 502.

H. The demonstration of Lemma 16

(1) This is a minor point. It is not a priori clear that each link contains some symbol not a σ_m-symbol, i.e., that all links of the chain of Lemma 16 can be written in the form ΣFZ. One could cite the obvious:

LEMMA 15.1. *Each member of an* F.R. *contains the same number of active symbols, namely, one.*

LEMMA 15.2. *Every link of a chain contains the same number of active symbols.*

LEMMA 15.3. *Chains with more than one link contain at least one active symbol in each link.*[9]

PROOFS. 15.1 implies 15.2; 15.1 and 15.2 imply 15.3. Then 15.3 implies that all links of the chain of Lemma 16 can be written in the form ΣFZ.

(2) Note that the void word is an allowable value for Σ_1 or Σ_2. (Pertinent to F_r and F_s in the proof of Lemma 17.)

I. The demonstration of Lemma 17

There are two lacunae here: (1) a citing of Lemma 13 so that one may assume the chain discussed has no change of direction; (2) a citing of a certain property of the F.R.'s to validate the sentence beginning in the third line of the proof. The observation made there does not hold for an arbitrary cancellation semi-group. Therefore, *first* precede Lemma 17 by

LEMMA 16.1. *There is no generator*, G_1, *and* F.R., (U, V), *of* $\mathfrak{S}_0$, *such that* (U, V) *is of the form* (G_1U', G_1V') *or* $(U'G_1, V'G_1)$.

Second, substitute for the first three sentences of the proof of Lemma 17, the following:

Let (GA, GB) be a given pair and H the chain for (GA, GB) which has

[9] Lemma 15.3 makes the proof of Lemma 18 immediate: For the chain (existence by Lemma 5) corresponding to the relation described in Lemma 18 must have one link.

the smallest number of links and denies the lemma. Then, by its definition, H can have no changes of direction, for should it have, by Lemma 13, there would exist, as a result of the presence of repetitious links in H, a chain with smaller number of links than H and denying the lemma.

We consider two cases:

Case I: Every link of H begins with G. Clearly it is sufficient to show that the existence of the F.R. (U, V) justifying (GA_i, GA_{i+1}) implies the existence of an F.R. justifying (A_i, A_{i+1}).[10] Now write (GA_i, GA_{i+1}) as (XUY, XVY). Then if X were void, U and V would both begin with G contradicting Lemma 16.1. Therefore the non-void X is of the form GX' and (GA_i, GA_{i+1}) of the form $(GX'UY, GX'VY)$. But then (A_i, A_{i+1}) is of the form $(X'UY, X'VY)$ which is also justified by (U, V).

Case II: Not every link of H begins with G. *Third*, now proceed in accordance with Turing's development from his fourth sentence on except that "from F_r to F_s" should be substituted for "from V to U'", line 23, page 503. (See Section H2.)

J. The demonstration of Lemma 19

(1) The following lemma, which is a kind of converse of Lemma 5, is needed here:

LEMMA 18.1. *If there is a normal chain for (A, B), then there is a normal proof, $\{A, B\}^u$, i.e., a proof using only normal words and whose length is $< \omega^2$.*

(We omit the straightforward proof.)

This lemma should be used by adding, after "by Lemma 17", line 5, page 504, "and hence a normal proof for (P, Q) by Lemma 18.1" and also as indicated in Section J2 which follows.

(2) The argument as given unjustifiably assumes that the shortest proof denying Lemma 19 which is being considered (we shall call this proof $\{P, Q\}^u$) contains at most a single application of (vii) or (viii).

If $[u] \geqq \omega^2$ we confine our considerations of changes necessary to the case wherein: (a) the first cancellation in u is an application of (viii), the premise (conclusion) of this operation being step K^v asserting (GA, GB) (step K^w asserting (A, B)); (b) "(A, B) is of form $(A'G_1G_2F, B'G_3G_4H)$ where G_1G_2 and G_3G_4 are barriers and $GA'G_1$, $GB'G_3$ without barriers" (page 504, line 13). Let $\{GA, GB\}^v$ ($\{A, B\}^w$) be the proof obtained from u by deleting all steps after K^v (K^w). Then, by Lemma 11, there exist proofs $\{GA'G_1, GB'G_3\}^m$ and $\{G_2F, G_4H\}^n$ such that $[m] + [n] + 1 \leqq [v]$. Since $[m] < [v] < [u]$, by the definition of u and Lemma 5 there is a chain

[10] GA normal implies A normal.

for $(GA'G_1, GB'G_2)$. By Lemma 8, Lemma 18, and (b) we have that either (α) $(GA'G_1, GB'G_2)$ is a normal relation or (β) $(GA'G_1, GB'G_2)$ is an identity. If (α), then there is a normal chain for $(GA'G_1, GB'G_2)$ by Lemma 6; hence there is a normal chain for $(A'G_1, B'G_2)$ by Lemma 17. Therefore, by Lemma 18.1, there is a normal proof, $\{A'G_1, B'G_2\}^f$. Combining f with n if (α), or making several applications of (vi) to n if (β), gives us in either case a proof $\{A, B\}^y$ which does not apply (vii) or (viii) and is therefore shorter than w. Let $\{P, Q\}^t$ be the proof consisting of v followed by y followed by that part of the proof u which comes after w with, however, the last step of y asserting (A, B) used as premise instead of K^w. Since $[y] < [w]$, $[t] < [u]$ and the definition of u is contradicted.

If $[u] < \omega^2$, then there is a chain for (P, Q) by Lemma 5, hence a normal chain by Lemma 6, hence there is a normal proof for (P, Q) by Lemma 18.1. Therefore u does not deny the lemma being shown.

K. Lemma 20 and statement following

(1) We must require in the statement of the lemma that W be normal. Lemma 2.1 should be cited to show $\varphi_1(C)$ normal. Lemma 19 (first sentence of demonstration) applies only to normal relations. ("Lemma 15" in this first sentence should be "Lemma 5"). Note that since Lemma 19 must be applied *before* Lemma 5, Lemma 6 is not applicable here.

(2) *Sixth line of proof*: "by Lemma 15" should be added after "$(\varphi_1(C), XUY)$".

(3) *Last sentence of page* 504: "and the fact that $\varphi_2(C)$ is normal by Lemma 2.1" should be added.

OXFORD UNIVERSITY

BIBLIOGRAPHY

1. W. W. BOONE, Review of [5], J. Symbolic Logic, vol. 17 (1952), pp. 74-76.
2. P. S. NOVIKOV, On the algorithmic unsolvability of the word problem in group theory (in Russian), Trudy Mat. Inst. Steklov, no. 44, 1 955.
3. E. L. POST, *Recursive unsolvability of a problem of Thue*, J. Symbolic Logic, vol. 12 (1947), pp. 1-11.
4. A. M. TURING, *On computable numbers, with an application to the Entscheidungsproblem*, Proc. London Math. Soc., vol. 42 (1936-1937), pp. 230-265. Corrections in the same periodical, vol. 43 (1937), pp. 544-546.
5. ——, *The word problem in semi-groups with cancellation*, Ann. of Math., vol. 52 (1950), pp. 491-505.

ON THE DIFFERENCE $\pi(x) - \operatorname{li} x$

By A. M. COHEN *and* M. J. E. MAYHEW

[Received 21 September 1965—Revised 4 January 1967]

Introduction

1. This paper is based on ideas of A. M. Turing, as recorded in an unpublished, and in places inaccurate, manuscript. We reproduce part of Turing's introduction and, in substance, his first few lemmas, but diverge from his work in Lemma 6.

'Let $\pi(x)$ be the number of primes less than x, and $\operatorname{li} x$ the "logarithmic integral" of x, defined by

$$\operatorname{li} x = \lim_{\eta \to +0} \left(\int_0^{1-\eta} + \int_{1+\eta}^{x} \right) \frac{du}{\log u}.$$

In this, as in other matters of notation, we follow Ingham (**1**). We propose to investigate where $\pi(x) - \operatorname{li} x$ is positive. If x is less than about 1·42 this quantity is positive, and from there up to 10^7 it is negative. The figures suggest that

$$\pi(x) - \operatorname{li} x \sim -x^{\frac{1}{2}}/\log x \quad \text{as } x \to \infty,$$

but Littlewood (**2**) has proved that $\pi(x) - \operatorname{li} x$ changes sign infinitely often. The argument proceeds by cases, according to whether the Riemann hypothesis is true or false. In the event of the Riemann hypothesis being true Skewes (**3**) has shown that

$$\pi(x) - \operatorname{li} x > 0 \quad \text{for some } x, \quad 2 < x < 10^{10^{10^{34}}},$$

while if the Riemann hypothesis is untrue he has shown (**4**) that

$$\pi(x) - \operatorname{li} x > 0 \quad \text{for some } x, \quad 2 < x < 10^{10^{10^{10^{3}}}}.\dagger$$

'In the present paper it is proposed to establish that $\pi(x) > \operatorname{li} x$ for some x, $2 < x < \exp\exp(1236)$,‡ without the restriction of assuming the Riemann hypothesis. We shall also prepare the ground for the possibility of improving the bound to

$$10^{10^5},$$

with the aid of extensive computations, essential to the whole method being the fact that the Riemann hypothesis has been tested in the region

† This result was not known to Turing and was published after his death. Turing only refers to the earlier result in his manuscript.

‡ Our figure. Turing has $\exp\exp(661)$.

Proc. London Math. Soc. (3) 18 (1968) 691–713

$|\tau| < 1468$ (Titchmarsh (**5**)).' The present authors have used the fact that much more is now known about the early errors of the Riemann zeta function (see Haselgrove and Miller (**9**)). Zeros computed by Haselgrove to almost double the precision of the Haselgrove-and-Miller tables are used in the ensuing calculations.

Turing, continuing his introduction, outlines his method as follows:

'The necessity of using the functions $\Pi(x)$ and $\log\zeta(s)$ somewhat complicates the argument. The general outline of the method may be illustrated by dealing with the analogous problem of finding where $\vartheta(x) > x$. Since

$$\psi(x) = \vartheta(x) + \vartheta(x^{\frac{1}{2}}) + \vartheta(x^{\frac{1}{3}}) + \ldots,$$

and

$$\vartheta(x) \sim x,$$

this is essentially the question as to where

$$\psi(x) > x + x^{\frac{1}{2}}.$$

Now

$$\psi(x) - x = -\sum_{\rho} \frac{x^{\rho}}{\rho} - \frac{\zeta'}{\zeta}(0) - \tfrac{1}{2}\log(1 - 1/x^2),$$

so that the problem reduces essentially to the question: For what value of t does the inequality

$$G(t) = -\sum_{\rho} \frac{e^{(\rho-\frac{1}{2})t}}{\rho} > 1$$

hold, summation being over the complex zeros?

'One may consider various expressions of the form

$$\int_0^{\infty} G(t) f(t)\, dt.$$

Putting

$$F(u) = \frac{1}{\sqrt{(2\pi)}} \int_0^{\infty} f(t) e^{iut}\, dt,$$

we have

$$I = \int_0^{\infty} G(t) f(t)\, dt = -\sqrt{(2\pi)} \sum_{\rho} \frac{1}{\rho} F\left(\frac{\rho - \frac{1}{2}}{i}\right).$$

If $f(t)$ is positive for positive t, and decreases to zero sufficiently quickly, it is possible to argue back from the value of I to the inequality $G(t) > 1$. For instance, it will suffice to have

$$\int_0^{\infty} f(t)\, dt = 1, \quad f(t) > 0, \quad I > \tfrac{5}{4}, \quad \int_A^{\infty} G(t) f(t)\, dt < \tfrac{1}{4},$$

to infer $G(t) > 1$, for some t satisfying $0 < t < A$. The value of t for which

I is sufficiently large and positive is to be obtained by Diophantine approximation. In carrying out the approximation we try to adjust the phases of the terms with small ρ so that these are maximized, and to arrange that the remaining, unadjusted, terms are small. We therefore wish $F((\rho - \frac{1}{2})/i)$ to be small for the larger values of ρ. By taking

$$f(t) = \{(\sin \beta t)/t\}^2,$$

Ingham (**6**) ensured that, if the Riemann hypothesis is true, only a finite number of terms were different from zero. In the present paper we use instead

$$f(t) = \{(\sin \beta t)/t\}^4 \exp(-\tfrac{1}{2}\alpha^2 t^2),$$

which is largely inspired by Ingham's argument. It does not result in the vanishing of any of the terms but, if α is small, the later terms are extremely small. The present function has various advantages (for the present purpose) over that used by Ingham. The factor $\exp(-\frac{1}{2}\alpha^2 t^2)$ encourages the quick convergence of the integral

$$\int_0^\infty G(t) f(t)\, dt,$$

facilitating the inference of inequalities about $G(t)$ from values of I. This factor also causes $F(u)$ to be an integral function (whereas, otherwise, it would only be regular in two squares and two right-angle segments). The use of higher powers of $(\sin \beta t)/t$ results in a rather sharper transition from large to small values of $f(u)$ leading to appreciable numerical improvement.'

Thanks are due to the late Dr C. B. Haselgrove who suggested the problem to the former of us, to Professor J. E. Littlewood for help with the writing of this paper and constant encouragement, to Mr A. E. Ingham and Professor P. Stein for some substantial corrections of Turing's manuscript, and, finally, to the referee for pointing out a howler in the first version and for many valuable suggestions.

2. Let $f(t)$ be an even function regular in the strip $-\frac{3}{2} \leqslant \mathscr{I}(t) \leqslant \frac{3}{2}$ and such that

$$f(t) = \begin{cases} O(e^{-(\lambda-\epsilon)\xi}), & \xi \to \infty, \\ O(e^{(\mu-\epsilon)\xi}), & \xi \to -\infty, \end{cases}$$

for every positive ϵ, where $\lambda > 0$, $\mu > 0$, and $\xi = \mathscr{R}(t)$. Define

$$F(u) = \mathscr{F}(-iu) = \frac{1}{\sqrt{(2\pi)}} \int_{-\infty}^{\infty} e^{-iut} f(t)\, dt, \tag{2.1}$$

$$\varphi(s) = \int_{0+}^{\infty} x^{-s}\, d\{h(x)\}, \tag{2.2}$$

where $h(x)$ is of bounded variation in every finite interval $(0, X)$, $s = \sigma + i\tau$, and the following conditions are satisfied:

(i) $\mathscr{F}(s)$ is regular and uniformly $O(e^{-\delta|\tau|})$ in $0 \leqslant \sigma \leqslant \frac{3}{2}$, for some fixed $\delta > 0$.

(ii) $h(x) = o(x^2)$ as $x \to 0$ and $x \to \infty$;

(iii) $\int_0^\infty x^{-3}|h(x)|\,dx$ is convergent.

Then we have, if t_0 is a real number,

LEMMA 1.

$$\int_{-\infty}^{\infty} \frac{h(e^t)f(t-t_0)}{e^{\frac{3}{2}t}}\,dt = \frac{1}{i\sqrt{(2\pi)}}\int_{2-i\infty}^{2+i\infty} \frac{\varphi(s)}{s} e^{(s-\frac{1}{2})t_0}\mathscr{F}(s-\tfrac{1}{2})\,ds.$$

For $\sigma = 2$ we have, on integrating (2.2) by parts and using condition (ii),

$$\varphi(s) = s\int_0^\infty h(x)\,x^{-s-1}\,dx.$$

Thus, substituting for $\varphi(s)/s$ in the right-hand expression of the lemma, we obtain

$$\frac{1}{i\sqrt{(2\pi)}}\int_{2-i\infty}^{2+i\infty}\left(\int_0^\infty h(x)x^{-s-1}\,dx\right)\exp\{(s-\tfrac{1}{2})t_0\}\mathscr{F}(s-\tfrac{1}{2})\,ds.$$

By virtue of (i) and (iii) the double integral is absolutely convergent and we may invert the order of integration. The integral then becomes

$$\int_0^\infty x^{-\frac{3}{2}}h(x)\chi(x)\,dx,$$

where

$$\chi(x) = \frac{1}{i\sqrt{(2\pi)}}\int_{\frac{3}{2}-i\infty}^{\frac{3}{2}+i\infty} x^{-s}\mathscr{F}(s)e^{st_0}\,ds.$$

If we now write $x = e^t$, then, in view of the order conditions on $\mathscr{F}$ arising from (2.1) and (i), we can displace our line of integration to $\sigma = 0$ and use the Fourier inversion formula to obtain

$$\chi(x) = f(t-t_0),$$

and Lemma 1 follows.

We define $f(t)$ by the relation

$$f(t) = f_1(t)f_2(t), \tag{2.3}$$

where

$$f_1(t) = \tfrac{3}{2}\{(\sin \mu t/\mu t)\}^4, \quad f_2(t) = \exp(-\tfrac{1}{2}\alpha^2t^2), \tag{2.4}$$

and $\mu > 0$, $\alpha > 0$.

Then

$$F(u) = \frac{1}{\sqrt{(2\pi)}}\int_{-\infty}^{\infty} F_1(\xi)F_2(u-\xi)\,d\xi, \tag{2.5}$$

where

$$\left.\begin{aligned} &F_1(\xi) = 3\sqrt{(2\pi)}K(\xi/2\mu)/4\mu, \quad F_2(\xi) = \alpha^{-1}\exp(-\xi^2/2\alpha^2),\\ &K(x) = \begin{cases} -\tfrac{2}{3}(1-|x|)^3+\tfrac{1}{6}(2-|x|)^3 & (0 \leqslant |x| \leqslant 1),\\ \tfrac{1}{6}(2-|x|)^3 & (1 \leqslant |x| \leqslant 2),\\ 0 & (2 \leqslant |x|).\end{cases}\end{aligned}\right\} \tag{2.6}$$

and

The result (2.5) and the formula for $F_2(\xi)$ can be found in Titchmarsh ((7) 51, 177), while the formula for $K(x)$ can be obtained by integrating around a semicircular contour indented at the origin. We note that $K(x) \geqslant 0$ and that $K(x)$ attains its maximum value, $\tfrac{2}{3}$, at $x = 0$. This implies that $F_1(\xi) \geqslant 0$ and the maximum value of $F_1(\xi)$ is $F_1(0) = \tfrac{1}{2}\sqrt{(2\pi)}/\mu$. We also require the maximum value of $|F_2(u-\xi)|$ as ξ varies. This is $\alpha^{-1}\exp(y^2/2\alpha^2) = M_1$, say, where $u = x+iy$. We now deduce some inequalities for $F(u)$ and $\mathscr{F}(u) = F(iu)$, where, in what follows, f and F refer to the functions defined above.

Lemma 2. *If $u = x+iy$ then*

(i) $|F(u)| \leqslant \tfrac{3}{2}\alpha^{-1}\exp(y^2/2\alpha^2)$,

(ii) $|F(u)| \leqslant \sqrt{(2\pi)}\exp(y^2/2\alpha^2)/2\mu$,

(iii) $|F'(u)| \leqslant (\alpha\mu)^{-1}\exp(y^2/2\alpha^2)$,

(iv) $\displaystyle\int_{c-i\infty}^{c+i\infty} |\mathscr{F}(u)\,du| \leqslant \tfrac{3}{2}\sqrt{(2\pi)}\exp(c^2/2\alpha^2)$ (*c real*),

(v) $|F(u)| \leqslant \tfrac{3}{2}\alpha^{-1}\exp[\{y^2-(|x|-4\mu)^2\}/2\alpha^2]$, $(|x| > 4\mu)$.

From (2.5),

$$\begin{aligned} |F(u)| &= (2\pi)^{-\frac{1}{2}}\left|\int_{-\infty}^{\infty} F_1(\xi)F_2(u-\xi)\,d\xi\right| \\ &\leqslant (2\pi)^{-\frac{1}{2}}M_1\int_{-\infty}^{\infty} |F_1(\xi)\,d\xi| = M_1 f_1(0).\end{aligned}$$

Since $f_1(0) = \tfrac{3}{2}$, we have (i).

Next, from (2.5),

$$\begin{aligned} |F(u)| &\leqslant (2\pi)^{-\frac{1}{2}}\int_{-\infty}^{\infty} |F_1(\xi)|\,|F_2(u-\xi)\,d\xi|, \\ &\leqslant (2\pi)^{-\frac{1}{2}}\tfrac{1}{2}\sqrt{(2\pi)}\mu^{-1}\alpha^{-1}\exp(y^2/2\alpha^2)\int_{-\infty}^{\infty}\exp\{-(\xi-x)^2/2\alpha^2\}\,d\xi, \\ &= \sqrt{(2\pi)}\exp(y^2/2\alpha^2)/2\mu,\end{aligned}$$

which proves (ii).

Let $\xi = u-\eta$ in (2.5). Then differentiation with respect to u gives

$$F'(u) = \frac{-1}{\sqrt{(2\pi)}}\int_{-\infty}^{\infty} F_1'(u-\eta)F_2(\eta)\,d\eta,$$

and thus

$$|F'(u)| = (2\pi)^{-\frac{1}{2}}\left|\int_{-\infty}^{\infty} F_2(u-\xi)F_1'(\xi)\,d\xi\right|,$$

$$\leqslant (2\pi)^{-\frac{1}{2}}\alpha^{-1}\exp(y^2/2\alpha^2)2\int_0^{4\mu}|F_1'(\xi)|\,d\xi,$$

$$|F'(u)| \leqslant (\alpha\mu)^{-1}\exp(y^2/2\alpha^2),$$

since $|F_1'(\xi)| = -F_1'(\xi)$, which gives (iii).

For (iv) we have

$$\int_{c-i\infty}^{c+i\infty}|\mathscr{F}(u)\,du| = \int_{c-i\infty}^{c+i\infty}\left|(2\pi)^{-\frac{1}{2}}\int_{-\infty}^{\infty} F_2(iu-\xi)F_1(\xi)\,d\xi\right||du|.$$

We use the fact that $F_1(\xi) \geqslant 0$ always; then the left-hand integral is at most

$$\int_{c-i\infty}^{c+i\infty}(2\pi)^{-\frac{1}{2}}\left\{\int_{-\infty}^{\infty}|F_2(iu-\xi)|\,F_1(\xi)\,d\xi\right\}du,$$

and, since it is permissible to change the order of integration here, this equals

$$\int_{-\infty}^{\infty}F_1(\xi)\left\{\int_{c-i\infty}^{c+i\infty}(2\pi)^{-\frac{1}{2}}|F_2(iu-\xi)|\,du\right\}d\xi.$$

Now in the range $(c-i\infty, c+i\infty)$,

$$|F_2(iu-\xi)| = \alpha^{-1}\exp Q,$$

where

$$Q = \mathscr{R}\{-(iu-\xi)^2/2\alpha^2\} = \{c^2-(\xi+t)^2\}/2\alpha^2, \quad u = c+it.$$

Thus

$$\int_{c-i\infty}^{c+i\infty}(2\pi)^{-\frac{1}{2}}|F_2(iu-\xi)\,du| = \exp(c^2/2\alpha^2),$$

so

$$\int_{c-i\infty}^{c+i\infty}|\mathscr{F}(u)\,du| \leqslant \exp(c^2/2\alpha^2)\int_{-\infty}^{\infty}F_1(\xi)\,d\xi = \tfrac{3}{2}\sqrt{(2\pi)}\exp(c^2/2\alpha^2),$$

which gives (iv).

Again from (2.5),

$$|F(u)| \leqslant (2\pi)^{-\frac{1}{2}}\int_{-\infty}^{\infty} |F_1(\xi)|\,|F_2(u-\xi)\,d\xi|$$

$$\leqslant (2\pi)^{-\frac{1}{2}}\int_{-4\mu}^{4\mu} |F_2(u-\xi)|\,|F_1(\xi)|\,d\xi$$

$$\leqslant (2\pi)^{-\frac{1}{2}}M_2\int_{-4\mu}^{4\mu} F_1(\xi)\,d\xi = \tfrac{3}{2}M_2,$$

where M_2 is the maximum value of $|F_2(u-\xi)|$ in the range $|\xi| \leqslant 4\mu$. Since $|x| > 4\mu$,

$$M_2 = \alpha^{-1}\exp[\{y^2-(|x|-4\mu)^2\}/2\alpha^2],$$

giving (v). This completes the proof of Lemma 2.

We may state at once that μ, α, t_0 satisfy

$$\mu = 175, \quad t_0 > 2500, \quad \alpha^2 t_0 = 400. \tag{2.7}$$

The last equation connects α and t_0, where t_0 is closely connected with the X of $\pi(x) > \operatorname{li}x$ for some $x < X$, and it is not determined until the last moment.

3. Define $g(s)$ by the relation

$$g(s) = \int_{a(s-1)}^{a(s-1)+\infty} \frac{e^{-t}}{t}\,dt = \int_{s}^{s+\infty} \frac{\exp\{-a(z-1)\}}{z-1}\,dz, \tag{3.1}$$

where $s = \sigma+i\tau$ and $a = 1/10$. The function $g(s)$ is well defined everywhere except on the real axis for $\sigma \leqslant 1$. We define $g(s)$ on the real axis for $\sigma < 1$ by

$$g(\sigma) = \tfrac{1}{2}\{g_+(\sigma)+g_-(\sigma)\},$$

where $g_+(\sigma)$ and $g_-(\sigma)$ are the limits as $\tau \to \pm 0$ of $g(\sigma+i\tau)$, and $g_+(\sigma)-g_-(\sigma) = -2\pi i$. Our definition of $g(s)$ now covers the whole plane except $s = 1$. We shall require later the following inequality for $g(s)$:

LEMMA 3.

$$|g(s)| \leqslant \{\pi+10/(|s-1|)\}\exp(a-a\sigma).$$

Proof. For all values of s apart from those on the real axis to the left of $s = 1$, we integrate the function $\exp\{-a(z-1)\}/(z-1)$ round the contour $\Gamma = \Gamma_1+\Gamma_2+\Gamma_3+\Gamma_4$, where Γ_1, Γ_2, Γ_3 are the straight lines joining s to $s+R$, $s+R$ to s_0+R, s_0+R to s_0 respectively, where we choose $s_0 = 1+|s-1|$ and we let $R \to \infty$. Γ_4 is the arc of the circle centre 1 and radius $|s-1|$ joining s to s_0 and not crossing the axis to the left of $s = 1$.

Since the function is regular in the region(s) enclosed by Γ,

$$\int_{\Gamma} \exp\{-a(z-1)\}/(z-1)\,dz = 0,$$

and, since the integral over Γ_2 tends to zero as $R \to \infty$, we have

$$g(s) = g(s_0) - \int_{\Gamma_4} \exp\{-a(z-1)\}/(z-1)\,dz.$$

If we now understand by logarithm the principal value we obtain an alternative definition for $g(s)$ to (3.1), namely

$$g(s)+\log(s-1) = g(s_0)+\log(s_0-1) - \int_{\Gamma_4} \frac{\exp\{-a(z-1)\}-1}{z-1}\,dz, \quad (3.2)$$

and we note that $g(s)+\log(s-1)$ is regular in the whole plane. Now

$$0 < g(s_0) = \int_{a|s-1|}^{\infty} \frac{e^{-t}}{t}\,dt \leqslant \frac{e^{-a|s-1|}}{a|s-1|},$$

and since

$$-|s-1| \leqslant (1-\sigma),$$

$$g(s_0) \leqslant 10\exp\{a(1-\sigma)\}/|s-1|.$$

Next, for z on Γ_4, let $z-1 = |s-1|e^{i\theta}$. We then have

$$|e^{-a(z-1)}| \leqslant e^{-a(\sigma-1)}, \quad |dz/(z-1)| = d\theta,$$

so that

$$\left|\int_{\Gamma_4} \frac{e^{-a(z-1)}}{z-1}\,dz\right| \leqslant \pi\exp\{a(1-\sigma)\},$$

and combining the above results gives Lemma 3.

When s lies on the real axis to the left of $s = 1$ we resort to our definition $g(\sigma) = \frac{1}{2}\{g_+(\sigma)+g_-(\sigma)\}$. Since, in the limit,

$$|g_+(\sigma)| \leqslant \{\pi+10/(|\sigma-1|)\}\exp(a-a\sigma),$$

$$|g_-(\sigma)| \leqslant \{\pi+10/(|\sigma-1|)\}\exp(a-a\sigma),$$

it follows that the inequality of Lemma 3 is also satisfied for real $s < 1$.

4. We define $\Pi(x)$ by the formula

$$\Pi(x) = \sum_{p^m \leqslant x} \frac{1}{m} = \sum_{m=1}^{M} \frac{1}{m}\pi(x^{1/m}), \quad M = [\log x/\log 2],$$

and $M(x)$ by

$$M(x) = \int_a^{\max(x,a)} \frac{e^t}{t}\,dt, \quad a = \tfrac{1}{10}.$$

We now have

LEMMA 4.

$$\int_{-\infty}^{\infty} \frac{\Pi(e^t)-M(t)}{e^{\frac{1}{2}t}} f(t-t_0)\,dt$$
$$= \frac{1}{i\sqrt{(2\pi)}} \int_{2-i\infty}^{2+i\infty} s^{-1}\{\log \zeta(s)-g(s)\}\mathscr{F}(s-\tfrac{1}{2})\exp\{(s-\tfrac{1}{2})t_0\}\,ds,$$

where the functions f and $\mathscr{F}$ are defined by (2.1), (2.3), (2.4), *and where t_0 is a number satisfying* (2.7) *which will be chosen later.*

Proof. We take first $h(x) = \Pi(x)$. Then, since $h(x)$ is of bounded variation and satisfies conditions (ii) and (iii) of § 2, and

$$\varphi(s) = \int_{0+}^{\infty} x^{-s}\,d\{\Pi(x)\} = s\int_{0}^{\infty} x^{-s-1}\Pi(x)\,dx = \log \zeta(s),$$

we have from Lemma 1,

$$\int_{-\infty}^{\infty} \frac{\Pi(e^t)}{e^{\frac{1}{2}t}} f(t-t_0)\,dt = \frac{1}{i\sqrt{(2\pi)}} \int_{2-i\infty}^{2+i\infty} s^{-1}\log \zeta(s)\mathscr{F}(s-\tfrac{1}{2})\exp\{(s-\tfrac{1}{2})t_0\}\,ds.$$

If we now take $h(x) = M(\log x)$, then $h(x)$ is of bounded variation and conditions (ii) and (iii) of § 2 are satisfied, so that

$$\varphi(s) = \int_{\exp a}^{\infty} \frac{x^{-s}}{\log x}\,dx = \int_{a}^{\infty} \frac{\exp\{(1-s)t\}}{t}\,dt,$$
$$= \int_{a(s-1)}^{(s-1)\infty} \frac{e^{-t}}{t}\,dt, \quad a = \tfrac{1}{10}.$$

Since $\sigma = 2$ and there are no singularities inside the triangular contour $[a(s-1),\, R(s-1),\, a(s-1)+R]$ and, further, the integral along the line $R(s-1)$ to $a(s-1)+R$ tends to zero as $R \to \infty$, we have

$$\varphi(s) = \int_{a(s-1)}^{a(s-1)+\infty} \frac{e^{-t}}{t}\,dt = g(s).$$

Hence

$$\int_{-\infty}^{\infty} \frac{M(t)}{e^{\frac{1}{2}t}} f(t-t_0)\,dt = \frac{1}{i\sqrt{(2\pi)}} \int_{2-i\infty}^{2+i\infty} s^{-1}g(s)\mathscr{F}(s-\tfrac{1}{2})\exp\{(s-\tfrac{1}{2})t_0\}\,ds,$$

and Lemma 4 follows.

5. Let ρ be a non-trivial zero of $\zeta(s)$, and define

$$I_1 = \frac{1}{i\sqrt{(2\pi)}} \int_{2-i\infty}^{2+i\infty} s^{-1}\log\{(s+1)(s+2)\zeta(s+2)\}\mathscr{F}(s-\tfrac{1}{2})\exp\{(s-\tfrac{1}{2})t_0\}\,ds, \quad (5.1)$$

$$I_2 = \frac{1}{i\sqrt{(2\pi)}} \int_{\frac{1}{2}-\Delta-i\infty}^{\frac{1}{2}-\Delta+i\infty} s^{-1}[\log\{\zeta(s)/(s+1)(s+2)\zeta(s+2)\}-g(s)]\mathscr{F}(s-\tfrac{1}{2})$$
$$\times \exp\{(s-\tfrac{1}{2})t_0\}\,ds, \quad (5.2)$$

where in I_1 the logarithm is real at $s = 2$ and continuous on $\sigma = 2$, and in I_2 the imaginary part of the logarithm changes from $i\pi$ to $-i\pi$ as s crosses the real axis from negative to positive along $\sigma = \frac{1}{2}-\Delta$, but is otherwise continuous. We choose Δ later to be 400. Define also

$$I_\rho = (2\pi)^{\frac{1}{2}} \int_{\rho-2}^{\rho} s^{-1}\mathscr{F}(s-\tfrac{1}{2})\exp\{(s-\tfrac{1}{2})t_0\}\,ds, \tag{5.3}$$

$$J = (2\pi)^{\frac{1}{2}}\{\log(|\zeta(0)|/2\zeta(2))-g(0)\}\mathscr{F}(-\tfrac{1}{2})\exp(-\tfrac{1}{2}t_0); \tag{5.4}$$

then we have

LEMMA 5.

$$\int_{-\infty}^{\infty} \frac{\Pi(e^t)-M(t)}{e^{\frac{1}{2}t}} f(t-t_0)\,dt = I_1+I_2-\sum_\rho I_\rho+J,$$

where f and $\mathscr{F}$ are defined by (2.1), (2.3), (2.4), *$g(s)$ is defined by* (3.1), *and t_0 is a number satisfying* (2.7) *which will not be determined until the last moment.*

Proof. We require the following results from Ingham ((**1**) 71 et seq.):

(A) *There exists a sequence of numbers $T_2, T_3, \ldots, T_r, \ldots$ such that*

$$r < T_r < r+1 \quad (r = 2, 3, \ldots)$$

and

$$|\zeta'(s)/\zeta(s)| < A\log^2 T_r \quad (-1 \leqslant \sigma \leqslant 2,\ t = T_r),$$

where A is a constant.

(B) *In the region obtained by removing from the half-plane $\sigma \leqslant -1$ the interiors of a set of circles of radius $\frac{1}{2}$ with centres at $s = -2, -4, -6, \ldots$, that is in the region defined by*

$$\sigma \leqslant -1, \quad |s-n| \geqslant \tfrac{1}{2} \quad (n = -2, -4, -6, \ldots),$$

we have

$$|\zeta'(s)/\zeta(s)| < A\log(|s|+1),$$

where A is a constant.

If now $\Delta \geqslant \frac{3}{2}$ we have on the segment $-\Delta+\frac{1}{2} \leqslant \sigma \leqslant -1$, $t = T_r$,

$$|\zeta'(s)/\zeta(s)| < A\log(|\sigma+iT_r|+1) < A'\log T_r.$$

This result combined with (A) gives

$$|\zeta'(s)/\zeta(s)| < A_1\log^2 T_r \quad (-\Delta+\tfrac{1}{2} \leqslant \sigma \leqslant 2),$$

where A_1 is a suitable constant. Integrating from $\sigma+iT_r$ to $2+iT_r$ we have

$$|\log\zeta(\sigma+iT_r)|-|\log\zeta(2+iT_r)| < (2-\sigma)A_1\log^2 T_r \leqslant (\Delta+\tfrac{3}{2})A_1\log^2 T_r,$$

where, for $\sigma < 2$, we define $\log\zeta(s)$ as the analytic continuation of

$$\log\zeta(s) = \log|\zeta(s)|+i\arg\zeta(s), \quad -\tfrac{1}{2}\pi < \arg\zeta(s) < \tfrac{1}{2}\pi, \quad \sigma \geqslant 2,$$

along the straight line $(\sigma+it,\ 2+it)$ provided that $\zeta(s) \neq 0$ on this segment of line and thus

$$|\log\zeta(\sigma+iT_r)| < 1+(\Delta+\tfrac{3}{2})A_1\log^2 T_r < A_2\log^2 T_r,$$

where A', A_1, A_2 in the above depend on Δ.

Now define $\chi(s)$ by

$$\chi(s) = \{is\sqrt{(2\pi)}\}^{-1}[\log\{\zeta(s)/(s+1)(s+2)\zeta(s+2)\}-g(s)] \times \mathscr{F}(s-\tfrac{1}{2})\exp\{(s-\tfrac{1}{2})t_0\}.$$

We integrate $\chi(s)$ round the rectangle whose vertices are

$$(2-iT_r,\ 2+iT_r,\ -\Delta+\tfrac{1}{2}+iT_r,\ -\Delta+\tfrac{1}{2}-iT_r)$$

with $r = 2, 3, 4, \ldots$. Defining I_3, I_4 to be the integrals

$$I_3 = \int_{2+iT_r}^{-\Delta+\frac{1}{2}+iT_r}\chi(s)\,ds, \quad I_4 = \int_{-\Delta+\frac{1}{2}-iT_r}^{2-iT_r}\chi(s)\,ds,$$

we see that I_3, I_4 are less in modulus than

$$T_r^{-1}(2\pi)^{-\frac{1}{2}}[2A_2\log^2 T_r+A\log T_r+(\pi+10T_r^{-1})\exp\{a(\Delta+\tfrac{1}{2})\}]M_3(\Delta+\tfrac{3}{2})e^{\frac{3}{2}t_0},$$

where M_3 is the upper bound of $|\mathscr{F}(s-\tfrac{1}{2})|$ in the region $-\Delta+\tfrac{1}{2} \leqslant \sigma \leqslant 2$. This implies that I_3, $I_4 \to 0$ as $T_r \to \infty$.

We next investigate the singularities of $\chi(s)$. Rewrite

$$\chi(s) = \{is\sqrt{(2\pi)}\}^{-1}\{\chi_1(s)-\chi_2(s)\}\mathscr{F}(s-\tfrac{1}{2})\exp\{(s-\tfrac{1}{2})t_0\},$$

where

$$\chi_1(s) = \log[\{(s-1)\zeta(s)\}/\{(s+1)(s+2)\zeta(s+2)\}],$$
$$\chi_2(s) = g(s)+\log(s-1).$$

The function $\chi_1(s)$ has logarithmic singularities at ρ and $\rho-2$ while, in view of (3.2), it follows that $\chi_2(s)$ is well defined and has no singularities. Thus $\chi(s)$ has logarithmic singularities at ρ and $\rho-2$, and a pole at $s = 0$. When we integrate round the given rectangle the former contribute $-\sum I_\rho$ to the integral, as can be deduced by modifying Theorem 1 in Littlewood (8) by replacing $\log\varphi(s)$ by $\psi(s)\log\varphi(s)$.

To find the contribution of the pole we need the residue at $s = 0$. Now in the interior of the rectangle $(2, 2+iT_r, -\Delta+\tfrac{1}{2}+iT_r, -\Delta+\tfrac{1}{2})$, to every zero of $\zeta(s)$ there is a zero of $\zeta(s+2)$, if $\Delta > \tfrac{5}{2}$. Hence as we describe the rectangle once, the logarithm of $\{(s-1)\zeta(s)\}/\{(s+1)(s+2)\zeta(s+2)\}$ returns to its original value. Note that because of the factor $(s+1)(s+2)$ the above expression is real, positive, and regular all along the real axis. We take the logarithm as real at $s = 2$. Also, the expression is positive for $s = 0$. Hence we may take its logarithm as $\log\{|\zeta(0)|/2\zeta(2)\}$. The contribution of $\chi_2(s)$ to the residue is

$$\chi_2(0) = \tfrac{1}{2}\{\log 1 + g_-(0) + g_+(0)\} = g(0).$$

Hence

$$-I_2 + \int_{2-i\infty}^{2+i\infty} \chi(s)\,ds = J - \sum_\rho I_\rho.$$

If we now add I_1 to both sides and rearrange we have

$$\frac{1}{i\sqrt{(2\pi)}}\int_{2-i\infty}^{2+i\infty} s^{-1}\{\log\zeta(s) - g(s)\}\mathscr{F}(s-\tfrac{1}{2})\exp\{(s-\tfrac{1}{2})t_0\}\,ds = I_1 + I_2 + J - \sum_\rho I_\rho,$$

and Lemma 5 follows from Lemma 4.

6. Let $\Delta = 400 = \alpha^2 t_0$ and $t_0 > 2500$. We have

LEMMA 6.

$$\left|(2\pi)^{-\frac{1}{2}}\int_{-\infty}^{\infty}\{\Pi(e^t) - M(t)\}e^{-\frac{1}{2}t}f(t-t_0)\,dt + \Sigma_1 \frac{F(\gamma)e^{i\gamma t_0}}{t_0(\frac{1}{2}+i\gamma)}\right|$$

$$< e^{-\frac{1}{8}t_0} + \frac{t_0^{-\frac{3}{2}}}{\mu}\Sigma_1 \frac{0{\cdot}051}{|\gamma|} + \frac{1}{\mu t_0}\Sigma_1 \frac{0{\cdot}93}{\gamma^2},$$

where Σ_1 denotes summation over all zeros satisfying $|\gamma| \leqslant 4\mu + 50$, where $10 \leqslant \mu \leqslant 250$, f and F are defined by (2.1), (2.3), *and* (2.4), *and $\Pi(x)$ and $M(x)$ are defined just prior to Lemma* 4. *Later we take $\mu = 175$ and determine a value for t_0.*

This is the point where we depart more from Turing's manuscript, our right-hand side being slightly larger than Turing's. The major departure, however, is in § 9.

Proof. We return to Lemma 5 and show first that

$$|I_1|,\ |I_2|,\ |J| < \tfrac{1}{5}\sqrt{(2\pi)}\exp(-\tfrac{1}{8}t_0). \tag{6.1}$$

We observe that the integrand in I_1 is regular in the strip $\frac{1}{4} \leqslant \sigma \leqslant 2$ and tends to zero uniformly in σ as $\tau \to \pm\infty$. Hence, integrating round the rectangle $(2-iR, 2+iR, \frac{1}{4}+iR, \frac{1}{4}-iR)$ and letting $R \to \infty$, we have

$$I_1 = \frac{1}{i\sqrt{(2\pi)}}\int_{\frac{1}{4}-i\infty}^{\frac{1}{4}+i\infty} s^{-1}\log\{(s+1)(s+2)\zeta(s+2)\}\mathscr{F}(s-\tfrac{1}{2})\exp\{(s-\tfrac{1}{2})t_0\}\,ds.$$

Now, for $\sigma \geqslant \frac{1}{4}$,

$$|s^{-1}\log\{(s+1)(s+2)\zeta(s+2)\}|$$
$$< \{\log|s+1| + \log|s+2| + |\log\zeta(s+2)| + \pi\}/|s|,$$

and

$$\log|s+a| \leqslant |s| + a - 1 \leqslant (4a-3)|s| \quad (a = 1, 2),$$

$|\log\zeta(s+2)| \leqslant \log\zeta(\sigma+2)$, since the coefficients in the Dirichlet series for $\log\zeta(s)$ are non-negative. Hence

$$|s^{-1}\log\{(s+1)(s+2)\zeta(s+2)\}| < 1+5+4\log 2+4\pi < 22. \tag{6.2}$$

By Lemma 2(iv),

$$\int_{\frac{1}{4}-i\infty}^{\frac{1}{4}+i\infty} |\mathscr{F}(s-\tfrac{1}{2})\,ds| \leqslant \int_{-\frac{1}{4}-i\infty}^{-\frac{1}{4}+i\infty} |\mathscr{F}(s)\,ds| \leqslant \tfrac{3}{2}\sqrt{(2\pi)}\exp(\tfrac{1}{32}\alpha^{-2}).$$

Hence from (6.2),

$$|I_1| < 22(2\pi)^{-\frac{1}{2}}\exp(-\tfrac{1}{4}t_0).\tfrac{3}{2}\sqrt{(2\pi)}\exp(t_0/12800) < \tfrac{1}{5}\sqrt{(2\pi)}\exp(-\tfrac{1}{8}t_0),$$

since $t_0 > 2500$.

We now call on the functional relationship between $\zeta(s)$ and $\zeta(1-s)$, which gives

$$\zeta(s)/(s+1)(s+2)\zeta(s+2) = -s\zeta(1-s)/4\pi^2(s+2)\zeta(-1-s),$$

so that if $\sigma < -3$,

$$\begin{aligned}|\log\{\zeta(s)/(s+1)(s+2)\zeta(s+2)\}| &\leqslant \pi+2\log 2\pi+\log|\{s/(s+2)\}| \\ &\quad +|\log\zeta(1-s)|+|\log\zeta(-1-s)|, \\ &< \pi+2\log 2\pi+\log 3+2\log\zeta(2) < 12,\end{aligned}$$

since, for $\sigma < -3$,

$$1 < |\{s/(s+2)\}| < 3$$

and, as above,

$$|\log\zeta(\pm 1-s)| \leqslant \log\zeta(\pm 1-\sigma) \leqslant \log\zeta(2).$$

We have, from Lemma 3 on the line of integration $s = -\Delta+\frac{1}{2}+i\tau$,

$$|g(s)| \leqslant (\pi+10/|-\Delta-\tfrac{1}{2}|)\exp\{a(\Delta+\tfrac{1}{2})\} \leqslant \tfrac{10}{3}e^{40},$$

since $\Delta = 400$, $a = \frac{1}{10}$; so that

$$|\log\{\zeta(s)/(s+1)(s+2)\zeta(s+2)\}-g(s)| \leqslant 12+\tfrac{10}{3}e^{40} < 4e^{40}. \tag{6.3}$$

By Lemma 2(iv),

$$\int_{-\Delta+\frac{1}{2}-i\infty}^{-\Delta+\frac{1}{2}+i\infty} |\mathscr{F}(s-\tfrac{1}{2})\,ds| = \int_{-\Delta-i\infty}^{-\Delta+i\infty} |\mathscr{F}(s)\,ds| \leqslant \tfrac{3}{2}\sqrt{(2\pi)}\exp(\Delta^2/2\alpha^2).$$

Hence from (6.3),

$$\begin{aligned}|I_2| &< (2\pi)^{-\frac{1}{2}}(\Delta-\tfrac{1}{2})^{-1}.4e^{40}\exp(-\Delta t_0).\tfrac{3}{2}\sqrt{(2\pi)}\exp(\Delta t_0/2), \\ &= 6(\Delta-\tfrac{1}{2})^{-1}.\exp(40-200t_0) \\ &< \tfrac{1}{5}\sqrt{(2\pi)}\exp(-\tfrac{1}{8}t_0),\end{aligned}$$

since $\Delta = 400 = \alpha^2 t_0$ and $t_0 > 2500$.

From (5.4),

$$|J| = \sqrt{(2\pi)}\,|\{\log(|\zeta(0)|/2\zeta(2)) - g(0)\}\mathscr{F}(-\tfrac{1}{2})\exp(-\tfrac{1}{2}t_0)|.$$

Now

$$|\log\{|\zeta(0)|/2\zeta(2)\}| = |\log(\tfrac{3}{2}\pi^{-2})| < 3,$$

and Lemma 3 gives

$$|g(0)| \leqslant |(\pi+10)\exp(0{\cdot}1)| \leqslant 15.$$

Hence from Lemma 2(ii),

$$|J| < (2\pi)^{\frac{1}{2}}(3+15)\sqrt{(2\pi)}(2\mu)^{-1}\exp\{(\tfrac{1}{8}\alpha^{-2}) - \tfrac{1}{2}t_0\},$$
$$< \tfrac{9}{5}\pi\exp\{t_0(-\tfrac{1}{2}+\tfrac{1}{3200})\} < \tfrac{1}{5}\sqrt{(2\pi)}\exp(-\tfrac{1}{8}t_0).$$

This completes the proof of (6.1).

The only term which has not been estimated in the right-hand side of Lemma 5 is the term $\sum I_\rho$ defined in (5.3). For $|\gamma| \leqslant 4\mu+50$ we estimate each individual I_ρ as follows. We write

$$K_\rho = \int_{\rho-2}^{\rho}\left(\frac{1}{s}-\frac{1}{\rho}\right)\mathscr{F}(s-\tfrac{1}{2})\exp\{(s-\tfrac{1}{2})t_0\}\,ds, \tag{6.4}$$

$$L_\rho = \int_{\rho-2}^{\rho}\rho^{-1}\{\mathscr{F}(s-\tfrac{1}{2}) - F(-\gamma)\}\exp\{(s-\tfrac{1}{2})t_0\}\,ds, \tag{6.5}$$

$$M_\rho = \int_{\rho-\infty}^{\rho}\rho^{-1}F(-\gamma)\exp\{(s-\tfrac{1}{2})t_0\}\,ds, \tag{6.6}$$

$$N_\rho = \int_{\rho-\infty}^{\rho-2}\rho^{-1}F(-\gamma)\exp\{(s-\tfrac{1}{2})t_0\}\,ds. \tag{6.7}$$

Since $\rho = \frac{1}{2}+i\gamma$ for $|\gamma| \leqslant 4\mu+50 \leqslant 1050$, we have for each ρ in the range the identity

$$(2\pi)^{-\frac{1}{2}}I_\rho = K_\rho + L_\rho + M_\rho - N_\rho. \tag{6.8}$$

We shall prove that

$$\left.\begin{aligned} &|K_\rho| \leqslant 0{\cdot}93\gamma^{-2}/(\mu t_0), \quad |L_\rho| < 0{\cdot}051 t_0^{-\frac{3}{2}}/\mu|\gamma|, \\ &M_\rho = F(\gamma)\exp(i\gamma t_0)/t_0(\tfrac{1}{2}+i\gamma), \quad |N_\rho| < \tfrac{1}{5}N^{-1}\exp(-\tfrac{1}{8}t_0), \end{aligned}\right\} \tag{6.9}$$

where N is the number of pairs of conjugate zeros satisfying $|\gamma| \leqslant 4\mu+50$, i.e. $25 \leqslant N \leqslant 690$.

From (6.4),

$$|K_\rho| = \left|\int_{\rho-2}^{\rho}\left(\frac{\rho-s}{\rho s}\right)\mathscr{F}(s-\tfrac{1}{2})\exp\{(s-\tfrac{1}{2})t_0\}\,ds\right|.$$

Here $s = \rho - x$, where $0 \leqslant x \leqslant 2$, and we have

$$|1/\rho s| \leqslant 1/\gamma^2; \quad |\exp\{(s-\tfrac{1}{2})t_0\}| = \exp(-xt_0);$$
$$|\mathscr{F}(s-\tfrac{1}{2})| \leqslant \tfrac{1}{2}\sqrt{(2\pi)}\mu^{-1}\exp(x^2/2\alpha^2), \quad \text{by Lemma 2(ii).}$$

Hence

$$|K_\rho| \leqslant \tfrac{1}{2}\sqrt{(2\pi)}\mu^{-1}\gamma^{-2}\int_0^2 x\exp\{(x^2/2\alpha^2)-xt_0\}\,dx.$$

Now in $0 \leqslant x \leqslant 2$, $(x^2/2\alpha^2)-xt_0 \leqslant -xt_0(1-\frac{1}{400})$, and, since

$$x\exp\{-xt_0(1-\tfrac{1}{400})\}$$

attains its maximum value for $x = 1/t_0(1-\frac{1}{400})$, we have

$$|K_\rho| \leqslant 1{\cdot}005\sqrt{(2\pi)}e^{-1}\gamma^{-2}/(\mu t_0) \leqslant 0{\cdot}93\gamma^{-2}/(\mu t_0).$$

From (6.5),

$$|L_\rho| = \left|\int_{\rho-2}^{\rho} \rho^{-1}\{\mathscr{F}(s-\tfrac{1}{2})-F(-\gamma)\}\exp\{(s-\tfrac{1}{2})t_0\}\,ds\right|.$$

Integrating by parts, we have

$$|L_\rho| \leqslant T_1+T_2, \quad \text{say,}$$

where

$$\begin{aligned} T_1 &= |[\exp\{(s-\tfrac{1}{2})t_0\}\{\mathscr{F}(s-\tfrac{1}{2})-F(-\gamma)\}/\rho t_0]_{\rho-2}^{\rho}| \\ &< \exp(-2t_0)[\tfrac{1}{2}\sqrt{(2\pi)}\mu^{-1}\{\exp(2/\alpha^2)+1)\}]/|\gamma|t_0 \\ &< 0{\cdot}00075t_0^{-\frac{3}{2}}/\mu|\gamma|, \end{aligned}$$

and

$$T_2 = \left|\frac{1}{\rho t_0}\int_{\rho-2}^{\rho}\exp\{(s-\tfrac{1}{2})t_0\}\mathscr{F}'(s-\tfrac{1}{2})\,ds\right|.$$

On putting $x = \rho - s$, we see from Lemma 2(iii) that

$$T_2 \leqslant (|\gamma|t_0\alpha\mu)^{-1}\int_0^2\exp\{(x^2/2\alpha^2)-xt_0\}\,dx,$$

in which

$$\int_0^2\exp\{(x^2/2\alpha^2)-xt_0\}\,dx < 1/t_0(1-\tfrac{1}{400}),$$

whence $T_2 < 0{\cdot}05025t_0^{-\frac{3}{2}}/\mu|\gamma|$, since $\alpha = 20t_0^{-\frac{1}{2}}$. Hence

$$|L_\rho| \leqslant T_1+T_2 < 0{\cdot}051t_0^{-\frac{3}{2}}/\mu|\gamma|.$$

Integrating (6.6), we have

$$M_\rho = F(-\gamma)\exp(i\gamma t_0)/\rho t_0 = F(\gamma)\exp(i\gamma t_0)/t_0(\tfrac{1}{2}+i\gamma).$$

Similarly, integrating (6.7), we have

$$\begin{aligned} |N_\rho| = |F(\gamma)\exp\{(i\gamma-2)t_0\}/\rho t_0| &< |F(\gamma)\exp(-2t_0)/t_0|\gamma|| \\ &< \tfrac{1}{5}N^{-1}\exp(-\tfrac{1}{8}t_0), \end{aligned}$$

by Lemma 2(ii).

This completes the proof of (6.9). To complete the proof of Lemma 6 it now remains to show that $\sum_2|I_\rho|$, where $\sum_2$ denotes summation over

$|\gamma| > 4\mu+50$, converges to something small. In fact, we show that

$$\Sigma_2|I_\rho| < \tfrac{1}{5}\sqrt{(2\pi)}\exp(-\tfrac{1}{8}t_0). \tag{6.10}$$

If $\rho = \beta+i\gamma$, and $|\gamma| > 4\mu+50$, then $0 < \beta < 1$ and

$$|(2\pi)^{-\frac{1}{2}}I_\rho| = \left|\int_{\rho-2}^{\rho} s^{-1}\mathscr{F}(s-\tfrac{1}{2})\exp\{(s-\tfrac{1}{2})t_0\}\,ds\right| = \left|\int_0^2 U\,dx\right|,$$

where $s = \rho-x$, and, by Lemma 2(v),

$$|U| \leqslant |\gamma|^{-1}\tfrac{3}{2}\alpha^{-1}\exp[\{(\beta-x-\tfrac{1}{2})^2-(|\gamma|-4\mu)^2\}/2\alpha^2]\exp\{(\beta-x-\tfrac{1}{2})t_0\}.$$

Now the term in square brackets is less than

$$\{(\tfrac{1}{2}+x)^2-(|\gamma|-4\mu)^2\}/2\alpha^2,$$

and the other exponent is less than $(\frac{1}{2}-x)t_0$, so that

$$|(2\pi)^{-\frac{1}{2}}I_\rho| < 3\alpha^{-1}|\gamma|^{-1}\exp[t_0\{\tfrac{25}{4}-(|\gamma|-4\mu)^2+400\}/800].$$

Now

$$(|\gamma|-4\mu)^2 \geqslant 50(|\gamma|-4\mu) \geqslant 25(50+|\gamma|-4\mu) = 1250+25(|\gamma|-4\mu),$$

so that

$$|(2\pi)^{-\frac{1}{2}}I_\rho| < 3\alpha^{-1}|\gamma|^{-1}\exp[-t_0\{1+\tfrac{1}{32}(|\gamma|-4\mu)\}].$$

Further, for $\mu \leqslant 250$ and $|\gamma| > 4\mu+50$,

$$|\gamma|-4\mu > |\gamma|/24, \tag{6.11}$$

so that

$$|(2\pi)^{-\frac{1}{2}}I_\rho| < 3\alpha^{-1}\gamma^{-2}\exp(-t_0)\{|\gamma|\exp(-t_0|\gamma|/768)\}.$$

The term in curly brackets is less than $\frac{1}{3}$, and $\alpha^{-1} < \exp(\frac{1}{2}t_0)$, giving

$$|I_\rho| < \sqrt{(2\pi)}\gamma^{-2}\exp(-\tfrac{1}{2}t_0).$$

Hence

$$\Sigma_2|I_\rho| < \sqrt{(2\pi)}\exp(-\tfrac{1}{2}t_0)\,\Sigma_2\frac{1}{\gamma^2} < \tfrac{1}{5}\sqrt{(2\pi)}\exp(-\tfrac{1}{8}t_0),$$

since $\Sigma_2(1/\gamma^2) \leqslant 0{\cdot}0466$ from ((4) Lemma 1(iii)).

From (6.8) and Lemma 5 we have

$$\int_{-\infty}^{\infty}(\Pi(e^t)-M(t))e^{-\frac{1}{2}t}f(t-t_0)\,dt+\sqrt{(2\pi)}\,\Sigma_1 M_\rho$$
$$= I_1+I_2+J-\sqrt{(2\pi)}\,\Sigma_1(K_\rho+L_\rho-N_\rho)-\Sigma_2 I_\rho,$$

and hence

$$\left|(2\pi)^{-\frac{1}{2}}\int_{-\infty}^{\infty}(\Pi(e^t)-M(t))e^{-\frac{1}{2}t}f(t-t_0)\,dt+\Sigma_1 M_\rho\right|$$
$$\leqslant (2\pi)^{-\frac{1}{2}}(|I_1|+|I_2|+|J|+\Sigma_2|I_\rho|)+\Sigma_1(|K_\rho|+|L_\rho|+|N_\rho|).$$

Collecting from this and (6.1), (6.9), (6.10), we get the result of Lemma 6.

7. Lemma 7. *If*

$$\int_{-\infty}^{\infty} \{\Pi(e^t) - M(t)\} e^{-\frac{1}{2}t} f(t - t_0)\, dt > 1{\cdot}0025\pi/\mu t_0, \tag{7.1}$$

then there exists a number t_1 *satisfying* $0{\cdot}950 t_0 \leqslant t_1 \leqslant 1{\cdot}052 t_0$, *such that*

$$\Pi(e^{t_1}) - \mathrm{li}(e^{t_1}) > 1{\cdot}002 t_1^{-1} \exp(\tfrac{1}{2}t_1),$$

where $\Pi(x)$ *and* $M(x)$ *are defined just prior to Lemma* 4, f *is defined by* (2.1), (2.3), *and* (2.4), *and* μ *and* t_0 *satisfy* (2.7).

Proof. Since

$$\frac{1}{\sqrt{(2\pi)}} \int_{-\infty}^{\infty} \{(\sin \mu t)/\mu t\}^4\, dt = \sqrt{(2\pi)}/3\mu,$$

we have, if (7.1) holds,

$$\int_{-\infty}^{\infty} \tfrac{3}{2}[\{\Pi(e^t) - M(t)\} e^{-\frac{1}{2}t} - 1{\cdot}0024 t_0^{-1} \exp\{\tfrac{1}{2}\alpha^2 (t - t_0)^2\}]$$
$$\times \exp\{-\tfrac{1}{2}\alpha^2(t - t_0)^2\}\{\sin \mu(t - t_0)/\mu(t - t_0)\}^4\, dt > 0{\cdot}0001\pi/\mu t_0.$$

This means that for some t, say $t = t_1$, we have

$$\Pi(e^{t_1}) - M(t_1) > 1{\cdot}0024 t_0^{-1} \exp\{\tfrac{1}{2}t_1 + \tfrac{1}{2}\alpha^2(t_1 - t_0)^2\}$$
$$> \exp\{\tfrac{1}{2}t_1 + 200 t_0^{-1}(t_1 - t_0)^2 - 0{\cdot}01 t_0\},$$

since $1{\cdot}0024 t_0^{-1} > \exp(-0{\cdot}01 t_0)$ for $t_0 > 2500$. If $t_1 < 0{\cdot}950 t_0$ then

$$200 t_0^{-1}(t_1 - t_0)^2 - 0{\cdot}01 t_0 > 0{\cdot}5 t_1,$$

which implies that $\Pi(e^{t_1}) > e^{t_1}$, and this contradicts the trivial fact that $\Pi(x) \leqslant x$. Similarly, if $t_1 > 1{\cdot}052 t_0$ we have a contradiction. Hence, for some t_1 satisfying $0{\cdot}950 t_0 \leqslant t_1 \leqslant 1{\cdot}052 t_0$, we have

$$\Pi(e^{t_1}) - M(t_1) > 1{\cdot}0024 t_0^{-1} \exp\{\tfrac{1}{2}t_1 + 200 t_0^{-1}(t_1 - t_0)^2\}.$$

However, $M(t_1) \geqslant \mathrm{li}(e^{t_1})$, and, for the quoted range of t_1,

$$1{\cdot}0024 t_0^{-1} \exp\{200 t_0^{-1}(t_1 - t_0)^2\} > 1{\cdot}002 t_1^{-1},$$

so that Lemma 7 follows.

8. Lemma 8. *If* $\{\Pi(x) - \mathrm{li}\, x\} x^{-\frac{1}{2}} \log x \geqslant 1{\cdot}002$ *and* $x > \exp(2000)$, *then either*

$$\pi(x) > \mathrm{li}\, x \quad \textit{or} \quad \pi(x^{\frac{1}{2}}) > \mathrm{li}\, x^{\frac{1}{2}}.$$

Proof. Let $E = \Pi(x) - \pi(x) - \frac{1}{2}\pi(x^{\frac{1}{2}}) = \sum_{r=3}^{[\log_2 x]} r^{-1}\pi(x^{1/r})$. If $r > \log_e x$ then $x^{1/r} < e$, so that $\pi(x^{1/r}) < 2$, and

$$E < \sum_{r=3}^{[\log_e x]} \frac{1}{r}\pi(x^{1/r}) + \sum_{r=[\log_e x]+1}^{[\log_2 x]} \frac{2}{r}.$$

However, for all x, $\pi(x) < 2+\frac{1}{3}x$, so that

$$E < \sum_{r=3}^{m} \tfrac{1}{3}r^{-1}x^{1/r} + \sum_{r=3}^{n} \frac{2}{r},$$

where $m = [\log_e x]$, $n = [\log_2 x]$. Since $x^{1/r} < x^{\frac{1}{3}}$ for $r \geqslant 3$ and

$$x > \exp(2000),$$

we have

$$E < \tfrac{1}{3}x^{\frac{1}{3}} \sum_{r=3}^{m} \frac{1}{r} + 2\sum_{r=3}^{n} \frac{1}{r} < \tfrac{1}{3}x^{\frac{1}{3}} \log(\tfrac{1}{2}m) + 2\log(\tfrac{1}{2}n)$$

$$< \tfrac{1}{3}x^{\frac{1}{3}} \log\log x + 2\log\log x. \tag{8.1}$$

Now

$$\{\Pi(x) - \mathrm{li}\, x\}x^{-\frac{1}{2}} \log x = \{E + \pi(x) + \tfrac{1}{2}\pi(x^{\frac{1}{2}}) - \mathrm{li}\, x\}x^{-\frac{1}{2}} \log x,$$

and if we assume that $\pi(x) \leqslant \mathrm{li}\, x$ and $\pi(x^{\frac{1}{2}}) \leqslant \mathrm{li}\, x^{\frac{1}{2}}$, then

$$\{\Pi(x) - \mathrm{li}\, x\}x^{-\frac{1}{2}} \log x \leqslant \{E + \tfrac{1}{2}\pi(x^{\frac{1}{2}})\}x^{-\frac{1}{2}} \log x,$$

$$\leqslant \tfrac{1}{3}x^{-\frac{1}{6}} \log x \log\log x + 2x^{-\frac{1}{2}} \log x \log\log x$$

$$+ \tfrac{1}{2}(\mathrm{li}\, x^{\frac{1}{2}})x^{-\frac{1}{2}} \log x,$$

by (8.1). By differentiating the function

$$\{x/\log x - 1{\cdot}5)\} - \mathrm{li}\, x$$

we see that this function is positive (and increasing) for $x > 1000$. Thus

$$\{\Pi(x) - \mathrm{li}\, x\}x^{-\frac{1}{2}} \log x \leqslant (\tfrac{1}{3}x^{-\frac{1}{6}} + 2x^{-\frac{1}{2}})\log x \log\log x + \log x/(\log x - 3)$$

$$< 1{\cdot}002, \quad \text{if } x > \exp(2000).$$

Hence, if $\{\Pi(x) - \mathrm{li}\, x\}x^{-\frac{1}{2}} \log x > 1{\cdot}002$ and $x > \exp(2000)$, our premise is false; i.e. either $\pi(x) > \mathrm{li}\, x$ or $\pi(x^{\frac{1}{2}}) > \mathrm{li}\, x^{\frac{1}{2}}$, which proves Lemma 8.

Combining Lemmas 6, 7, and 8 we have

THEOREM 1. *If there exists a number $t_0 > 2500$ such that*

$$-\sum\nolimits_1 \frac{F(\gamma)e^{i\gamma t_0}}{\frac{1}{2} + i\gamma} > 0{\cdot}521\sqrt{(2\pi)}/\mu, \tag{8.2}$$

then there exists a number t_1, with $0{\cdot}950t_0 \leqslant t_1 \leqslant 1{\cdot}052t_0$, such that either $\pi(e^{t_1}) > \mathrm{li}\, e^{t_1}$ or $\pi(e^{\frac{1}{2}t_1}) > \mathrm{li}\, e^{\frac{1}{2}t_1}$, where F satisfies (2.1) and (2.5), μ and t_0 satisfy (2.7), and $\sum_1$ denotes summation over all zeros satisfying $|\gamma| \leqslant 4\mu + 50$.

Proof. Applying condition (8.2) in Lemma 6, we have

$$\int_{-\infty}^{\infty} \{\Pi(e^t) - M(t)\}e^{-\frac{1}{2}t}f(t-t_0)\, dt > 1{\cdot}042\pi/\mu t_0 - \sqrt{(2\pi)}\exp(-\tfrac{1}{8}t_0)$$

$$-\sqrt{(2\pi)}\mu^{-1}t_0^{-\frac{3}{2}} \sum\nolimits_1 \frac{0{\cdot}051}{|\gamma|} - \sqrt{(2\pi)}\mu^{-1}t_0^{-1} \sum\nolimits_1 \frac{0{\cdot}93}{\gamma^2}.$$

Now, for t_0 in the given range,

$$\sqrt{(2\pi)}\exp(-\tfrac{1}{8}t_0) < 0{\cdot}0001\pi/\mu t_0, \quad t_0^{-\frac{1}{2}} < 0{\cdot}02,$$

$$\textstyle\sum_1 1/\gamma^2 < 0{\cdot}0466, \quad \text{and} \quad \sum_1 1/|\gamma| < 5$$

by direct enumeration, so that

$$\int_{-\infty}^{\infty} \{\Pi(e^t) - M(t)\}e^{-\frac{1}{2}t}f(t-t_0)\,dt > \pi\mu^{-1}t_0^{-1}(1{\cdot}042 - 0{\cdot}0001 - 0{\cdot}0042 - 0{\cdot}0346)$$

$$> 1{\cdot}0025\pi/\mu t_0,$$

and the theorem is a consequence of Lemmas 7 and 8.

9. Our problem now is to find a suitable t_0 satisfying (8.2). Since $F(\gamma) = F(-\gamma)$, and the zeros of $\zeta(s)$ on $s = \frac{1}{2}$ occur in conjugate pairs, the sum on the left-hand side of (8.2) can be expressed as

$$-\sum\nolimits_1 \frac{F(\gamma)e^{i\gamma t_0}}{\frac{1}{2}+i\gamma} = 2\sum{}' \frac{F(\gamma)}{|\rho|}\sin(\gamma t_0 + \vartheta + \pi),$$

where $\sum'$ denotes summation over γ for $0 < \gamma \leqslant 4\mu + 50$, and $\tan\vartheta = \frac{1}{2}\gamma^{-1}$. If we order the zeros so that $\gamma_1 < \gamma_2 < \ldots$, the expression can be written as

$$2\sum{}' \frac{F(\gamma_n)}{|\rho_n|}\sin(\gamma_n t_0 + \vartheta_n + \pi), \tag{9.1}$$

where $\vartheta_n = \arctan(\frac{1}{2}\gamma_n^{-1})$ and $0 < \gamma_n \leqslant 4\mu + 50$. To evaluate (9.1) we require some idea of the terms $F(\gamma_\eta)$. From (2.5) and (2.6) we have

$$F(\gamma_n) = \tfrac{3}{4}\mu^{-1}\alpha^{-1}\left\{\int_{-2\mu}^{2\mu}\Phi_1(u)\,du + \int_{-4\mu}^{-2\mu}\Phi_2(u)\,du + \int_{2\mu}^{4\mu}\Phi_2(u)\,du\right\},$$

where

$$\Phi_1(u) = \{-\tfrac{2}{3}(1-|u|/2\mu)^3 + \tfrac{1}{6}(2-|u|/2\mu)^3\}\exp\{-(\gamma_n - u)^2/2\alpha^2\},$$

$$\Phi_2(u) = \tfrac{1}{6}(2-|u|/2\mu)^3\exp\{-(\gamma_n - u)^2/2\alpha^2\},$$

i.e.

$$F(\gamma_n) = \tfrac{3}{4}\mu^{-1}\alpha^{-1}\left\{\int_0^{4\mu}\tfrac{1}{6}(2-u/2\mu)^3\Phi(u)\,du - \int_0^{2\mu}\tfrac{2}{3}(1-u/2\mu)^3\Phi(u)\,du\right\},$$

where

$$\Phi(u) = \exp\{-(\gamma_n - u)^2/2\alpha^2\} + \exp\{-(\gamma_n + u)^2/2\alpha^2\}.$$

This gives, by appropriate substitution,

$$F(\gamma_n) = X_n + Y_n + Z_n + T_n,$$

where

$$X_n = \frac{1}{8\mu}\int_{-e'}^{a'} e^{-\frac{1}{2}x^2}(a-qx)^3\,dx,$$

$$Y_n = \frac{1}{8\mu}\int_{e'}^{b'} e^{-\frac{1}{2}x^2}(b-qx)^3\,dx,$$

$$Z_n = -\frac{1}{2\mu}\int_{-e'}^{c'} e^{-\frac{1}{2}x^2}(c-qx)^3\,dx,$$

$$T_n = -\frac{1}{2\mu}\int_{e'}^{d'} e^{-\frac{1}{2}x^2}(d-qx)^3\,dx,$$

and a, b, c, d, q denote, respectively, the terms $2-\gamma_n/2\mu$, $2+\gamma_n/2\mu$, $1-\gamma_n/2\mu$, $1+\gamma_n/2\mu$, $\alpha/2\mu$, and where a', b', c', d', e' denote the terms $\alpha^{-1}(4\mu-\gamma_n)$, $\alpha^{-1}(4\mu+\gamma_n)$, $\alpha^{-1}(2\mu-\gamma_n)$, $\alpha^{-1}(2\mu+\gamma_n)$, $\alpha^{-1}\gamma_n$ respectively. Now for $t_0 > 2500$, $\alpha^{-1} > 2{\cdot}5$, $e' > 2{\cdot}5\gamma_n > 30$, with the assistance of tables of

$$\int_{-\infty}^{t} e^{-\frac{1}{2}x^2}\,dx,$$

we find that $|Y_n|$, $|T_n| < 10^{-10}\mu^{-1}$. We remark here that our $F(\gamma_n)$ has been expressed in terms of X_n, Y_n, Z_n, and T_n to assist the numerical computation. These terms involve linear combinations of $x^r e^{-\frac{1}{2}x^2}$ $(r = 0, 1, 2, 3)$, and when $r = 1$, 3, $x^r e^{-\frac{1}{2}x^2}$ is directly integrable, while for $r = 0$, 2 we can obtain an asymptotic expansion provided the upper limit of integration is sufficiently large (say 7). We mention a particularly useful expansion which we have used extensively, namely

$$\int_{x}^{\infty} e^{-\frac{1}{2}t^2}\,dt \sim e^{-\frac{1}{2}x^2}\left(\frac{1}{x}-\frac{1}{x^3}+\frac{3}{x^5}-\ldots+(-1)^n\frac{(2n)!}{2^n n!}\frac{1}{x^{2n+1}}+\ldots\right).$$

In the instances where the upper limits of integration a' and c' are near zero we find that a, c, q are, fortuitously, very small. Thus if we take $\mu = 175$, we find that X_n and Z_n are negligibly different from

$$X'_n = \frac{1}{8\mu}\int_{-2\cdot5\gamma_n}^{2\cdot5(4\mu-\gamma_n)} e^{-\frac{1}{2}x^2}(a-qx)^3\,dx,$$

and

$$Z'_n = -\frac{1}{2\mu}\int_{-2\cdot5\gamma_n}^{2\cdot5(2\mu-\gamma_n)} e^{-\frac{1}{2}x^2}(c-qx)^3\,dx,$$

respectively. More precisely, we can show that

$$|X_n - X'_n| < 10^{-9}\mu^{-1},$$

$$|Z'_n - Z_n| < 10^{-8}\mu^{-1}.$$

Before evaluating further the terms of (9.1) we prove the following lemma.

LEMMA 9. *Given real numbers $a_1, \ldots, a_k$, a positive real number τ, and positive integers $m_1, m_2, \ldots, m_k$, we can find a number t, an integral multiple of τ, satisfying $\tau \leqslant t \leqslant \tau \prod m_n$, and such that, for each n, $1 \leqslant n \leqslant k$, ta_n differs from an integer by at most $1/m_n$.*

Proof. The proof is basically the same as that for Dirichlet's theorem, Ingham ((**1**) Theorem J, p. 94), except that out 'unit cube' is divided into $\prod m_n$ 'rectangular parallelepipeds', the length of the nth edge being $1/m_n$.

By the lemma, there exists a t in the range

$$2500 \leqslant t \leqslant 2500 \prod_n (1+2R\gamma_N/\gamma_n),$$

where $R = 60$, $N = 100$, $\gamma_N = 236{\cdot}524\ldots$, such that

$$\|\gamma_n t\| \leqslant 1/m_n, \quad n = 1, 2, \ldots, 241,$$

where $m_n = [2R\gamma_N/\gamma_n]+1$, and where $\|x\|$ denotes the non-negative distance of x from the nearest integer to x. Now take $t_0 = 2\pi t - \pi/2\gamma_N$. Then

$$\|\gamma_n(t_0+\pi/2\gamma_N)/2\pi\| \leqslant 1/m_n,$$

$$\|\gamma_n\{t_0+(\vartheta_n/\gamma_n)+(\pi/2\gamma_n)\}/2\pi\| \leqslant \|\gamma_n(t_0+\pi/2\gamma_N)/2\pi\|+\|X\|,$$

where

$$X = \tfrac{1}{4}+(\vartheta_n/2\pi)-\tfrac{1}{4}(\gamma_n/\gamma_N).$$

Since

$$(\tfrac{1}{2}\gamma_n^{-1})-\tfrac{1}{3}(\tfrac{1}{2}\gamma_n^{-1})^3 < \vartheta_n < \tfrac{1}{2}\gamma_n^{-1},$$

we have for $\gamma_1 \leqslant \gamma_n \leqslant \gamma_{241}$,

$$-\tfrac{1}{4}+1/m_n < X < \tfrac{1}{4}-1/m_n, \quad n = 1, \ldots, 241,$$

and hence

$$\|X\| < \tfrac{1}{4}-1/m_n.$$

R has been specially chosen to achieve this result, which ensures that the first 241 terms of (9.1) are positive. This follows for the chosen t_0 because now

$$\|(\gamma_n t_0+\vartheta_n+\tfrac{1}{2}\pi)/2\pi\| < 1/m_n+(\tfrac{1}{4}-1/m_n) = \tfrac{1}{4},$$

and, consequently, $\gamma_n t_0+\vartheta_n+\frac{1}{2}\pi$ differs from an integral multiple of 2π by at most $\frac{1}{2}\pi$, i.e. for the chosen t_0 we have

$$\sin(\gamma_n t_0+\vartheta_n+\pi) > 0, \quad n = 1, 2, \ldots, 241.$$

Moreover, we have

$$\sin(\gamma_n t_0 + \vartheta_n + \pi)\begin{cases} \geqslant \sin\{(\tfrac{1}{2}\pi\gamma_n/\gamma_N) - (2\pi/m_n) - \vartheta_n\}, & X > 0, \\ \geqslant \sin\{(\tfrac{1}{2}\pi\gamma_n/\gamma_N) + (2\pi/m_n) - \vartheta_n\}, & X < 0, \end{cases}$$

$X = 0$ being precluded by our formulation of X. With $\mu = 175$, we are using 452 zeros of the ζ-function in our sum (9.1) which will thus be greater than

$$2\sum_{n=1}^{241} \frac{(X'_n + Z'_n)}{|\rho_n|} \sin\{(\tfrac{1}{2}\pi\gamma_n/\gamma_N) \pm (2\pi/m_n) - \vartheta_n\}$$

$$+2\sum_{n=242}^{452} \frac{F(\gamma_n)}{|\rho_n|} \sin(\gamma_n t_0 + \vartheta_n + \pi) + \text{small term},$$

where, in the first term, the plus sign is appropriate when $X < 0$ and the minus sign is appropriate when $X > 0$. The small term is less in modulus than $10^{-6}\mu^{-1}$, so that (9.1) is greater than

$$2\sum_{n=1}^{241} \frac{(X'_n + Z'_n)}{|\rho_n|} \sin\{(\tfrac{1}{2}\pi\gamma_n/\gamma_N) + (2\pi/m_n) - \vartheta_n\}$$

$$-2\sum_{n=232}^{452} \frac{F(\gamma_n)}{|\rho_n|} - 10^{-6}\mu^{-1}.$$

Evaluating these sums numerically (it may interest readers to know the above calculations took approximately 20 minutes on an Elliott 803 computer, although an additional hour or two must be added to cover the time for programming and tape preparation), and making the necessary allowances for rounding errors, we find that (9.1) is larger than

$$(1{\cdot}3280 - 0{\cdot}0171)\mu^{-1} = 1{\cdot}3109\mu^{-1} > 1{\cdot}3062\mu^{-1} > 0{\cdot}521\surd(2\pi)/\mu.$$

We have thus shown the existence of a t_0 satisfying

$$2\pi.2500 \leqslant t_0 + (\pi/2\gamma_N) \leqslant 2\pi.2500 \prod_1^{241} (1 + 2R\gamma_N/\gamma_n),$$

such that (8.2) is true. By direct calculation,

$$\sum_1^{241} \log\{1 + (2R\gamma_N/\gamma_n)\} < 1207,$$

i.e.

$$2\pi.2500 \leqslant t_0 + (\pi/2\gamma_N) \leqslant 2\pi.2500.\exp(1207),$$

or

$$2500 < t_0 < 10^{530}.$$

It follows from Theorem 1 that there exists a number t, with $t_1 \leqslant 1{\cdot}052 t_0$, for which *either* $\pi(e^{t_1}) > \operatorname{li} e^{t_1}$ *or* $\pi(e^{\frac{1}{2}t_1}) > \operatorname{li} e^{\frac{1}{2}t_1}$. Thus we have

THEOREM 2. *There exists a number x, with $2 \leqslant x \leqslant 10^{10^{529.7}}$, for which*

$$\pi(x) > \operatorname{li} x.$$

Note added 26 *June* 1967. The referee has drawn our attention to a paper by R. Sherman Lehman, *Acta Arithmetica* 11 (1966) 397–410, where it is shown that there are more than 10^{500} successive integers x, in the range

$$1{\cdot}53 \times 10^{1165} < x < 1{\cdot}65 \times 10^{1165},$$

for which $\pi(x) > \operatorname{li} x$.

REFERENCES

1. A. E. INGHAM, *The distribution of prime numbers* (Cambridge Mathematical Tracts, No. 30, 1932).
2. J. E. LITTLEWOOD, 'Sur la distribution des nombres premiers', *Comptes Rendus* 158 (1914) 1869–72.
3. S. SKEWES, 'On the difference $\pi(x) - \operatorname{li}(x)$ (I)', *J. London Math. Soc.* 8 (1933) 277–83.
4. —— 'On the difference $\pi(x) - \operatorname{li}(x)$ (II)', *Proc. London Math. Soc.* (3)5 (1955) 48–70.
5. E. C. TITCHMARSH, 'The zeros of the zeta-function', *Proc. Royal Soc.* (A) 157 (1936) 261–63.
6. A. E. INGHAM, 'A note on the distribution of primes', *Acta Arithmetica* 1 (1936) 201–11.
7. E. C. TITCHMARSH, *Introduction to the theory of Fourier integrals* (Oxford, 1937).
8. J. E. LITTLEWOOD, 'On the zeros of the Riemann zeta-function', *Proc. Cambridge Phil. Soc.* 22 (1924) 295–318.
9. C. B. HASELGROVE and J. C. P. MILLER, *Tables of the Riemann zeta function*, Roy. Soc. Math. Tables, Vol. 6 (1960).

University of Wales Institute of Science and Technology
Cardiff

Rugby College of Engineering Technology
Rugby

Biometrika (1979), **66**, 2, *pp.* 393-6
Printed in Great Britain

Studies in the History of Probability and Statistics. XXXVII
A. M. Turing's statistical work in World War II

By I. J. GOOD

Department of Statistics, Virginia Polytechnic Institute & State University, Blacksburg

SUMMARY

An account is given of A. M. Turing's unpublished contributions to statistics during 1941 or 1940.

Some key words: Bayes factors; Cryptology; Decibans; Diversity; Empirical Bayes; History of statistics; Information; Repeat rate; Sequential analysis; Weight of evidence; World War II.

1. Preamble

Alan Mathison Turing (1912–54) is best known for the concept of the Turing Machine (Turing, 1936–7). He introduced this concept in his proof that no finite decision procedure can solve all mathematical problems. Owing to a security curtain that lifted only a few years ago, it is less well known that he made important contributions to cryptanalysis during World War II. I was familiar with much of this work because of being his main statistical assistant in 1941. During the war, part of his work related to electromagnetic and electronic machinery, but I shall deal here only with his statistical ideas. All of these date from 1941 or 1940. These statistical ideas are not treated in the biography of Turing by his mother (Turing, 1959).

2. Bayes factors

In practical affairs and in philosophy it is useful to introduce intuitively appealing terminology. Turing introduced the expression '(Bayes) factor in favour of a hypothesis', without the qualification 'Bayes'. The (Bayes) factor in favour of a hypothesis H, provided by evidence E, is $O(H \mid E)/O(H)$, the factor by which the initial odds of H must be multiplied to get the final odds. It is an easy but important theorem that the Bayes factor is equal to $\mathrm{pr}(E \mid H)/\mathrm{pr}(E \mid \bar{H})$, where $\bar{H}$ denotes the negation of H. Perhaps it is fair to say that Bayes only got half-way to the Bayes factor. This theorem was already familiar to Jeffreys (1939), but without Turing's appealing terminology. The result is especially 'Bayesian' if either H or $\bar{H}$ is composite.

3. Sequential analysis and log factors

Turing was one of the independent inventors of sequential analysis for which he naturally made use of the logarithm of the Bayes factor. He did not know that the logarithm of a Bayes factor occurred in a paper by the famous philosopher Charles Saunders Peirce (1878), who had called it weight of evidence.

To show the relationship to Shannon information it is convenient to write $W(H : E)$ for the 'weight of evidence, or log factor, in favour of H provided by E'. The colon, meaning provided by, must be distinguished from a vertical stroke, meaning given. As a very slight generalization we naturally define the weight of evidence concerning H as against H', provided by E, by

$$W(H/H' : E) = \log \frac{O(H/H' \mid E)}{O(H/H')} = \log \frac{\text{pr}(E \mid H)}{\text{pr}(E \mid H')} = W(H : E \mid H \text{ or } H').$$

We then see that weight of evidence is closely related to amount of information concerning H provided by E, $I(H : E)$, defined as $\log\{\text{pr}(E \mid H)/\text{pr}(E)\}$. In fact

$$W(H/H' : E) = I(H : E) - I(H' : E).$$

The expectation of $I(H : E)$ with respect to H and E, when H and E have a joint probability distribution, is prominent in Shannon's mathematical theory of communication (Shannon, 1948). One can also regard amount of information as a special case of weight of evidence $W(H/H' : E)$ in which H' is replaced by a tautology. In fact weight of evidence is a more intuitive concept than amount of information; and expected weight of evidence, which is an expression of the form $\Sigma p_i \log (p_i/q_i)$, is more fundamental than entropy. It even seems to be advantageous to replace entropy by expected weight of evidence in the proof of Shannon's coding theorems: see Good & Toulmin (1968). Turing's interest in expected weight of evidence will be explained below.

4. The deciban

Turing was the first to recognize the value of naming the units in terms of which weight of evidence is measured. When the base of logarithms was e he called the unit a natural ban, and simply a ban when the base was 10. It was much later that a unit of information for base 2 was called a bit and the same units can be used for information as for weight of evidence. Turing introduced the name deciban in the self-explanatory sense of one-tenth of a ban, by analogy with the decibel. The reason for the name ban was that tens of thousands of sheets were printed in the town of Banbury on which weights of evidence were entered in decibans for carrying out an important classified process called Banburismus.

A deciban or half-deciban is about the smallest change in weight of evidence that is directly perceptible to human intuition. I feel that it is an important aid to human reasoning and will eventually improve the judgements of doctors, lawyers and other citizens.

The main application of the deciban was to sequential analysis, not for quality control but for discriminating between hypotheses, just as in clinical trials or in medical diagnosis.

5. The weighted average of factors

The main application of weights of evidence in 1941 was in situations where H and $\bar{H}$, or H', were simple statistical hypotheses, so that the Bayes factor then reduced to a likelihood ratio. But this was not always so, and sometimes a theorem of 'weighted averages of factors' was relevant (Good, 1950, pp. 68, 71). Turing had noticed a special case of this theorem and the generalization was straightforward.

6. The design of experiments and expected weights of evidence

For evaluating Banburismus in advance Turing calculated the expected weight of evidence. In other words, for this application, he recognized that expected weight of evidence was a criterion for the value of an experimental design. In view of the close relationship between weight of evidence and amount of information, it should be recognized that he partially anticipated unpublished work by L. J. Cronbach, in a College of Education, University of Illinois, Urbana report in 1953, Good (1955–6) and Lindley (1956), all of whom proposed the use of expected amount of information in the Shannon sense. Of course, Fisher (1925) had used the same philosophical concept much earlier, but with his different definition of amount of information.

Turing remarked that the expected weight of evidence in favour of a true hypothesis is nonnegative, as one would intuitively require. As a mathematical inequality this is a simple result, previously known, for example, to Gibbs (1902, pp. 136–7), but the application to statistical inference is of interest.

7. The variance of weight of evidence

Also while evaluating Banburismus in advance, Turing considered a model in which the weight of evidence W in favour of the true hypothesis H had a normal distribution, say with mean μ and variance σ^2. He found, under this assumption, (i) that if H is false W must again have a normal distribution with mean $-\mu$ and variance σ^2, and (ii) that $\sigma^2 = 2\mu$ when natural bans are used; it follows that σ is about $3\sqrt{\mu}$ when decibans are used. This result was published by Birdsall (1955) in connection with radar, and was generalized by Good (1961) to the case where the distribution of W is only approximately normal. In radar applications the variance is disconcertingly large and the same was true of Banburismus.

8. Expected Bayes factors

Turing noticed a simple and curious property of Bayes factors, namely that the expectation of the Bayes factor against a true hypothesis is equal to unity. This is equivalent to the fundamental identity of sequential analysis (Wald, 1944, p. 285). Wald gave it useful applications that were not anticipated by Turing.

9. Search trees

Closely related to sequential analysis is the concept of a search tree, now familiar in most expositions of decision theory. Such trees occurred centuries ago in games such as chess, and they form part of the technique of cryptanalysis. Certainly Turing made use of search trees, though not with the full explicit apparatus of expected utilities. I do not know whether he had the idea independently of other people or whether it was obvious to many cryptanalysts.

10. The repeat rate

One cryptanalytic idea that I believe Turing had for himself, but which had been anticipated, was that of a repeat rate. If $p_1, \ldots, p_t$ are the mutually exclusive and exhaustive probabilities of the symbols or letters of a t-letter alphabet, occurring in a random sequence, then the probability that two letters in different places will be the same letter of the alphabet is of couse Σp_i^2. Since this is the probability of a 'repeat', Turing called it the repeat rate ρ, an almost self-explanatory term. If, in a sample of N letters, letter i occurs ν_i times, then Turing knew that an unbiased estimate of ρ is $\Sigma\nu_i(\nu_i - 1)/\{N(N-1)\}$. Friedman (1922) had previously in effect called $t\rho$ the index of coincidence. See also Saccho (1951, p. 185). The repeat rate had also been used as a measure of diversity by Gini (1912), according to Bhargava & Uppulari (1975). E. H. Simpson and I both obtained the notion from Turing.

11. Empirical Bayes

Suppose that a random sample is drawn from an infinite population of animals of various species, or from a population of words. Let the sample size be N and let n_r distinct species be each represented exactly r times in the sample, so that $\Sigma r n_r = N$, and n_r can be called 'the frequency of the frequency r'. Turing, using an urn model, showed that the expected population frequency of a species represented r times is about $(r+1)\,n_{r+1}/(Nn_r)$. For a more exact statement, including the need for smoothing the n_r's, and for numerous elaborations and deductions see Good (1953, 1969) and Good & Toulmin (1956). This work was an example of the empirical Bayes method which method now of course has an extensive literature both with hyperparameterized families of priors and with general priors.

This work was supported in part by a grant from the National Institutes of Health.

References

Bhargava, T. N. & Uppulari, V. R. R. (1975). On an axiomatic derivation of Gini diversity, with applications. *Metron* **33**, 41–53.

Birdsall, T. G. (1955). The theory of signal detectability. In *Information Theory in Psychology*, Ed. H. Quastler, pp. 391–402. Glencoe, Illinois: The Free Press.

Fisher, R. A. (1925). Theory of statistical estimation. *Proc. Camb. Phil. Soc.* **22**, 700–25.

Friedman, W. F. (1922). *The Index of Coincidence and its Applications in Cryptography*. Geneva, Illinois: Riverbank Laboratories.

Gibbs, J. W. (1902). *Elementary Principles in Statistical Mechanics*. Reprinted (1960), New York: Dover.

Gini, C. (1912). Variabilità e mutabilità. *Studi Economico-giuridici della Facoltà di Giurisprodenza dell' 'Università' di Cagliari*, III, part II.

Good, I. J. (1950). *Probability and the Weighing of Evidence*. London: Griffin.

Good, I. J. (1953). The population frequencies of species and the estimation of population parameters. *Biometrika* **40**, 237–64.

Good, I. J. (1955–6). Some terminology and notation in information theory. IEE Monograph 155R (1955). *Proc. Inst. Elec. Eng.* C **103**, 200–4.

Good, I. J. (1961). Weight of evidence, causality and false-alarm probabilities. In *Information Theory, Fourth London Symposium* (1960), Ed. C. Cherry, pp. 125–36. London: Butterworth.

Good, I. J. (1969). Statistics of language. In *Encyclopaedia of Linguistics, Information and Control*, Ed. A. R. Meetham, pp. 567–81. London: Pergamon.

Good, I. J. & Toulmin, G. H. (1956). The number of new species, and the increase in population coverage, when a sample is increased. *Biometrika* **43**, 45–63.

Good, I. J. & Toulmin, G. H. (1968). Coding theorems and weight of evidence. *J. Inst. Math. & Applic.* **4**, 94–105.

Jeffreys, H. (1939). *Theory of Probability*. Oxford: Clarendon.

Lindley, D. V. (1956). On a measure of the information provided by an experiment. *Ann. Math. Statist.* **27**, 986–1005.

Peirce, C. S. (1878). The probability of induction. *Pop. Sci. Monthly*. Reprinted (1956) in *The World of Mathematics*, Vol. 2, Ed. J. R. Newman, pp. 1341–54. New York: Simon and Schuster.

Saccho, L. (1951). *Manuel de Cryptographie*. French edn by J. Bres from the Italian. Paris: Payot.

Shannon, E. C. (1948). A mathematical theory of communication. *Bell System Tech. J.* **27**, 379–423, 623–56.

Turing, A. M. (1936–7). On computable numbers, with an application to the Entscheidungsproblem. *Proc. Lond. Math. Soc.* 2, **42**, 230–65; **43**, 544–6.

Turing, S. (1959). *Alan M. Turing*. Cambridge: Heffer.

Wald, A. (1944). On cumulative sums of random variables. *Ann. Math. Statist.* **15**, 283–96.

[*Received October* 1978. *Revised January* 1979]

INTRODUCTORY REMARKS FOR THE ARTICLE IN BIOMETRIKA 66 (1979), "A. M. Turing's Statistical Work in World War II"

I.J. Good

(i) For a fairly full appreciation of Turing's statistical ideas, in historical perspective, related to the cryptanalysis of the Enigma, during World War II, it is necessary to know some of the background and also some of the later developments of his ideas. It would take up too much space to give full details so I shall make a liberal use of citations.

Turing did not publish these war-time statistical ideas because, after the war, he was too busy working on the ground floor of computer science and artificial intelligence. I was impressed by the importance of his statistical ideas, for other applications, and developed and published some of them in various places. Much of my delay was caused by the wartime attitude that everything was classified, from Hollerith cards to sequential statistics, to empirical Bayes, to Markov chains, to decision theory, to electronic computers. These extreme standards of secrecy only gradually abated after the war.

I shall begin this Introduction with a few comments concerning the Enigma.

(ii) The Enigma is a cryptographic (not cryptanalytic) machine used for enciphering messages in a 26-letter alphabet. To encipher a letter one presses a knob on a 26-letter keyboard and the cipher letter is indicated by a small electric light bulb lighting up. There are 26 of these bulbs and they also have the geometrical arrangement of a keyboard. The machine contains a number of internally and permanently wired wheels, usually three, at least one of which advances each time a letter is enciphered, in fact the "right-hand" wheel advances one step each time, and the others only occasionally, somewhat like an odometer. Owing to this motion of the wheels the state of the machine constantly changes in the course of encipherment of a message. For any given state of the machine it produces a simple substitution so if the wheels did not move, the machine would have no security. If a letter ξ enciphers as η, for a given state of the machine, then η enciphers as ξ for the machine in the same state. This reciprocal property is convenient for the legitimate users of the machine, because it means that deciphering is done by the same procedure as enciphering, and this decreases the chance of enciphering and "legitimate" deciphering errors, but the property is also useful for the cryptanalyst. For a more complete description of the Enigma see, for example, REJEWSKI (1981).

The Enigma had been used by various commercial institutions before World War II. By a curious coincidence I first heard this in 1941 from a retired banker named Burbury who used to sit at my table in a hotel near Bletchley where I was temporarily billeted. He didn't use the name "Enigma". (Burbury was a nephew of S.H. Burbury of the Burbury-Boltzmann "molecular disorder" hypothesis: see BRUSH (1983, pp.89, 92).) A more secure version of the Enigma was adopted by the German army in 1929 (GARLIŃSKI 1979, p.12). Of course the wheels (or rotors) did not have the same wiring as in the commercial models. Much of the further security was achieved by the addition of a plugboard (or steckerboard) which was replugged periodically (every day in Naval usage at least by 1941). The effect of any given plugging was to define a simple substitution that applied both to the plain-language letters entering the wheels and to the letters emerging from the wheels before becoming cipher letters. The plugboard was such that letters could be paired off, for example, if *Q* was plugged (steckered) to *X*, then *X* was plugged to *Q*. Thus the simple substitution of the plugboard was reciprocal or idempotent and did not destroy the reciprocal property of the Enigma.

For the "Home Waters" traffic (as distinct from the Mediterranean traffic) the operator would first choose a trigraph, say *XQV*, as a system discriminator, from a table called the Kennbuch. Next he would set the three wheels at positions $G_1G_2G_3$, known as the Grundstelling, which was part of the daily keys, and, at this initial position of the wheels, he would encipher his selection $M_1M_2M_3$ (the setting for the real message) and obtain the enciphment *LRP*, say. The six letters *XQV* and *LRP* would be further encrypted by the following procedure which does not use the Enigma. (That's why, to avoid confusion, I've here used the word "encrypted" in place of "enciphered".) First the six letters would be written one under the other at a stagger as in (a). Then two letters would be chosen haphazardly to fill a two by four rectangle as in (b). Then the four vertical digraphs *XL*, *QL*, *VR* and *AP* would be encrypted with the help of a secret printed digraph table, giving, say, (c). Finally *PTOW XUBN* would be the first two groups, the "indicator groups", of the enciphered message. There were ten digraph tables and which one was to be used would be part of the daily keys.

(a)	(b)	(c)
XQV	*XQVA*	*PTOW*
LRP	*LLRP*	*XUBN*

Each digraph table was reciprocal; for example, if *XL* became *PX*, then *PX* would become *XL*. This again was helpful both for the encrypter and the cryptanalyst.

For the Mediterranean traffic the indicator system was much the same as the one described by Rejewski (1981, p.217).

At a certain stage in the war we had partial reconstructions of the digraph tables, based on previous successes at breaking daily keys.

On one night shift I discovered that one could identify, by deciban scoring, which digraph table was in use, on the basis of about thirty messages, by taking advantage of the "human random" choices of the two dummy letters.

The original cryptanalytic success against the German military use of the Enigma was achieved by three Polish mathematicians, and especially by the brilliant work of Rejewski in 1932 or 1933. For a description of their efforts, see Rejewski (1981). At the time of this success the Enigma had three wheels that could be inserted in any of the six possible orders, the wheel order being fixed for three months at a time. (For each wheel order there were 26^3 possible settings of the wheels.) The Polish cryptanalysis was aided enormously by "intelligence" material supplied by Gustave Bertrand of the French Secret Service. Rejewski (1981, p.221) says that this intelligence was "the decisive factor in breaking the machine's secrets". Bertrand had bought the material from Hans-Thilo Schmidt (code-name Asché), a member of the German cryptographic agency. Kahn (1983) describes Hans-Thilo as "the spy who most affected World War II". Hans-Thilo was shot in 1943 by the Nazis for his service to humanity. He had been betrayed by R.S. Lemoine, a member of the French Secret Service, to save his own life (Kahn 1983, pp. 76–88). Hans-Thilo's brother was Rudolf Schmidt, a high-ranking general who was demoted by Hitler after Hans-Thilo was betrayed. Apparently Hitler suspected that helping civilization might be a familial trait.

In September and December of 1938 (according to Calvocoressi 1980, p.38) the Germans introduced two changes, one of which was an increase in the number of wheels in the "library" from three to five. There were therefore now $5 \times 4 \times 3 = 60$ possible wheel orders. The method for indicating the settings of the wheels for individual messages was also changed. At about the same time the number of unplugged letters (or self-steckers) was decreased from 12 to 6. The number of possible pluggings was now $26!/(10!\,6!\,2^{10}) = 1.51 \times 10^{14}$ which is greater than if there were no self-steckers. The Poles were able to discover what changes had been made, and to discover the wirings of the two new wheels, but they didn't have the facilities to cope with these changes on a regular daily basis and accordingly decided to give all their methods and results to the British and French. They managed to do so, late in July 1939, only weeks before Poland was overrun. If the Germans had delayed the improvements in security perhaps the Polish gift would not have been made in time!

(iii) When I arrived at the Government Code and Cypher School in Bletchley Park in the town of Bletchley on 1941 May 27 (the day the Bismarck was sunk), Turing was in charge of the cryptanalysis of the Naval Enigma. It then had eight wheels in its library, of known permanent wirings, and therefore $8 \times 7 \times 6 = 336$ possible wheel orders. The wheel order was changed every two days. One of the methods of attack was the "clock method" of Różycki (REJEWSKI 1981, p.223; GOOD 1981) which Turing had improved by using respectable probabilistic methods. (One of his ideas, involving regenerative Markov chains, is developed by GOOD 1973, esp. p.936.) The elaborated clock method was called *Banburismus* and is mentioned in Section 4 of the following article. When I wrote that article I was unaware that the Poles had used the clock method. The game of Banburismus involved putting together large numbers of pieces of probabilistic information somewhat like the reconstruction of DNA sequences. The best player, out of about ten, was Hugh Alexander, the British chess champion, who became head of the Naval Enigma section when Turing started work on speech secrecy.

The security principle of "need to know" was applied fairly rigorously in Bletchley, but on one occasion my curiosity got the better of me and I asked Turing "How on earth did we find the wirings of the wheels?" although my work on Banburismus did not require this knowledge. (It should be held in mind that cryptographers always have to assume that the permanent features of a cryptographic machine are known to the cryptanalyst.) He replied "Well, I suppose, the Poles" and I added "And a pinch?" and then the conversation fizzled out. (By a "pinch" I meant the capture of an Enigma machine.) That was the only time, until many years after the war, that I had any inkling of the Polish contribution.

Without the help of powerful cryptanalytic machines it would not have been practicable for us to read the Naval Enigma at all regularly. The main such machine used for this purpose was called the Bombe which was also used on all other Enigma traffic. It was a greatly improved form of a cryptanalytic machine that the Poles had invented. It was electromagnetic rather than electronic and should not be confused with the electronic machine the Colossus. For information concerning the Colussus see, for example, RANDELL (1980) and GOOD (1980). The Colossus had much more in common with modern electronic computers. It was used against a German cryptographic teleprinter machine which we called *Fish* (Sägefisch, Geheimschreiber, Schlüsselzusatz, SZ40, SZ42, made by Lorenz). There were various links or "species" of Fish, such as Tunny, Bream and Jellyfish. Fish was entirely different from the Enigma and was used for even higher-level communications (HINSLEY 1984, Vol. 3, Part I, pp.477–482, who says, for example, that, for the importance of its effects, the solution of Jellyfish was the most significant cryptanalytic achievement of the Government Code and Cypher School in 1944). The Colossus was not used against the Enigma, contrary to the impression conveyed by the penultimate paragraph of REJEWSKI (1981). Turing made an early contribution to the cryptanalysis of Fish but I believe he did not contribute directly to the design of the Colossus apart from rightly suggesting that Thomas H. Flowers, of the Dollis Hill Research Station of the Post Office, would be a good man to head the engineering effort. Turing was not one of the users of Colossus after it was built. But his influence on the design of the Bombe was great and was perhaps his most important contribution to the cryptanalysis of the Enigma.

In February 1942, the U-boats began to use a four-wheel Enigma and we read no more U-boat messages until 13th December of that year. From that date we were again able to read the U-boat messages (HINSLEY 1981, Vol. 2, p.548). According to CALVOCORESSI (1980, pp.104, 126), Turing played a major part in the new break-in, but in fact Turing was not involved in that exercise. I obtained that information recently from Shaun Wylie who remained in Hut 8 until September 1943 whereas I moved out in June or July of that year. We were both transferred to M.H.A. Newman's section to work on Fish.

For much more detail about the U-boat traffic see HINSLEY (1981, Vol. 2, pp.547–572 and pp.747–752). The value of reading U-boat traffic was reduced because, for much of the time up to June 1943, the German Naval Cryptanalytic Service (B-Dienst) was reading the British Naval Cypher No. 3, known to B-Dienst as the convoy cipher. When the cryptanalysts on both sides were having successes, the Atlantic battle was to some extent like chess instead of Kriegspiel. But neither side knew how much the other side knew.

Turing's idea for the Bombe was to use a modification of the principle that from a logical contradiction all propositions can be deduced whether they are true or false, a proposition that Wittgenstein had thought was unimportant in an argument with Turing (HODGES 1983, p.154). Perhaps one reason Turing had his idea was to win his argument with Wittgenstein retrospectively, but that's just an amusing speculation. The modification was that, in an appropriate context, from a false assumption one could very probably deduce a large number of mutually contradictory consequences. This idea made it possible to cut the running time of the Bombe by a factor of 26. This fact is mentioned in the play "Breaking the Code" which is about the life of Turing and is based on the book by HODGES (1983). Turing's idea helped to win the war but would have been unnecessary if modern electronics had been available.

The efficiency of the Bombe was greatly increased by the use of the "diagonal board". This device, which takes advantage of the reciprocal property of the Enigma and of its plugboard, was suggested by Welchman and is described in WELCHMAN (1982).

For further information about the Enigma, and the effects of cryptanalysis on the war, see the additional books listed in the References. I believe LEWIN (1978) gives one of the most accurate accounts of the effects in a book of moderate length. KAHN (1983, p.218), in a reprint of a book review published in 1979, says "Michael Howard, Chichele professor of the history of war at Oxford, is right when he says that this [LEWIN's book] is 'perhaps the most important book to have appeared on the Second World War since Chester Wilmot wrote *The Struggle for Europe* a quarter of a century ago'." (But Lewin erred when he gave my address as the University of West Virginia.)

(iv) Most of the ideas described in the following article reprinted from *Biometrika* were invented, discovered or rediscovered by Turing in relation to Banburismus. (Please read Sections 2, 3 and 4 of the *Biometrika* article before reading the rest of this Introduction.)

From a mathematical point of view, the concept of a Bayes factor is only a small modification of concepts familiar to Laplace and Poisson, but, when combined with the terminology of odds, log-odds, and weights of evidence, the appeal to the untrained intuition is immediate. The idea of a Bayes factor could be readily explained to the woman in the street.

Turing's name for weight of evidence was "decibannage" or "score". In the following article I mention that C. S. Peirce had used the expression *weight of evidence* in 1878 in its technical sense. This was only in a short comment, and a careful reading of Peirce's obscurely written article shows that he was dealing with the special case in which the prior probability of the hypothesis H is equal to $\frac{1}{2}$. For a fuller explanation see my response to Barnard in GOOD (1988).

Turing did not have a notation for weight of evidence although he once remarked that mathematics is more in need of good notations than of new theorems. He probably meant that any given proved theorem would have been discovered eventually whereas a bad notation (like bad terminology) is liable to become entrenched. At any rate I took to heart what Turing had said, and in GOOD (1960 A, B, C), or perhaps earlier, I introduced the notations $W(H:E)$ and $W(H:E|G)$, together with the almost obvious additive property

$$W(H:E\&F)=W(H:E)+W(H:F|E),$$

where $W(H:E|G)$ denotes *the weight of evidence concerning H provided by E assuming G all along* (*or given G*).

The successful use of weights of evidence against the Enigma, and the way that the modest deciban, and the doubly modest half-deciban, helped to rid the world of Criminal Lunatic #1, caused me to become somewhat obsessed with the topic, and I have returned to it repeatedly like a dog returneth to his or her vomit (Proverbs 26:11). (See "Weight of evidence" in both indexes of GOOD 1983 A.) I am totally convinced that the technical concept of weight of evidence precisely and uniquely captures the corresponding intuitive concept and it appeals to physicians. The expression was independently proposed in the same sense by MINSKY and SELFRIDGE (1961). The concept is logically fundamental in medicine, in the law, and in statistics (for example, GOOD and CARD 1971; SPIEGELHALTER and KNILL-JONES 1984; GOOD 1986; BERNSTEIN et al. 1989). That the concept follows uniquely from natural desiderata was shown, in various ways, and with increasing simplicity, by GOOD (1968, Appendix A; 1984; 1989 A, B). But it is an uphill struggle to get the philosophers to listen. Here is the simplest proof. It is based on the assumption that $W(H:E)$ depends only on $x=P(E|H)$ and $y=P(E|\bar{H})$, say $W(H:E)=f(x,y)$. (I am taking the background information for granted, that is, omitting it from the notation.) Let F be unrelated to H and E. Then we must have $W(H:E\&F)=W(H:E)$. Hence $f(x,y)=f(\lambda x,\lambda y)$ for all λ and therefore $f(x,y)$ is a function of x/y. We must take the logarithm to justify the use of the word *weight*. Q.E.D.

I conjecture that theoretical scientists, doctors, magistrates and detectives could be trained to make useful subjective estimates of Bayes factors or weights of evidence, but careful experiments have not yet been done to test this conjecture as far as I know. It might be found that some people usually overestimate and others underestimate weights of evidence, and of course wishful thinking might also affect the judgement. (Compare GOOD and CARD 1971, p.187.) Perhaps Alexander's ability at Banburismus was partly because he had unusually objective judgement as well as outstanding energy, determination and intelligence.

In more general terms, the judgements used, when applying a subjective theory of probability, do not need to be restricted to (inequality) judgements of individual probabilities (Good 1983 A, p.76). Judgements can be made of (inequalities between) ratios of probabilities, including Bayes factors, odds, weights of evidence, utilities, ratios of utilities, expected utilities and "causal tendencies" (mentioned below).

When discriminating between two hypotheses, it is natural to think of weight of evidence as a "quasi-utility" owing to its additive property. Expected weight of evidence can then be taken as a substitute for expected utility when the true expected utility cannot be readily estimated. (Donald Michie points out the analogy with an investor who can't make money now but acquires knowledge that might pay off later.) Expected weight of evidence has many names, one being cross-entropy. Entropy and cross-entropy have had applications too numerous to be mentioned here. For some of them, and for citations to E.T. Jaynes, H. Jeffreys, S. Kullback, D.V. Lindley, J. Rothstein, S. Watanabe and others, see the indexes of Good (1983 A).

The theorem in Section 7 of the *Biometrika* paper, concerning the variance of a normally distributed weight of evidence shows that its spread is unexpectedly wide. As pointed out, the result is "disconcerting" in application to radar, but *terrifying* might be a more apt description.

Turing's surprising-sounding little theorem, in Section 8, that the expected Bayes factor in favour of a false hypothesis is 1 (Good 1950, p.72, generalized on p.74) is trivial to prove once it is pointed out. One simple consequence is that *if it is possible, by means of some experiment, to undermine a hypothesis, then it is also possible to obtain evidence in its favour from the same experiment.* This almost obvious proposition (which can also be readily proved without mentioning the little theorem) has sometimes been implicitly denied by philosophers and statisticians who say you can refute a theory or null hypothesis, but you can't support it. It is clear to a Bayesian that it is often possible to get an enormous weight of evidence against a false ("simple statistical") hypothesis (when it is embedded in a composite hypothesis), and often difficult to get *much* support in favour when it is true.

The terminology and notation for weight of evidence is suggestive for the philosophical problem of defining, in terms of probability, the tendency of one event F to cause another later one, E. (This is not at all the same as the extent to which F actually caused E.) For any one familiar with the concept of weight of evidence, just four feasible definitions might spring to mind, $W(E:F|U)$, $W(F:E|U)$, $W(\bar{E}:\bar{F}|U)$ and $W(\bar{F}:\bar{E}|U)$, where the bar denotes negation, and U stands for the state of the universe just before F occurred. (Some people might forget to mention U.) Of these, the first three can be easily eliminated (Good 1988) and we are left with the weight of evidence against F if E does not occur, an explicandum first reached by arguments depending on causal networks (Good 1961/1962). For example, in the extreme case in which F is a sufficient cause of E, the tendency is infinite because then $P(\bar{E}|\bar{F}\cdot U)/P(\bar{E}|F\cdot U)=\infty$ unless E is almost certain whether F occurs or not (in which case the premise that F is a sufficient cause of E would be like assuming that eating popcorn will cause the sun to rise tomorrow).

In my work on explicativity or explanatory power (Good 1977) I again found it convenient to use the notation $W(H:E|F)$. For many other references to weight of evidence see the two indexes of Good (1983 A). For a survey, which needs updating, see Good (1983 B).

(v) The empirical Bayes method can be classified as either the (fairly obvious) parametric method or the (ingenious) nonparametric method. Turing is not always given credit for his anticipation, in an interesting special case, of the nonparametric empirical Bayes idea. (See also von Mises 1942.) This work is mentioned in Section 11 of the *Biometrika* paper. The formula given there is almost equivalent to formula (19) in Robbins (1956). I found it philosophically fascinating that one could use a Bayesian argument while making only weak qualitative assumptions about the priors. Robbins was certainly one of the pioneers in this field although the basic idea had been anticipated.

The following simple deduction from Turing's work is interesting (Good 1953, 1969). The probability that the next "animal" (or word) sampled will belong to a new "species" (or will be a new word) is close to n_1/N if $n_1>20$. In other words the *coverage* of the existing sample of N animals or words is about $1-n_1/N$. Also interesting are estimates of the expected coverage of future samples, see Good and Toulmin (1956).

On one occasion in 1941 or 1942 I met George A. Barnard in London and told him that we were using Bayes factors, and their logarithms, sequentially, to discriminate between two hypotheses but of course I did not mention the application. Barnard said that curiously enough a similar method was being used for quality control in the Ministry of Supply for discriminating between lots rather than hypotheses. It was really the same method because the selection of a lot can be regarded as the acceptance of a hypothesis. Barnard no longer remembers our meeting but I remember it with great clarity, possibly because I was worried in case the discussion might be regarded, by people of poor judgement concerning statistical theory, as a breach of security. If Barnard had not already been using the idea my remark would have fallen on fertile ground! Turing, Barnard and Wald all deserve credit for the concept of sequential analysis, and for using it in important applications.

Additional References*

BEESLY, Patrick
1977 *Very Special Intelligence*
(Hamish Hamilton, London)

BENNETT, Ralph
1979 *Ultra in the West*
(Charles Scribner's Sons, New York)

BERNSTEIN, L.H., I.J. GOOD, G.I. HOLTZMAN, M.L. DEATON and J. BABB
1989 Diagnosis of acute myocardial infarction from two measurements of creatine kinase isoenzyme MB with use of nonparametric probability estimation
Clinical Chemistry (USA) **35** (3), 444–447

BRUSH, S.G.
1983 *Statistical Physics and the Atomic Theory of Matter, from Boyle and Newton to Landau and Onsager*
(Princeton Univ. Press, Princeton, NJ)

CALVOCORESSI, Peter
1980 *Top Secret Ultra*
(Ballantine Books, New York)

CAVE-BROWN, A.
1975 *Bodyguard of Lies*
(Harper & Row, New York)

GARLIŃSKI, J.
1979 *The Enigma War*
(Charles Scribner's Sons, New York)

* I have not repeated references that were given in the *Biometrika* article.

GOOD, I.J.
1960A The paradox of confirmation
British J. Philos. Sci. **11**, 145–149
1960B Weight of evidence, corroboration, explanatory power, information, and the utility of experiments
J. Roy. Statist. Soc. Ser. B **22**, 319–331
1960C Effective sampling rates for signal detection: or can the Gaussian model be salvaged?
Inform. and Control **3**, 116–140
1961/1962 A causal calculus
British J. Philos. Sci. **11** (1961) 305–318; **12** (1961) 43–51; **13** (1962) 88
(reprinted in GOOD 1983 A)
1968 Corroboration, explanation, evolving probability, simplicity, and a sharpened razor
British J. Philos. Sci. **19**, 123–143
1973 The joint probability generating function for run-lengths in regenerative binary Markov chains, with applications
Ann. Statist. **1**, 933–939
1977 Explicativity: a mathematical theory of explanation with statistical applications
Proc. Roy. Soc. London Ser. A **354**, 303–330
(largely reprinted in GOOD 1983 A)
1980 Pioneering work on computers at Bletchley
In: N. METROPOLIS, J. HOWLETT and G.-C. ROTA (Eds.), *A History of Computing in the Twentieth Century*
(Academic Press, New York) 31–45.
1981 Contribution to the discussion of Rejewski (1981)
Ann. Hist. Comput. **3**, 232–234
1983A *Good Thinking: The Foundations of Probability and its Applications*
(Univ. of Minnesota Press, Minneapolis, MN)
1983B Weight of evidence: a brief survey
In: J.M. BERNARDO, M.H. DEGROOT, D.V. LINDLEY and A.F.M. SMITH (Eds.), *Bayesian Statistics 2: Proceedings of the Second Valencia International Meeting September* 6/10, 1983
(North-Holland, New York, 1985) 249–269 (including discussion)
1984 The best explicatum for weight of evidence
J. Statist. Comput. Simulation **19** (C197), 294–299; **20**, 89
1986 The whole truth
Inst. Math. Statist. Bull. **15**, 366–373

1988 The interface between statistics and philosophy of science
Statist. Sci. **3**, 386–412 (with discussion)

1989A Yet another argument for the explication of weight of evidence
J. Statist. Comput. Simulation **31** (C312), 58–59

1989B Weight of evidence and a compelling metaprinciple
J. Statist. Comput. Simulation **31** (C319) 121–123

GOOD, I.J. and W.I. CARD

1971 The diagnostic process with special reference to errors
Methods Inform. Medicine **10**, 176–188

HINSLEY, F.H.

1979/ *British Intelligence in the Second World War*, three volumes
1981/ (Her Majesty's Stationary Office, London)
1984 (this is the "official" history)

HODGES, A.

1983 *Alan Turing: the Enigma*
(Burnett Books, London)

JOHNSON, Brian

1978 *The Secret War*
(British Broadcasting Corporation, London) Chapter 6

KAHN, David

1983 *Kahn on Codes: Secrets of the New Cryptology*
(Macmillan, New York)

KOZACZUK, W.

1984 *Enigma: how the German Machine Cipher was Broken and how it was Read by the Allies in World War Two*
(Arms and Armour Press, London)

LEWIN, Ronald

1978 *Ultra Goes to War*
(McGraw-Hill, New York)

MINSKY, M. and O.G. SELFRIDGE

1961 Learning in random nets
In: Collin CHERRY (Ed.), *Information Theory*
(Butterworths, London) 335–347

RANDELL, Brian
1980 The Colossus
In: N. METROPOLIS, J. HOWLETT and G.-C. ROTA (Eds.), *A History of Computing in the Twentieth Century*
(Academic Press, New York) 47–92

REJEWSKI, M.
1981 How Polish mathematicians deciphered the Enigma
Ann. Hist. Comput. **3**, 213–234 (with discussion)
(this is a translation by Joan STEPENSKE from the Polish; another translation, by C. KASPAREK appears at Appendix D in KOZACZUK (1984))

ROBBINS, H.E.
1956 An empirical Bayes approach to statistics
In: *Proceedings of the Third Berkeley Symposium Math. Statist. Probab.* **1**, 157–163

SPIEGELHALTER, D.J. and R.P. KNILL-JONES
1984 Statistical and knowledge-based approaches to clinical decision-support systems, with an application in gastroenterology
J. Roy. Statist. Soc. Ser. A **147**, 35–77

STEVENSON, William
1976 *A Man Called Intrepid*
(Harcourt Brace Jovanovich, New York)

VON MISES, R.
1942 On the correct use of Bayes's formula
Ann. Math. Statist. **13**, 156–165

WELCHMAN, G.
1982 *The Hut Six Story*
(McGraw-Hill, New York)

WINTERBOTHAM, F.W.
1974 *Very Special Intelligence*
(Hamish Hamilton, London)

NOTES AND SUMMARIES

1935 *Equivalence of Left and Right Almost Periodicity*

Notes

⟦1⟧ Thus f is r.a.p. if and only if the set of translates

$$x \to f(xa), \quad a \in \mathfrak{G},$$

is totally bounded with respect to the norm $\|g\| = \sup_{x \in \mathfrak{G}} |g(x)|$.

⟦2⟧ (a) If f is l.a.p., the left mean $L(f)$ will satisfy the following condition. For all $\varepsilon > 0$ there exist n, c_i, a_i, $i = 1, \ldots, n$, such that $c_i > 0$, $\sum c_i = 1$ and, for all x in G,

$$\left| \sum_{i=1}^{n} c_i f(a_i x) - L(f) \right| < \varepsilon.$$

(b) For a discussion of almost periodic functions and the Von Neumann mean when $\mathfrak{G}$ is a compact group, see LOOMIS (1953) Chap. VIII.

1938 A *Finite Approximations to Lie Groups*

Summary

Let G be an abstract group and a metric space such that

$$D(ax, ay) = D(x, y) = D(xa, ya)$$

for all x, y, a in G. (Then G is a topological group.) For example, any compact Hausdorff topological group such that each point has a countable base of neighbourhoods has such a metric.

Let $\varepsilon > 0$. A finite group H_ε with product denoted by $x \circ y$ is called an *ε-approximation* to G if H_ε is a subset of G and

(i) each $x \in G$ is within distance ε of some $r(x) \in H_\varepsilon$;

(ii) $a, b \in H_\varepsilon$ implies $D(a \circ b, ab) < \varepsilon$.

G is *approximable* if it has an ε-approximation for each $\varepsilon > 0$.

Note that an approximable group is totally bounded.

Theorem 1. *Let G be an approximable group with a faithful continuous representation ϱ_1 by matrices over $\mathbb{C}$. Then there exists a faithful continuous representation ϱ_2 by complex matrices, of degree n say, such that G may be approximated by finite groups with faithful representations of degree n.*

Theorem 2. *Let G be a connected Lie group. If G is approximable, then G is compact and Abelian.*

Notes

⟦1⟧ A *metrical group* is presumably an abstract group and a metric space such that the product and inverse are continuous. It will appear shortly that in this paper the only metrical groups of interest are those satisfying $D(xa, ya) = D(x, y) = D(ax, ay)$.
⟦2⟧ Instead of 'Let H_ε' read 'For $\varepsilon > 0$ let H_ε'.
⟦3⟧ For '$>$' read '$<$'.
⟦4⟧ Recall that a metric space is *totally bounded* if for every $\varepsilon > 0$ there exists a finite number of open balls of radius ε which cover the space. A metric space is compact if and only if it is complete and totally bounded.

Actually, in view of the results to be proved, it would seem that little is lost by assuming at the outset that G is compact.
⟦5⟧ Insert 'for all x in G' after 'such that'.
⟦6⟧ For later note that $D(e_\varepsilon, e) < \varepsilon$.
⟦7⟧ Matrices are over $\mathbb{C}$. 'True' means 'faithful'.

Given any character χ of a compact or abstract finite group G we may write $\chi = \sum_{i=1}^{s} n_i \chi^{(i)}$ where $n_i > 0$ is an integer and $\chi^{(i)}$, $i = 1, \dots, s$, are distinct irreducible characters of G (but not in general the full set of irreducible characters). Let us say that χ is *normalized* if, for each i, n_i equals the degree d_i of $\chi^{(i)}$ and that the *normalization* of the character χ expressed as above is $\sum_{i=1}^{s} d_i \chi^{(i)}$. If χ is faithful, then so is its normalization. What will be proved is not Theorem 1 but the following modification of it.

Theorem 1′. *Let G be an approximable group with a true continuous representation ϱ. Let ϱ' be a normalization of ϱ and have degree n; thus ϱ' is normalized and true. Then G may be approximated by finite groups with true representations of degree n.*

⟦8⟧ This is slightly misleading; the argument required is as follows. $D(ca_i, c\,r(a_i)) < \eta$ so $|f(ca_i) - f(c\,r(a_i))| < \Delta$; $D(c\,r(a_i), c \circ r(a_i)) < \eta$ so $|f(c\,r(a_i) - f(c \circ r(a_i))| < \Delta$. Hence $|f(ca_i) - f(c \circ r(a_i))| < 2\Delta$.
⟦9⟧ Let χ be any character of a compact group G, written as a sum as in ⟦7⟧. Let $W(x) = \int_G \chi(xy)\overline{\chi(y)}\, dy$. Then

$$W(x) = \sum n_i^2 \int \chi^{(i)}(xy)\overline{\chi^{(i)}(y)}\, dy = \sum n_i^2 \frac{\chi^{(i)}(x)}{d_i},$$

where d_i is the degree of $\chi^{(i)}$. This is not in general equal to $\chi(x)$ but we do have equality if $n_i = d_i$ for all i, that is, if χ is normalized. Thus in order that (7) should hold we must replace the given representation by a normalization of it; see ⟦7⟧ above.

⟦10⟧ For later we need that α is less than $\frac{50}{51}$.

⟦11⟧ This is by the uniform continuity of the function $g(x, y) = \chi(xy)\overline{\chi(y)}$ on the compact space $G \times G$.

⟦12⟧ Let $f(y) = \chi(ay)\overline{\chi(y)}$. If $D(y, y') < \eta$, then $D(y, y') < 2\eta$ so by (11) $|f(y) - f(y')| < \alpha/(50n)$. Hence by the lemma

$$\left| h_\eta^{-1} \sum_{b \in H_\eta} f(b) - \int_G f(y)\,dy \right| \leqslant 2\alpha/(50n).$$

⟦13⟧ Since $D(a \circ b, ab) < \eta < 4\eta$ we have by (10) $|\chi(a \circ b) - \chi(ab)| < \alpha/(50n^2)$.

⟦14⟧ $|\varphi(b)| \leqslant n$ by the definition (12).

⟦15⟧ The right-hand side should be divided by the degree d_λ of $\chi^{(\lambda)}$.

⟦16⟧ From this point to the inequality (21) the argument given should be corrected as follows:

$$h_\eta^{-1} \sum_{b \in H_\eta} \varphi(a \circ b)\overline{\varphi(b)} = h_\eta^{-1} \sum_b \sum_\lambda \alpha_\lambda \chi^{(\lambda)}(a \circ b) \sum_\mu \overline{\alpha_\mu \chi^{(\mu)}(b)}$$

$$= \sum_\lambda \alpha_\lambda \bar{\alpha}_\lambda \frac{\chi^{(\lambda)}(a)}{d_\lambda}.$$

Let $J'_\lambda = (\alpha_\lambda \bar{\alpha}_\lambda / d_\lambda) - \alpha_\lambda$ and $J_\lambda = |J'_\lambda|$. Then $|(\alpha_\lambda/d_\lambda)|\,|\alpha_\lambda - d_\lambda| = J_\lambda$. (19) becomes $|\sum J'_\lambda \chi^{(\lambda)}(a)| < \alpha/(8n)$, so

$$h_\eta^{-1} \sum_{a \in H_\eta} \sum_\lambda J'_\lambda \chi^{(\lambda)}(a) \sum_\mu \overline{J'_\mu \chi^{(\mu)}(a)} < \frac{\alpha^2}{64n^2}$$

and therefore $\sum J_\lambda^2 < \alpha^2/(64n^2)$.

Define $\xi(a)$ to be $\sum c_\lambda \chi^{(\lambda)}(a)$ where for each λ, c_λ is 0 or d_λ (the choice to be determined below). Then ξ is a character of H_η. We have

$$h_\eta^{-1} \sum_{b \in H_\eta} \xi(a \circ b)\overline{\xi(b)} = \sum c_\lambda \bar{c}_\lambda \frac{\chi^{(\lambda)}(a)}{d_\lambda}$$

$$= \sum c_\lambda \chi^{(\lambda)}(a) = \xi(a),$$

which is (20), apart from the obvious misprint. Next,

$$h_\eta^{-1} \sum_a |\xi(a) - \varphi(a)|^2 = h_\eta^{-1} \sum_a |\sum (c_\lambda - \alpha_\lambda)\chi^{(\lambda)}(a)|^2 = \sum |c_\lambda - \alpha_\lambda|^2.$$

If $|\alpha_\lambda / d_\lambda| \leqslant \frac{1}{2}$, then $|(\alpha_\lambda/d_\lambda) - 1| \geqslant \frac{1}{2}$ so $J_\lambda \geqslant \frac{1}{2}|\alpha_\lambda|$; define $c_\lambda = 0$ in this case.

If $|\alpha_\lambda/d_\lambda|>\frac{1}{2}$, then $J_\lambda \geqslant \frac{1}{2}|\alpha_\lambda - d_\lambda|$, so $|\alpha_\lambda - d_\lambda| \leqslant 2J_\lambda$; define $c_\lambda = d_\lambda$ here. Then we have

$$h_\eta^{-1} \sum_a |\xi(a)-\varphi(a)|^2 = \sum_{c_\lambda=0} |\alpha_\lambda|^2 + \sum_{c_\lambda=d_\lambda} |d_\lambda - \alpha_\lambda|^2 \leqslant \sum 4J_\lambda^2 < \frac{\alpha^2}{16n^2},$$

which is (21).

⟦17⟧ Although $|\varphi(b)| \leqslant n$ for each b in H_η, it is certainly not obvious at this stage that $|\xi(b)| \leqslant n$.

Thanks are due to the late J. Frank Adams for rescuing the argument at this point, as follows.

$$\left(h_\eta^{-1} \sum_{b\in H_\eta} |\xi(b)|^2\right)^{1/2} \leqslant \left(h_\eta^{-1} \sum |\xi(b)-\varphi(b)|^2\right)^{1/2} + \left(h_\eta^{-1} \sum |\varphi(b)|^2\right)^{1/2}$$
$$< \frac{\alpha}{4n} + n.$$

Hence

$$\left| h_\eta^{-1} \sum_{b\in H_\eta} (\xi(a\circ b)\overline{\xi(b)} - \varphi(a\circ b)\overline{\varphi(b)}) \right|$$
$$\leqslant \frac{\alpha}{4n}\left(\left(\frac{\alpha}{4n}+n\right)+n\right) = \frac{\alpha^2}{16n^2} + \tfrac{1}{2}\alpha.$$

Also $|\xi(a)-\chi(a)| < (\alpha^2/16n^2 + \frac{1}{2}\alpha) + \alpha/(10n) < \alpha$ since $n \geqslant 1$ and $0<\alpha<1$.

⟦18⟧ $D(e_\eta, e) < \eta$, so $|\chi(e_\eta)-\chi(e)| < \alpha/(50n^2)$. Now $|\xi(e_\eta)-\chi(e_\eta)|<\alpha$ so $|\xi(e_\eta)-\chi(e)| < \alpha + \alpha/(50n^2) \leqslant \frac{51}{50}\alpha < 1$. $\xi(e_\eta)$ is an integer and $\chi(e)=n$, hence $\xi(e_\eta)=\chi(e)=n$. Let $a\in H_\eta$ and $D(a,e)\geqslant \frac{1}{4}\varepsilon$. Then $|\chi(a)-n|>\alpha$ and $|\xi(a)-\chi(a)|<\alpha$ so $\xi(a)\neq n = \xi(e_\eta)$; thus a is not in N.

⟦19⟧ $N=\{a\in H_\eta : \xi(a)=\xi(e_\eta)\}$ is a self-conjugate subgroup of H_η since ξ, being a sum of irreducible characters, is a character of H_η.

⟦20⟧ For 'coset of N' read 'coset of N in H_η'.

⟦21⟧ K is isomorphic to H_η/N.

⟦22⟧ Recall that if $x\in G$, there exists $r(x)\in H_\eta$ such that $D(x, r(x))<\eta$.

⟦23⟧ $D(ab, v(a)v(b)) \leqslant D(ab, av(b)) + D(av(b), v(a)v(b)) < 2(\frac{1}{4}\varepsilon+\eta)$.

⟦24⟧ From the sequel it would appear that connectedness is assumed as part of the definition of a Lie group.

⟦25⟧ We have $D(xy,yx) \leqslant D(ada'd', aa') + D(aa', a'a) + D(a'a, a'd'ad)$. Since $a'\circ a = a\circ a'$, the middle term is less than 2ε. Now in general $D(r_1r_2r_3, s_1s_2s_3) \leqslant \sum_{i=1}^3 D(r_i, s_i)$. Hence

$$D(xy,yx) < 4\varepsilon + 2\varepsilon + 4\varepsilon = 10\varepsilon.$$

⟦26⟧ There seems to be a gap here. Let $J=\bigcup_{x\in G} \{x\}\times N_x$. Then $E\subset J\subset$

$G \times G$. If J is measurable, then the inequality follows from the Fubini theorem; for

$$m(E) = \int\int \chi_E \leqslant \int\int \chi_J = \int_G \left(\int_G \chi_{N_x} \, dy \right) dx = \int_G m(N_x) \, dx.$$

The measurability of J or some appropriate modification of J needs to be proved.

1938 *The Extensions of a Group*

Summary

In this summary, $\mathfrak{N}$ is a group, $\mathfrak{A}$ is its automorphism group and $\mathfrak{J}$ is the group of inner automorphisms. $\mathfrak{N}$ is a normal subgroup of a group $\mathfrak{F}$. $\chi : \mathfrak{F} \to \mathfrak{A}$, $a \mapsto \chi_a$, is a homomorphism such that $\chi(\mathfrak{N}) \subset \mathfrak{J}$; thus χ induces a homomorphism $X : \mathfrak{F}/\mathfrak{N} \to \mathfrak{A}/\mathfrak{J}$. If $\mathfrak{G}$ is an extension of $\mathfrak{N}$ by $\mathfrak{F}/\mathfrak{N}$, then we have a homomorphism $H : \mathfrak{G}/\mathfrak{N} \to \mathfrak{A}/\mathfrak{J}$ given by $g\mathfrak{N} \mapsto \hat{g}\mathfrak{J}$, where $\hat{g}$ is defined by $\hat{g}(n) = g^{-1}ng$, $n \in \mathfrak{N}$. We say that this extension *realizes* X if there is an isomorphism $\lambda : \mathfrak{G}/\mathfrak{N} \to \mathfrak{F}/\mathfrak{N}$ such that $H = X \circ \lambda$.

The first main result is the Corollary to Theorem 1.

Let $\mathfrak{F}$ be a free group. Then an extension $\mathfrak{G}$ of $\mathfrak{N}$ by $\mathfrak{F}/\mathfrak{N}$ realizing X exists if and only if there is a homomorphism $\mathfrak{a} : \mathfrak{N} \to \mathfrak{N}$ such that $\chi_a(\mathfrak{a}(r)) = \mathfrak{a}(a^{-1}ra)$ and $\chi_r(\mathfrak{b}) = \mathfrak{a}(r)^{-1}\mathfrak{b}\mathfrak{a}(r)$ for all $a \in \mathfrak{F}$, $r \in \mathfrak{N}$, $\mathfrak{b} \in \mathfrak{N}$.

Next we have the following theorem.

Theorem 2. *Let $\mathfrak{F}$ be the free group on $e_1, \ldots, e_n$. Let there exist elements $r_1, \ldots, r_l$ of $\mathfrak{F}$ such that the normal subgroup of $\mathfrak{F}$ which they generate is $\mathfrak{N}$. Let $\mathfrak{r}_i^* \in \mathfrak{N}$ satisfy $\chi_{r_i}(\mathfrak{b}) = \mathfrak{r}_i^{*-1}\mathfrak{b}\mathfrak{r}_i^*$ for all $\mathfrak{b}$ in $\mathfrak{N}$, $i = 1, \ldots, l$. Then there is an extension $\mathfrak{G}$ of $\mathfrak{N}$ by $\mathfrak{F}/\mathfrak{N}$ realizing X if and only if there exist elements $\mathfrak{z}_1, \ldots, \mathfrak{z}_l$ in the centre $Z(\mathfrak{N})$ of $\mathfrak{N}$ such that the following condition holds*:

(*) *If* $\prod_{i=1}^{N} a_i^{-1} r_{l_i}^{t_i} a_i = 1$ *in* $\mathfrak{F}$, *where each* t_i *is* ± 1,

then $\prod_{i=1}^{N} \chi_{a_i}(\mathfrak{r}_{l_i}^* \mathfrak{z}_{l_i}^{-1})^{t_i} = 1.$

The author next simplifies Theorem 2 by making use of some work of Reidemeister on generators and relations for subgroups.

Let $\mathfrak{F}$ and $\mathfrak{N}$ be as in Theorem 2. In each coset $a\mathfrak{N}$ of $\mathfrak{N}$ in $\mathfrak{F}$ pick a representative which may be denoted by v_a or $v(a)$; take $v_1 = 1$. For $a \in \mathfrak{F}$ let $r_a = v_a^{-1}a$. Let $E_{i,a}$ denote the ordered pair (i, a), $i = 1, \ldots, l$, $a \in \mathfrak{F}$. Let E_i

mean $E_{i,1}$. Let Φ be the free group on the $E_{i,a}$ and define the homomorphism $\tau:\Phi\to\mathfrak{N}$ by $\tau(E_{i,a})=a^{-1}r_ia$.

It is possible to exhibit a construction which for each $c\in\mathfrak{F}$ gives an element R_c of Φ such that $\tau(R_c)=r_c$. Since $r_{v_br_i}=r_i$ we have $\tau(R_{v_br_i}E_i^{-1})=1$.

The main result of this paper is the next theorem.

Theorem 4. *In Theorem 2 we may replace the condition (∗) by the condition that, where $\theta:\Phi\to\mathfrak{N}$ is the homomorphism such that*

$$\theta(E_{i,a})=\chi_a(\mathfrak{r}_i^*\mathfrak{z}_i^{-1}),$$

we must have $\theta(X)=1$ for all X of the form $R_{v_br_i}E_i^{-1}$.

Finally, a simple application is given:

Theorem 5. *Let $\mathfrak{J}$ be the group of inner automorphisms of a group $\mathfrak{N}$ and let $A=\mathrm{Aut}\,\mathfrak{N}$. Let $n\geqslant 2$ be an integer and let $\xi\in A$. Assume that the coset $\xi\mathfrak{J}$ has order n, so that $\mathfrak{r}^*\in\mathfrak{N}$ exists such that $\xi^n(b)=\mathfrak{r}^{*-1}b\mathfrak{r}^*$ for all $b\in\mathfrak{N}$. Then there is an extension $\mathfrak{G}$ of $\mathfrak{N}$ by the cyclic group $\langle a\rangle$ of order n realizing the homomorphism $\langle a\rangle\to A/\mathfrak{J}$, in which $a\to\xi\mathfrak{J}$, if and only if there exists $\mathfrak{z}$ in $Z(\mathfrak{N})$ such that*

$$\xi(\mathfrak{z})\mathfrak{z}^{-1}=\xi(\mathfrak{r}^*)\mathfrak{r}^{*-1}.$$

Note. There is doubt about the statement of Theorem 4. It may be correct but the proof given is inadequate. One can prove a weaker version, obtained from Theorem 4 by replacing 'for all X of the form $R_{v_br_i}E_i^{-1}$' by 'for all X of the form $R_{a^{-1}r_ia}E_{i,a}^{-1}$'.

Notes

⟦1⟧ Turing's convention is that the product of two automorphisms α, β is given by first applying α and then β. Consequently the coset $g\mathfrak{N}$ corresponds to the coset $\hat{g}\mathfrak{J}$, where $\hat{g}(n)=g^{-1}ng$, $g\in\mathfrak{G}$, $n\in\mathfrak{N}$.

⟦2⟧ Nowadays we would perhaps say rather that X (or $\gamma\mapsto X(\gamma)$ or $\lambda\gamma.X(\gamma)$) is a homomorphism (of $\mathfrak{G}'$ in $\mathfrak{A}/\mathfrak{J}$).

⟦3⟧ For 'χ_a' read '$a\mapsto\chi_a$'. Similarly, for '$\mathfrak{a}(r)$', '$\mathfrak{w}(r)$' read '$r\mapsto\mathfrak{a}(r)$', '$r\mapsto\mathfrak{w}(r)$', respectively.

⟦4⟧ That is, $\gamma\mapsto X(\gamma)$ is a homomorphism $\mathfrak{F}/\mathfrak{N}\to\mathfrak{A}(\mathfrak{N})/\mathfrak{J}(\mathfrak{N})$, where $\mathfrak{A}(\mathfrak{N})$, $\mathfrak{J}(\mathfrak{N})$ are the groups of automorphisms, inner automorphisms of $\mathfrak{N}$.

⟦5⟧ By 'an extension $\mathfrak{G}$ of $\mathfrak{N}$ by $\mathfrak{G}'$ in which the coset α of $\mathfrak{N}$ induces the class $X(\alpha)$' is meant an extension $\mathfrak{G}$ of $\mathfrak{N}$ by $\mathfrak{G}'$ for which there is an isomorphism $\mu:\mathfrak{G}/\mathfrak{N}\to\mathfrak{F}/\mathfrak{N}$ such that, for all $g\in\mathfrak{G}$, $\hat{g}\mathfrak{J}=X(\mu(g\mathfrak{N}))$, (where $\hat{g}(n)=g^{-1}ng$ for all $n\in\mathfrak{N}$).

⟦6⟧ It is worth noting for later that the proof of necessity does not use (b) or (c).
⟦7⟧ The subscripts b', b should be interchanged.
⟦8⟧ $\mathfrak{N}$ may be identified with a normal subgroup of $\mathfrak{G}$ by the mapping $n \mapsto (1,n)$ $(n \in \mathfrak{N})$.
⟦9⟧ An additional conclusion is that there is an isomorphism $\mathfrak{F}/\mathfrak{R} \to \mathfrak{G}/\mathfrak{N}$ given by $f\mathfrak{R} \mapsto \mathfrak{w}(f)\mathfrak{N}$; this is a homomorphism since $\mathfrak{w}(\mathfrak{R}) \subset \mathfrak{N}$, it is onto by condition (b) and one-one by the result proved in the previous line that if $a \in \mathfrak{F}$ and $\mathfrak{w}(a) \in \mathfrak{N}$, then $a \in \mathfrak{R}$.
⟦10⟧ To see this, choose $\mathfrak{e}_i' \in \mathfrak{G}$ such that $\mathfrak{e}_i\mathfrak{R} = \mu(\mathfrak{e}_i'\mathfrak{N})$, where μ is the isomorphism of note ⟦5⟧. Then

$$\widehat{\mathfrak{e}_i'}\mathfrak{J} = X(\mu(\mathfrak{e}_i'\mathfrak{N})) = X(\mathfrak{e}_i\mathfrak{R}),$$

which contains $\chi_{\mathfrak{e}_i}$. Hence $\chi_{\mathfrak{e}_i}$ is $\widehat{\mathfrak{e}_i'}$ multiplied by an inner automorphism $n \mapsto n_i^{-1}nn_i$ depending on i; that is, $\chi_{\mathfrak{e}_i}(n) = n_i^{-1}\mathfrak{e}_i'^{-1}n\mathfrak{e}_i'n_i$. Without loss of generality we may replace $\mathfrak{e}_i'$ by $\mathfrak{e}_i'n_i$, so $\chi_{\mathfrak{e}_i}(n) = \mathfrak{e}_i'^{-1}n\mathfrak{e}_i'$.
⟦11⟧ Using that $\mathfrak{F}$ is free.
⟦12⟧ Any coset α of $\mathfrak{N}$ may be written $\mathfrak{w}(f)\mathfrak{N}$ where $f \in \mathfrak{F}$. Let $\alpha' = f\mathfrak{R}$. Then $\chi_f \in X(\alpha')$ and, by (a), $\chi_f = \widehat{\mathfrak{w}(f)}$. Hence $X(\alpha') = \widehat{\mathfrak{w}(f)}\mathfrak{J}$, which is the coset induced by α.
⟦13⟧ As before, X is assumed to be a homomorphism $\mathfrak{F}/\mathfrak{R} \to \mathfrak{A}/\mathfrak{J}$.
⟦14⟧ We easily see that $\mathfrak{r}_i^*\mathfrak{r}_i^{-1}$ $(=h_i$, say) lies in the centre of $\mathfrak{N}$. But $\mathfrak{z}_i = \mathfrak{r}_i^{-1}h_i\mathfrak{r}_i$, so $\mathfrak{z}_i$ is in the centre.
⟦15⟧ τ_i should be replaced by σ_i.
⟦16⟧ There is no difficulty if $\mathfrak{N}$ is Abelian. In the case when the centre of $\mathfrak{N}$ is 1, we may argue as follows. Let $a \in \mathfrak{F}$, $\mathfrak{b} \in \mathfrak{N}$. Then

$$\begin{aligned}\chi_{a^{-1}\mathfrak{r}_i a}(\mathfrak{b}) &= \chi_a(\chi_{\mathfrak{r}_i}(\chi_{a^{-1}}(\mathfrak{b}))) \\ &= \chi_a(\mathfrak{r}_i^{*-1}\chi_a^{-1}(\mathfrak{b})\mathfrak{r}_i^*) \\ &= \chi_a(\mathfrak{r}_i^*)^{-1}\mathfrak{b}\,\chi_a(\mathfrak{r}_i^*),\end{aligned}$$

so $\chi_{a^{-1}\mathfrak{r}_i a} = (\chi_a(\mathfrak{r}_i^*))^{\wedge}$. Now let $\prod a_i^{-1}\mathfrak{r}_{l_i}^{t_i}a_i = 1$ and let $\beta = \prod \chi_{a_i}(\mathfrak{r}_{l_i}^{*t_i})$. Then $\chi_1 = \hat{\beta}$, hence β lies in the centre of $\mathfrak{N}$, so $\beta = 1$. Hence (7) implies (6), if each $\mathfrak{z}_i$ is defined to be 1, and therefore the required extension exists.
⟦17⟧ Thus Φ may be regarded as the free group with generating set $\{1,2,\ldots,l\} \times \mathfrak{F}$.
⟦18⟧ The term $r_{ce_i^{-1},e_i}$ should be replaced by its inverse.
⟦19⟧ Note that $r_{a,e_i} = r_{\upsilon(a),e_i}$ and for later note that $r_{\upsilon(a),e_i} = r_{\upsilon(a)e_i}$. Thus the second equation of (12) is an analogue of the first equation of (10). The analogue of the corrected second equation of (10) is

$$R_{ce_i^{-1}} = \chi_{e_i^{-1}}(R^{-1}_{\upsilon(ce_i^{-1}),e_i}R_c)$$

and the third equation of (12) should be replaced by this.

At this point the equations $R_{(ce_i)e_i^{-1}} = R_{(ce_i^{-1})e_i} = R_c$ are easily verified.

⟦20⟧ Actually, what the equations (12) imply is that when $k = e_i$ or e_i^{-1}

$$R_{ck} = R_{\upsilon(c)k}\chi_k(R^{-1}_{\upsilon(c)}R_c) \tag{13'}$$

I see no reason why $R_{\upsilon(c)} = 1$. (For example, (13′) with $k = e_i$ follows from the second equation of (12) and the equation arising from it when c is replaced by υ_c (noting that $\upsilon_{\upsilon_c} = \upsilon_c$).)

However, we do have $r_{\upsilon_c} = 1$ so $R_{\upsilon_c} \in \mathbf{P}$, that is, $R_{\upsilon_c} = 1$ in $\Phi/\mathbf{P}$. If we regard R_{υ_c} as included in (15), then we may regard (13) as valid in the author's subsequent discussion. In any case, (13) and (13′) are the same if $c \in \mathfrak{R}$.

⟦21⟧ Precisely, $\tau(R_c) = r_c$. This may be proved by induction using (12). (If $\psi_a : \mathfrak{R} \to \mathfrak{R}$ is defined by $\psi_a(r) = a^{-1}ra$, then we have $\psi_a \circ \tau = \tau \circ \chi_a$.)

⟦22⟧ Note also that χ_c applied to an expression of the form (9) also has the form (9). (Write $A = ac$, $B = bc$; then $ab^{-1} = AB^{-1}$.)

⟦23⟧ Without assuming $R_{\upsilon_c} = 1$ for all c in $\mathfrak{F}$ we have

$$R_{ax} = R_{\upsilon_a x}\chi_x(R^{-1}_{\upsilon_a}R_a) \quad \text{for all } a, x \text{ in } \mathfrak{F}. \tag{$*$}$$

To see this, first note that it is true when x is e_i or e_i^{-1}. Assume (∗) is true for x and true for y. Then

$$R_{ax \cdot y} = R_{\upsilon(ax)y}\chi_y(R^{-1}_{\upsilon(ax)}R_{ax}).$$

Now $\upsilon(\upsilon_a x) = \upsilon(ax)$, so

$$R_{\upsilon_a x \cdot y} = R_{\upsilon(ax)y}\chi_y(R^{-1}_{\upsilon(ax)}R_{\upsilon_a x}).$$

From these two equations we get

$$R_{ax \cdot y} = R_{\upsilon_a x \cdot y}\chi_y(R^{-1}_{\upsilon_a x}) \cdot \chi_y(R_{ax})$$

and since $R^{-1}_{\upsilon_a x}R_{ax} = \chi_x(R^{-1}_{\upsilon_a}R_a)$ we get that (∗) is true for xy.

⟦24⟧ (a) A weaker theorem is as follows.

Theorem 3′. *$\mathfrak{R}$ is isomorphic to $\Phi/\mathbf{P}'$, where $\mathbf{P}'$ is the least normal subgroup of Φ containing all elements of the form* (9) *and all elements of the form $\chi_b(R_{a^{-1}r_i a}E^{-1}_{i,a})$. (We do not need to include R_{υ_c} (see note ⟦20⟧).)*

To prove this, first note that each χ_c leaves $\mathbf{P}'$ invariant. Working in $\Phi/\mathbf{P}'$ we have the following.

(1) $\chi_b(R_\alpha) = R_{b^{-1}\alpha b}$ where α is $a^{-1}r_i a$. (For $\chi_b(R_\alpha) = \chi_b(E_{i,a}) = E_{i,ab} = R_{b^{-1}a^{-1}r_i ab}$.)

(2) For $r \in \mathfrak{R}$, $x \in \mathfrak{F}$ we have $R_{rx} = R_x \chi_x(R_r)$. (See note ⟦20⟧.)

(3) For $r \in \mathfrak{R}$, $\chi_r(R_{r^{-1}}) = R_r^{-1}$.

(4) If $\alpha = a^{-1} r_i a$, then $R_{\alpha^{-1}} = R_\alpha^{-1}$. ($R_{\alpha^{-1}}^{-1} = \chi_{\alpha^{-1}}(R_\alpha) = R_\alpha$ by (1).)

(5) If $\alpha = a^{-1} r_i a$, $\beta = b^{-1} r_j b$, then $\chi_\beta(R_\alpha) = R_{\beta^{-1}\alpha\beta} = R_\beta^{-1} R_\alpha R_\beta$. (This follows from (9).)

(6) The conclusion of (5) remains true if β is replaced by β^{-1}. ($R_\alpha = R_\beta \chi_\beta(R_\alpha) R_{\beta^{-1}} = \chi_\beta(R_\beta R_\alpha R_{\beta^{-1}})$. Hence $\chi_{\beta^{-1}}(R_\alpha) = R_{\beta^{-1}}^{-1} R_\alpha R_{\beta^{-1}}$.)

(7) The conclusions of (5), (6) remain true if α is replaced by α^{-1}. (This follows by taking the inverse of each side.)

(8) If $\alpha_1, \ldots, \alpha_k$ have the form $a^{-1} r_i^\varepsilon a$, $\varepsilon = \pm 1$, then $R_{\alpha_1 \alpha_2 \cdots \alpha_k} = R_{\alpha_1} R_{\alpha_2} \cdots R_{\alpha_k}$. Hence $R_{rs} = R_r R_s$, $r \in R$, $s \in R$. (By induction on k, using that $R_{\alpha_1 \cdots \alpha_{k+1}} = R_{\alpha_{k+1}} \chi_{\alpha_{k+1}}(R_{\alpha_1 \cdots \alpha_k})$.)

By (8) we have a homomorphism $\alpha : \mathfrak{R} \to \Phi/\mathbf{P}'$ given by $r \to R_r \mathbf{P}'$. Now τ induces a homomorphism $\tau' : \Phi/\mathbf{P}' \to \mathfrak{R}$ and $\tau'(\alpha(r)) = \tau'(R_r \mathbf{P}') = \tau(R_r) = r_r = r$. Also $\alpha\tau'(E_{i,a}\mathbf{P}') = \alpha(a^{-1} r_i a) = R_{a^{-1} r_i a} \mathbf{P}' = E_{i,a}\mathbf{P}'$. Hence $\mathfrak{R} \cong \Phi/\mathbf{P}'$.

(b) In the editor's opinion, the author is led into difficulties by not adhering exactly to the conventions of Reidemeister (1926). Had he done so, he would have considered right cosets Ra and written $a = r_a \upsilon_a$ and $\upsilon_a e_i = r_{a,e_i} \upsilon_{ae_i}$. (As an alternative to Reidemeister's paper, the reader may consult Magnus, Karras and Solitar (1966), Theorem 2.8.)

We now establish a presentation for $\mathfrak{R}$ using this approach.

Let $\mathfrak{F}$ be the free group on symbols e_i, $i \in I$, and let R be the subgroup generated by the elements g_j, $j \in J'$. R need not be normal. Let representatives $\bar{a}$ of the right cosets Ra be taken, where the representative of R is 1. Let P be the set of representatives. If p is a representative, let $s_{p,e_i} = pe_i \overline{pe_i}^{-1}$. Let S_{p,e_i} denote the ordered pair (p, e_i). The well-known Reidemeister rewriting gives for each word A in the e_i which represents an element of R a word $t(A)$ in the S_{p,e_i} such that the corresponding word in the s_{p,e_i} equals A in R. A special case of Reidemeister's theorem is that a presentation for R is

$$\langle S_{p,e_i}\ (p \in P, i \in I) : S_{p,e_i} = t(s_{p,e_i})\ (p \in P, i \in I)\rangle$$

Corollary. *There is a subset J of J', which is finite if I and P are finite, such that if Φ is the free group on symbols G_j, $j \in J$, then a presentation for R is*

$$\langle G_j\ (j \in J) : G_j = \upsilon t(g_j)\ (j \in J)\rangle,$$

where υ is defined as follows: for each s_{p,e_i} choose a word $w(g_j)$ equal to it in R and put $\upsilon(S_{p,e_i}) = w(G_j)$.

Proof of the Corollary. Let B be the free group on the S_{p,e_i}. By the known properties of t, t can be regarded as a homomorphism $R \to B$. We have a homomorphism $\varphi: B \to R$ given by $S_{p,e_i} \to s_{p,e_i}$ and $\varphi t(r) = r$ $(r \in R)$. Let the generators appearing in the words $w(g_j)$ above be $g_j, j \in J$. This defines J; let Φ be as in the statement of the corollary. There is a homomorphism $\tau: \Phi \to R$ in which $G_j \to g_j$. Also, υ extends to a homomorphism $B \to \Phi$ and we have $\tau\upsilon = \varphi$. Since $\tau\upsilon t = \varphi t = 1$, τ is onto. Let N be the normal subgroup of Φ generated by $G_j^{-1}\upsilon t\tau(G_j)$ $(= n_j$, say) where j runs through J. Then $N \subset \operatorname{Ker} \tau$.

In fact $N = \operatorname{Ker} \tau$, since if $\xi \in \operatorname{Ker} \tau$, then $\xi = w(G_j)$, say, and $\tau(\xi) = 1$, so $1 = \upsilon t\tau(\xi) = w(\upsilon t\tau(G_j)) = w(G_j n_j) = w(G_j)n = \xi n$, for some $n \in N$.

Hence $R \cong \Phi/N$ and the corollary follows.

(c) We have not been able to verify Theorem 3 (with or without the relations $R_{\upsilon_c} = 1$ (see note [[20]])). By Theorem 3′, it would be sufficient to prove that

$$R_{a^{-1}r_i a} E_{i,a}^{-1} \qquad (**)$$

belongs to $\bar{\mathbf{P}}$.

If we consider the application (see Theorem 4), Theorem 3′ is not as good as Theorem 3 because the number of expressions (**) may be infinite (if G' is a finite group and the free group $\mathfrak{F}$ is finitely generated).

In contrast, the presentation of R in (b) (the g_j being the $a^{-1}r_i a$) is finite so it does not have this defect.

[[25]] If we replace Theorem 3 by Theorem 3′ (see note [[24]]), we obtain the weaker Theorem 4′.

Theorem 4′. *The same statement as Theorem* 4 *but with 'for all X of form* (15)*' replaced by 'for all X of the form* $R_{a^{-1}r_i a} E_{i,a}^{-1}$*'.*

[[26]] It is sufficient to prove this when X is a generator $E_{i,b}$, that is, to prove $\theta(\chi_a^{(2)}(E_{i,b})) = \chi_a^{(1)}(\theta(E_{i,b}))$. But this is easily checked.

[[27]] Once this is established, θ may be viewed as a homomorphism from $\mathfrak{R}$ to $\mathfrak{N}$. We want to prove that (7) implies (6). But (7) implies that $\prod \theta(a_i^{-1} r_{l_i}^{\tau_i} a_i) = 1$, that is, $\prod \theta(E_{l_i,a_i}^{\tau_i}) = 1$ which by (18) is equivalent to (6).

[[28]] $\chi_{ab^{-1}r_j}(\mathfrak{r}_i) = \chi_{r_j}(\chi_{ab^{-1}}(\mathfrak{r}_i)) = \mathfrak{r}_j^{*-1}\chi_{ab^{-1}}(\mathfrak{r}_i)\mathfrak{r}_j^{*} = \mathfrak{r}_j^{-1}\chi_{ab^{-1}}(\mathfrak{r}_i)\mathfrak{r}_j$, by using that $\mathfrak{r}_j^* = \mathfrak{z}_j \mathfrak{r}_j$; $(\mathfrak{r}_j = \mathfrak{r}_j^* \mathfrak{z}_j^{-1} = \mathfrak{z}_j^{-1}\mathfrak{r}_j^*)$.

[[29]] To prove the converse, assume (7) implies (6). We want to prove that $\theta(R_{\upsilon_b r_i} E_i^{-1}) = 1$. Now $R_{\upsilon_b r_i} E_i^{-1}$ is a word in the E's, say $w(E_{i,a})$. It equals 1 in F so $w(a^{-1}r_i a) = 1$. Hence $w(\chi_a(r_i)) = 1$. Thus $\theta(w(E_{i,a})) = w(\theta(E_{i,a})) = 1$, as required.

[[30]] Proof of Theorem 5: Let $\mathfrak{F}$, $\mathfrak{R}$ be as at the beginning of §3. Then we have a homomorphism $X: \mathfrak{F}/\mathfrak{R} \to \mathfrak{A}/\mathfrak{Z}$ in which $a\mathfrak{R} \to \xi\mathfrak{Z}$. We also have a homomorphism $\chi: \mathfrak{F} \to \mathfrak{A}$ in which $a \to \xi$. For $b \in \mathfrak{N}$, $\chi_{a^n}(b) = \xi^n(b) = \mathfrak{r}^{*-1} b \mathfrak{r}^*$. Thus the hypothesis of Theorem 2 and hence of Theorem 4 holds. If the extension exists we have by Theorem 4 a central element ζ such that, where $\theta(E_{1,a}) = \chi_a(\mathfrak{r}^*\zeta^{-1})$, we have $\theta(Q_a Q^{-1}) = 1$; thus (21) holds. Conversely, if we have a central element ζ satisfying (21) and we put $\theta(E_{1,a}) = \chi_a(\mathfrak{r}^*\zeta^{-1})$, then we find $\theta(Q_{a^p}) = \xi^p(\mathfrak{r}^*\zeta^{-1})$. Hence for example

$$\theta(Q_{a^2}) = \xi^2(\mathfrak{r}^*\zeta^{-1}) = \xi(\mathfrak{r}^*\zeta^{-1}) = \mathfrak{r}^*\zeta^{-1} = \theta(Q).$$

Thus θ maps the elements (19) to 1. Hence the extension exists.

⟦31⟧ A^w should be A^n.

⟦32⟧ '$A, A^2, \ldots$' means '$A, A^2, \ldots, A^n$'.

⟦33⟧ As we have seen, the author deduces Theorem 5 as a special case of Theorem 4, which involves the machinery of §2. In fact Theorem 5 can be deduced directly from Theorem 2.

As shown in note ⟦30⟧, the hypothesis of Theorem 2 holds. Assume (21) and let $\prod a^{-m_i} r^{s_i} a^{m_i} = 1$ (where r is a^n). Then $\prod r^{s_i} = 1$ so $\sum s_i = 0$. Hence $\prod \chi_{a^{m_i}}((\mathfrak{r}^*\zeta^{-1})^{s_i}) = \prod \xi^m((\mathfrak{r}^*\zeta^{-1})^{s_i}) = \prod (\mathfrak{r}^*\zeta^{-1})^{s_i}$ (by repeated use of (21)) $= 1$, since $\sum s_i = 0$. By Theorem 2 the extension exists.

Conversely, if the extension exists, then since $a^{-1}rar^{-1} = 1$ we have $\chi_a(\mathfrak{r}^*\zeta^{-1})(\mathfrak{r}^*\zeta^{-1})^{-1} = 1$, by Theorem 2. Hence $\xi(\mathfrak{r}^*\zeta^{-1}) = \mathfrak{r}^*\zeta^{-1}$.

1943 *A Method for the Calculation of the Zeta-function*

Summary

The author first gives an integral representation of the zeta-function, as follows. Let $I(s) = \int_L h(z)\,dz$, where

$$h(z) = \frac{\exp(i\pi z^2) z^{-s}}{\exp(i\pi z) - \exp(-i\pi z)},$$

$\arg z = 0$ on the positive real axis and L is the line $z = \frac{1}{2} + \lambda\varepsilon$, with λ real and running from $-\infty$ to ∞ and $\varepsilon = \exp(\frac{1}{4}\pi i)$. Then

$$\zeta(s)\Gamma(\tfrac{1}{2}s)\pi^{-s/2} = -\Gamma(\tfrac{1}{2}s)\pi^{-s/2}\overline{I(\bar{s})} - \Gamma(\tfrac{1}{2} - \tfrac{1}{2}s)\pi^{s/2-1/2} I(1-s).$$

On the critical line $\sigma = \frac{1}{2}$ $(s = \sigma + it)$, we obtain

$$\zeta(s)\Gamma(\tfrac{1}{2}s)\pi^{-s/2} = -2\Re(\Gamma(\tfrac{1}{2}s)\pi^{-s/2} I(s)).$$

Most of the paper is devoted to giving an estimate for $I(s)$ in terms of positive real parameters κ, μ. A major step to this end is the result

$$I(s) = \sum_{k=-\infty}^{\infty} \left(\frac{\varepsilon}{\kappa}\right) h(p_k) - \sum_{r=1}^{\infty} \frac{r^{-s}}{1 - \exp(2\pi\kappa\varepsilon(r-\mu))} + R \qquad (1)$$

where $p_k = \mu + \varepsilon k/\kappa$ and $|R|$ is less than a very complicated expression (namely 5.18)).

[In (5.18), $\tau=t/2\pi$ while ζ_0, ζ_0' and y are functions of ϱ, where $\varrho=\kappa\tau^{-1/2}$, and y_0, z_0 are defined by $\zeta_0=z_0\tau^{-1/2}$, $z_0=x_0+\mathrm{i}y_0$ (and similarly for y_0', z_0').]

Turing now wants to restrict the choice of the parameters with the aims of simplifying and making small the bound for $|R|$ and replacing the infinite sums in (1) by finite sums, with a small error. As a first step he chooses μ to be close to a specific function μ_0 of τ, κ; if $\kappa \geqslant 2$ we may choose any μ in the interval $(\mu_0-\frac{1}{2}, \mu_0+\frac{1}{2})$.

Then, after some discussion of the general case, he deals with the special cases when $\varrho=\kappa\tau^{-1/2}\to 0$ and when $\varrho\to\infty$. In the first case $\mu_0\to\tau^{1/2}$ and in the second case $\mu_0\to\kappa/2\sqrt{2}$.

It turns out that the series $\sum_{k=-\infty}^{\infty}(\varepsilon/\kappa)h(p_k)$ can be approximated by $\sum_{k=-K}^{K}\varepsilon/\kappa h(p_k)$ with error of the order of magnitude $\exp(-\frac{1}{2}\pi(2K+1))$ if $\varrho\to 0$; $\exp(-\frac{1}{3}\pi(2K+1))$ if $\varrho\to\infty$.

The other infinite series in (1) is easier to approximate by a finite sum; for if r is not too close to μ, the rth term of the series is approximately equal to r^{-s} if $r<\mu$; 0 if $r>\mu$.

One particular specialization of the parameter in the case $\varrho\to 0$ leads to the theorem below. No explicit specialization is given in the case $\varrho\to\infty$. The theorem evidently gives a practical method of calculating $\zeta(\frac{1}{2}+\mathrm{i}t)$.

Theorem. *There exists a function $I(s)$, $s=\sigma+\mathrm{i}t$, which satisfies the equation*

$$\zeta(s)\Gamma(\tfrac{1}{2}s)\pi^{-s/2} = -2\Re(\Gamma(\tfrac{1}{2}s)\pi^{-s/2}I(s))$$

for s on the critical line and which satisfies the following condition. Let $\sigma=\frac{1}{2}$, $t>350$. Let $\kappa=1.6\sqrt{2}$, $\tau=t/2\pi$, $\varepsilon=\exp(\frac{1}{4}\mathrm{i}\pi)$, $\mu=\tau^{1/2}$ and $p_k=\mu+\varepsilon k/\kappa$. Then

$$I(s)=\varepsilon\kappa^{-1}(h(p_{-1})+h(p_0)+h(p_1))-\sum_{r=1}^{m-1} r^{-s}-m^{-s}F_m-(m+1)^{-s}F_{m+1}+E,$$

where h is the function defined previously, $F_r=(1-\exp(2\pi\kappa\varepsilon(r-\mu)))^{-1}$, $m=[\mu]$ and $|E|\leqslant 0.0044\,\tau^{-1/4}$; (if for example $t<1000$, then $m-1\leqslant 11$).

If μ is an integer, so $m=\mu$, then $\varepsilon h(p_0)/\kappa=\varepsilon h(\mu)/\kappa$ and $m^{-s}F_m$ are both infinite. But, writing $F_r(\mu)$ for F_r,

$$\lim_{d\to 0}\left(\frac{\varepsilon}{\kappa}h(m+d)-m^{-s}F_m(m+d)\right)$$

exists so the two terms may be replaced by this limit.

The condition $\mu=\tau^{1/2}$ may be replaced by $\tau^{1/2}-\frac{1}{4}\leqslant\mu\leqslant\tau^{1/2}+\frac{1}{4}$ at the expense of increasing the number 0.0044. This would allow us to take μ to be either an integer or half an odd integer in any calculation.

Notes

⟦1⟧ The required change of variables is $t = z - 1$.
⟦2⟧ To evaluate $\int \exp(\mathrm{i}\pi z^2 + 2\pi \mathrm{i}uz - \mathrm{i}\pi z)\,\mathrm{d}z$, complete the square to obtain $\exp(-\mathrm{i}\pi(u-\frac{1}{2})^2)\int \exp(\mathrm{i}\pi z^2)\,\mathrm{d}z$. Then put $z = (1+\mathrm{i})t/(2\pi)^{1/2}$ and use that $\int_{-\infty}^{\infty} \exp(-\frac{1}{2}t^2)\,\mathrm{d}t = (2\pi)^{1/2}$.
⟦3⟧ Appropriate values are $k + \frac{1}{2}$, $k = 1, 2, 3, \ldots$.
If $z = k + \xi$, where ξ is small but $|\xi| \geqslant e > 0$, then

$$|1 - \exp(-2\pi \mathrm{i}z)| = |1 - \exp(-2\pi \mathrm{i}\xi)| \simeq |2\pi \mathrm{i}\xi| \geqslant 2\pi e.$$

⟦4⟧ (i) $1 + \exp(\mathrm{i}\pi s) = 0$ for the *odd* integers.
(ii) The integral converges uniformly in the region $|s| < R$, for any R; for if $z = \lambda \mathrm{i}$, where λ is large and positive, then $|z^{-s}| < K\lambda^L$, where K and L are constants, $|1 - \exp(-2\pi \mathrm{i}z)| \sim \exp(2\pi\lambda)$ and $K\lambda^L/\exp(2\pi\lambda) < K'\exp(-\frac{1}{2}\lambda)$. Hence the integral is analytic in s for all s.
⟦5⟧ In the integral over L, first assume z is positive real and make the transformation $t = -2\pi \mathrm{i}uz$. L becomes $-C$, that is, C with the arrows reversed, where C is the contour in Hankel's expression of $\Gamma(z)$. Hence the integral equals $-2(2\pi)^{s-1}\sin(s\pi)\Gamma(1-s)\exp(\frac{1}{2}\mathrm{i}s\pi)z^{s-1}$. We may now remove the restriction on z.
⟦6⟧ That is, the lines $x = \frac{1}{2}$, $x = -\frac{1}{2}$.
⟦7⟧ $\exp(\mathrm{i}\pi u^2)$ should be $\exp(-\mathrm{i}\pi u^2)$.
⟦8⟧ In the integral on the previous line it is immaterial whether we require $-\frac{3}{2}\pi < \arg u < \frac{1}{2}\pi$ or $-\frac{1}{2}\pi < \arg u < \frac{3}{2}\pi$ but for the change to $0 \nwarrow 1$ we require the latter.
⟦9⟧ In general,

$$\overline{\int_{0 \nearrow 1} f(z)\,\mathrm{d}z} = -\int_{0 \nwarrow 1} \overline{f(\bar{z})}\,\mathrm{d}z.$$

⟦10⟧ In the formula $2^{2z-1}\Gamma(z)\Gamma(z+\frac{1}{2}) = \pi^{1/2}\Gamma(2z)$, take $z = \frac{1}{2} - \frac{1}{2}s$. In the formula $\Gamma(z)\Gamma(1-z) = \pi/\sin(\pi z)$, take $z = \frac{1}{2}s$. The equation now follows.
⟦11⟧ On the right-hand side, replace $\Gamma(\frac{1}{2}s)$ by $-\Gamma(\frac{1}{2}s)$.
⟦12⟧ Replace $I(\bar{s})$ by $I(s)$. (If in (2.3) we replace s by $\bar{s}$ and then conjugate the equation, $\overline{I(\bar{s})}$ becomes $I(s)$, $I(1-s)$ becomes $\overline{I(1-\bar{s})}$ and all the other terms are unchanged. On the critical line, $\overline{I(1-\bar{s})} = \overline{I(s)}$.)
⟦13⟧ This should be $-\frac{1}{2}\pi - \alpha > \arg z > -\pi + \alpha$.
⟦14⟧ It is important that P here should be replaced by J.

⟦15⟧ For any real λ and for $z \in J \cup J'$,

$$|2\pi\kappa\varepsilon(z-\mu-\lambda\varepsilon)| > 2\pi\kappa \cdot \tfrac{1}{4}\kappa^{-1} = \tfrac{1}{2}\pi.$$

Choose λ to make the imaginary part vanish. Then $|A| > \frac{1}{2}\pi$, where $A = \Re 2\pi\kappa\varepsilon(z-\mu) = 2\pi\kappa(x-\mu-y)/\sqrt{2}$. Now on the left of P, $A<0$ so $A < -\frac{1}{2}\pi$ and the expression is less than $(1-\exp-\frac{1}{2}\pi)^{-1} < 1.27$. There is a similar argument for the right of P.

⟦16⟧ Define $R_0 = \int_{J'} H - \int_J H$, where $H = h(z)e^{\xi}/(1-e^{\xi})$ and $\xi = 2\pi\kappa\varepsilon(z-\mu)\mathrm{sg}(z)$. Write $\int_{J+J'} g = \sum_1 + \sum_2$ and $V = \int_J h - \int_{m \nearrow m+1} h$. Then $V = \sum r^{-s}$ summed over r between P and J, and

$$I(s) = -\sum_{r=1}^{m} r^{-s} + \sum_1 + \sum_2 - V + R_0.$$

The equation for $I(s)$ in the text follows from this.

⟦17⟧ The integrand of (3.2) equals $1.27|h(z)\exp\xi|$, where ξ is as in note ⟦16⟧.

⟦18⟧ We have $R_1 \underset{\text{def}}{=} \sum_{r=1}^{\infty} D_r r^{-s}$, where $D_r = (1-\exp(2\pi\kappa\varepsilon(r-\mu)))^{-1} - \theta_r$. Writing $\eta = 2\pi\kappa\varepsilon(r-\mu)$, the values of D_r are

$$\begin{cases} \dfrac{\exp\eta}{1-\exp\eta} & \text{for } r \text{ left of } J', \\ 0 & \text{for } r \text{ between } J \text{ and } J', \\ \dfrac{\exp(-\eta)}{\exp(-\eta)-1} & \text{otherwise (that is, right of } J). \end{cases}$$

We have seen previously that if $d(z,P) > \frac{1}{4}\kappa^{-1}$, then

$$|1-\exp[2\pi\kappa\varepsilon\mathrm{sg}(z)(z-\mu)]|^{-1} < 1.27,$$

so if r is left of J', then $|1-\exp\eta|^{-1} < 1.27$, where $\eta = 2\pi\kappa\varepsilon(r-\mu)$, while if r is right of J, then $|1-\exp(-\eta)|^{-1} < 1.27$; here we have to assume that $d(m+1,P) > \frac{1}{4}\kappa^{-1}$. Hence for r not between J' and J, $|D_r| < 1.27\exp[-\sqrt{2}\pi\kappa|r-\mu|]$. This gives the bound for $|R_1|$ in the text.

⟦19⟧ This follows by integrating from 0 to l.

⟦20⟧ Here we use that $\int_{-\infty}^{\infty} \exp(-\frac{1}{2}t^2)\,dt = \sqrt{2\pi}$.

⟦21⟧ Assuming that $t>0$ we find that one of z_0, z_1 is in the first quadrant and the other is in the second quadrant; one of z_0', z_1' is in the third quadrant and the other is in the fourth. Requiring z_0, z_0' to be in the right half-plane implies that z_0, z_1, z_1', z_0' are in the first, second, third and fourth quadrants, respectively.

Also $z_1' = -z_0$, $z_0' = -z_1$, $0 \leqslant \arg z_0 \leqslant \frac{1}{4}\pi$ and the angle between the vectors z_0, z_1 is bisected by the vector i.

⟦22⟧ Turing is assuming $t>0$.

⟦23⟧ (i) The last statement should be $\frac{1}{4}\pi>\arg\zeta_0>0$.

(ii) *From this point onwards there is a change of notation as follows.* (This is confirmed later at the top of page 192.)

Present notation	*New notation*
$z_0,\ z_1,\ z_0',\ z_1',$	$z_0',\ z_1',\ z_0,\ z_1,$

and similarly for ζ_i, ζ_i'. From now on we shall use the new notation in these notes unless otherwise specified.

⟦24⟧ In Fig. 2, b is in the fourth quadrant. Hence in the old notation we have $b=z_0'-\frac{1}{2}y_0'(1+\mathrm{i})$.

⟦25⟧ In proving that the contribution to the remainder R_0 from that part of J_3 in the lower half-plane tends to zero, we need that, in the old notation, $|y_0'|$ should not be not too large compared with κ. For this it is enough to note that the modulus of the imaginary part of $z_0'+z_1'$ is greater than that of z_0', that is, $\kappa>|y_0'|$.

⟦26⟧ In the old notation b', which is in the first quadrant, equals $z_0-\frac{1}{2}y_0(1+\mathrm{i})$. Similarly the z_0' of Fig. 2 is z_0 in the original notation.

⟦27⟧ These facts follow once we notice that, since $\arg z_0\leqslant\frac{1}{4}\pi$, $|b'|\geqslant\frac{1}{2}|z_0|$ (old notation).

⟦28⟧ First, ζ should be ζ_0. Now $\tau^{1/2}\Re(\varepsilon\zeta_0)=\Re(\varepsilon z_0)=(x_0-y_0)/\sqrt{2}$ which is the perpendicular distance of 0 from the line J_1, so the result follows.

⟦29⟧ This is because $|y|\geqslant|\frac{1}{2}y_0|$ on J_1 and J_2.

⟦30⟧ An alternative approach to proving the inequalities (5.2), (5.3) is as follows.

Let l denote arclength from z_0 on J_1. Then on J_1 we have $z-z_0=\pm l\varepsilon$ and $\varphi'(z)\mathrm{d}z/\mathrm{d}l = 2\pi\mathrm{i}(z-z_0)(z-z_1)(\pm\varepsilon)/z = 2\pi\mathrm{i}l\varepsilon^2(z-z_1)/z = -2\pi l(z-z_1)/z$. Now suppose we have, for some d, $\Re(z-z_1)/z\geqslant d>0$. Then $\Re\varphi'(z)\mathrm{d}z/\mathrm{d}l\leqslant-2\pi ld$. By (a) of §4,

$$\int_{J_1}\exp\Re\varphi(z)\,|\mathrm{d}z|\leqslant(4d)^{-1/2}\exp\Re\varphi(z_0) \tag{5.2'}$$

and, since $|b-z_0|=|y_0|/\sqrt{2}$,

$$\Re\varphi(b)\leqslant-\tfrac{1}{2}(2\pi d)\tfrac{1}{2}y_0^2+\Re\varphi(z_0). \tag{5.3'}$$

We now show that we may take $d=\frac{3}{4}$.

On J_1, $x_0+y_0=-x_1-y_1$ and $x_0-y_0=x-y$ (new notation). Hence $y_1/x_1=y_0/x_0=(-x_1-y_1-x+y)/(x-y-x_1-y_1)$ and $(y-x)(x_1-y_1)=(x_1+y_1)^2=|z_1|^2+2x_1y_1$.

Now $\Re z_1/z=(x_1x+yy_1)/|z|^2$ and $|zz_1|\geqslant\Im zz_1=yx_1+xy_1$.

Thus $|z|^2\Re z_1/z = yx_1 + xy_1 - |z_1|^2 - 2x_1y_1 \leqslant |zz_1| - |z_1|^2$. Hence $\Re z_1/z \leqslant |z_1/z| - |z_1/z|^2 \leqslant \frac{1}{4}$ so that $\Re((z-z_1)/z) \geqslant \frac{3}{4}$.

⟦31⟧ In the old notation: $z_0'z_1' = -\tau$, so we may write $z_0' = a\exp(-\mathrm{i}\alpha)$ and $-z_1' = b\exp(\mathrm{i}\alpha)$, where $ab=\tau$ and $0<\alpha<\frac{1}{4}\pi$. Now $z_0'+z_1' = \kappa\bar{\varepsilon}$ so $a\cos\alpha - b\cos\alpha - a\sin\alpha - b\sin\alpha = 0$. Hence $a^2\cos^2\alpha/\tau = \cos^2\alpha(\cos\alpha + \sin\alpha)/(\cos\alpha - \sin\alpha) > 1$, that is, $x_0'^2 > \tau$.

⟦32⟧ In the new notation, $\tau < x_0^2 < x^2 < x^2 + \frac{1}{4}y_0^2$ so we have $\Re\varphi'(z) < -2\pi y_0 - \sqrt{2}\pi\kappa$, which equals $-2\pi y_0'$, since $z_0 - z_0' = -z_1' - z_0' = \kappa\bar{\varepsilon}$ and hence $y_0 - y_0' = \kappa/\sqrt{2}$.

⟦33⟧ The second occurrence of $\Re\varphi(z)$ should be $\Re\varphi(b)$.

⟦34⟧ This follows from the product formula $\sin(\pi z) = \pi z \prod(1 - z^2/n^2)$ and the fact that $y \leqslant x$ on J_1' and J_2'.

⟦35⟧ Another approach to (5.7), (5.8) is as follows; (in (5.8), b should be b'). In the old notation: $x_0 + x_1 + y_0 + y_1 = 0$. The line J_1' has equation $y - y_0 = x - x_0$. Also since z_0z_1 is real, $x_0y_1 + x_1y_0 = 0$. Thus we have the same three equations as in note ⟦30⟧ and we therefore get

$$|z|^2\frac{\Re z_1}{z} = yx_1 + xy_1 - |z_1|^2 - 2x_1y_1.$$

Writing $x = x_0 + \lambda$, $y = y_0 + \lambda$, $|z|^2\Re z_1/z = (x_1+y_1)(\lambda - x_1 - y_1)$. When $z=b$, we have $x = x_0 - \frac{1}{2}y_0$ so $\lambda = -\frac{1}{2}y_0$,

$$\lambda - x_1 - y_1 = \lambda + x_0 + y_0 \geqslant x_0 + \tfrac{1}{2}y_0 > 0$$

and $x_1 + y_1 = -x_0 - y_0 < 0$. Hence $\Re z_1/z < 0$ and $\Re(z - z_1)/z > 1$.

As before, we deduce that, in the new notation,

$$\int_{J'} \exp\Re\varphi(z)\,|\mathrm{d}z| \leqslant \tfrac{1}{2}\exp\Re\varphi(z_0') \tag{5.7'}$$

and also

$$\Re\varphi(b') \leqslant -\tfrac{1}{2}\pi y_0'^2 + \Re\varphi(z_0'). \tag{5.8'}$$

⟦36⟧ For z on J_2' we have $z = b' - l\exp(\mathrm{i}\theta)$, where $\theta = \arg z$. Hence $\mathrm{d}z/\mathrm{d}l = -z/|z|$.

⟦37⟧ Write $b' = x' + \mathrm{i}y'$ and let $F(z) = \Re(-2\pi\mathrm{i}z(z - \tau/z)/|z|)$. Now $z/|z| = b'/|b'|$ and $F(z) = 4\pi xy/|z| \leqslant 4\pi x'y'/|b| = F(b)$.

⟦38⟧ An alternative approach to (5.10) is as follows. Using the old notation, put $E = -2\pi\mathrm{i}(b' - z_0)(b' - z_1)/|b'|$. Now $x_1 + y_1 = -x_0 - y_0$, $y_0x_1 + x_0y_1 = 0$ and $b' = z_0 - \frac{1}{2}y_0(1+\mathrm{i})$. Hence $\Re(E) = -\pi y_0(1+2s)(s^2 - s + \frac{1}{2})^{-1/2}$, where $s = x_0/y_0$. Since $s>0$ we have $\Re(E) \leqslant -\pi y_0\sqrt{2}$ or in the new notation

$$\Re(E) \leqslant -\sqrt{2}\pi y_0'. \tag{5.10'}$$

⟦39⟧ J_5' could be replaced by J_2' but, as we shall see on page 191, J_5' is all we need.
⟦40⟧ On J_2' we have $\Re\varphi(z) \leqslant -\alpha l + \Re\varphi(b')$, where $\alpha = \sqrt{2}\pi y_0' \sin\gamma$. In particular, take $z = \frac{2}{3}b'$. Equation (5.12) follows, recalling that $|b'| \geqslant \frac{1}{2}|z_0'|$ (by note ⟦27⟧).
⟦41⟧ $J_6' \subset J_2'$ and (5.10′) holds on J_2'. On J_3', $z = r\exp(\mathrm{i}(\alpha - \theta))$, where $0 \leqslant \theta \leqslant 2\alpha + \frac{1}{2}\pi$ and $c = r\exp(\mathrm{i}\alpha)$. Hence $\mathrm{d}z/\mathrm{d}l = -z\mathrm{i}/r$. Using that $x_0 + x_1 + y_0 + y_1 = 0$, $x_0 y_1 + x_1 y_0 = 0$ in the old notation we find $\Re\varphi'(z)\mathrm{d}z/\mathrm{d}l = 2\pi[(x_0 - y_0)(x^2 - y^2) + 2x_0 y_0(x+y) - (x_0^2 + y_0^2)(x_0 + y_0)]/r(x_0 - y_0)$. Now $r = |c| \leqslant \frac{1}{3}|z_0|$, $x^2 - y^2 \leqslant r^2$ and $x + y \leqslant \sqrt{2}r$, so $\Re\varphi'(z)\mathrm{d}z/\mathrm{d}l \leqslant (x_0^2 + y_0^2)2\pi \cdot (-8(x_0 - y_0) - (18 - 6\sqrt{2})y_0)/9r(x_0 - y_0) \leqslant -\frac{16}{3}\pi|z_0|$ (or $-\frac{16}{3}\pi|z_0'|$ in the new notation).
⟦42⟧ On J_1', $y \geqslant \Im b' = \frac{1}{2}y_0'$ and $|z| \geqslant |b'| \geqslant \frac{1}{2}|z_0'| = \tau^{1/2}|\frac{1}{2}\zeta_0'|$ (see note ⟦27⟧).
⟦43⟧ On J_5', $z = \lambda b'$, $\frac{2}{3} \leqslant \lambda \leqslant 1$, so $|z| \geqslant \frac{1}{3}|z_0'|$. Also $y > \frac{1}{3}y_0'$ by definition.
⟦44⟧ On J_6', $|z| \geqslant |c|$ while $|\sin(\pi z)| \geqslant \pi y \geqslant \pi|c|\frac{1}{2}y_0'/|b'| \geqslant \frac{1}{2}\pi|c| \sin\arg\zeta_0'$, since $|b'| \leqslant |z_0'|$.

On the part of J_3' from c to $|c|\exp(-\frac{1}{4}\mathrm{i}\pi)$ we have $\frac{1}{2} \geqslant |c| \geqslant x \geqslant |c|/\sqrt{2}$ while on the remainder of J_3', $y \leqslant -|c|/\sqrt{2}$. On the first part, $|\sin(\pi z)| \geqslant |\sin(\pi x)| \geqslant \sin(\pi|c|/\sqrt{2}) \geqslant \pi|c|/\sqrt{2} - \frac{1}{6}(\pi|c|/\sqrt{2})^3 \geqslant \pi|c|/\sqrt{2}(1 - \frac{1}{48}\pi^2) \sin\arg\zeta_0'$.

On the other part, $|\sin(\pi z)| \geqslant |\sinh(\pi y)| = \sinh(-\pi y) \geqslant \sinh(\pi|c|/\sqrt{2}) \geqslant \pi|c|/\sqrt{2}\sin\arg\zeta_0'$.
⟦45⟧ The expression $-\pi\kappa\Re\varepsilon(z_0 - 2\mu) + 2\pi\tau\arg\zeta_0$ equals $\sqrt{2}\pi\kappa\mu + \Re\varphi(z_0)$.
⟦46⟧ Similarly, $-\pi\kappa\Re\varepsilon(2\mu - z_0') + 2\pi\tau\arg\zeta_0' = -\sqrt{2}\pi\kappa\mu + \Re\varphi(z_0')$.
⟦47⟧ The first statement should be $\sqrt{2}\Re\varepsilon\zeta_0 > 1$.

To obtain these results, simply use that ζ_0' satisfies $\zeta^2 + \bar{\varepsilon}\varrho\zeta - 1 = 0$ so that when ϱ is small $\xi_0' = 1 - \varrho/2\sqrt{2}$ and $\eta_0' = \varrho/2\sqrt{2} - \frac{1}{8}\varrho^2$, approximately. Similarly, $\xi_0 = 1 + \varrho/2\sqrt{2}$ and $\eta_0 = -\varrho/2\sqrt{2} - \frac{1}{8}\varrho^2$, approximately.

For later note the following. $|\zeta_0'| \leqslant 1$, $|z_0'| = \tau^{1/2}|\zeta_0'| > 0.81\tau^{1/2}$; $y_0 = \tau^{1/2}\eta_0$ so $|y_0| \geqslant \kappa/2\sqrt{2}$; $y_0' = \tau^{1/2}\eta_0' > \tau^{1/2}(0.29)\varrho = 0.29\kappa$.

$\tan\arg\zeta_0' = \varrho/2\sqrt{2} + \mathrm{O}(\varrho^3)$ so $\arg\zeta_0' \simeq \varrho/2\sqrt{2}$ and $\operatorname{cosech}\arg\zeta_0' \simeq 2\sqrt{2}/\varrho = (\kappa/\sqrt{2})^{-1}2\tau^{1/2}$.
⟦48⟧ In the last line of (5.20), the first occurrence of $\kappa/\sqrt{2}$ should be $(\kappa/\sqrt{2})^{-1}$.

To obtain (5.20), use the results of the previous note. Also use that $\exp(-\frac{1}{2}\pi t)$ is negligible and $\frac{1}{2} = \min(\frac{1}{2}, \dots)$.
⟦49⟧ $z_0' + z_0 = z_0 - z_1 = \tau^{1/2}(\zeta_0 - \zeta_1) = \tau^{1/2}(-\mathrm{i}\varrho^2 + 4)^{1/2}$.
⟦50⟧ Since $0 < \arg\zeta_0' < \frac{1}{4}\pi$, the first statement is clear.

When $\varrho \to 0$, $2\zeta_0' = -\varrho\bar{\varepsilon} + (\varrho^2\bar{\varepsilon}^2 + 4)^{1/2}$ so $\zeta_0' \sim 1 - \frac{1}{2}\varrho\bar{\varepsilon}$. Thus $\varrho^{-1}\arg\zeta_0' \to 1/(2\sqrt{2})$ and $\mu_0 \sim \tau^{1/2}$.

⟦51⟧ As $\varrho \to 0$, $\frac{1}{4} + \mathrm{i}/\varrho^2 \sim \mathrm{i}/\varrho^2$, $\varrho^{-1}\arg\zeta_0' \to 1/(2\sqrt{2})$ by note ⟦50⟧ and $1 - \Re\varepsilon/\varrho + 1/(\sqrt{2}\varrho) \sim 1$.

⟦52⟧ To prove that the expression is not less than $\frac{1}{2}$, we may proceed as follows. From the quadratic equation we have $\zeta_0' = \frac{1}{2}\varrho\bar{\varepsilon}E$, where $E = -1 + (1 + 4\mathrm{i}/\varrho^2)^{1/2}$. Put $\lambda = 4/\varrho^2$, $a + \mathrm{i}b = (1 + \lambda\mathrm{i})^{1/2}$, $D = (1 + \lambda^2)^{1/2}$, $\sigma = D - 1$. Then we find that $\tan\arg E = ((\sigma + 2)^{1/2} + \sqrt{2})/\sigma^{1/2} = W$, say, and $\Re(\frac{1}{4} + \mathrm{i}/\varrho^2)^{1/2} = \frac{1}{2}(1 + \frac{1}{2}\sigma)^{1/2}$. We want to prove that

$$1 - \tfrac{1}{2}(1 + \tfrac{1}{2}\sigma)^{1/2} + \tfrac{1}{2}(\sigma^2 + 2\sigma)^{1/2}(-\tfrac{1}{4}\pi + \arg E) \geqslant \tfrac{1}{2}.$$

Now $\tan(\arg E - \frac{1}{4}\pi) = (W - 1)/(W + 1)$ and $\sigma = 8W^2/(W^2 - 1)^2$ so we have to prove $\tan^{-1}((W-1)/(W+1)) - (W^2 - 1)/(2W^3 + 2W) \geqslant 0$.

This follows since the derivative of this expression is positive (because $W > 1$) and it has the value 0 when $W = 1$.

⟦53⟧ $\frac{1}{2}\kappa \geqslant \Re\varepsilon(\mu_0 - z_0') = \kappa - \Re\varepsilon(z_0 - \mu_0)$.

⟦54⟧ Recall that $A(\mu_0) = B(\mu_0)$. To see that $A(\mu_0) = \frac{1}{2}\pi\kappa^2$, we use that $\arg\zeta_0 + \arg\zeta_0' = 0$, so $2A(\mu_0) = A(\mu_0) + B(\mu_0) = \pi\kappa\Re\varepsilon(z_0 - z_0') = \pi\kappa\Re\varepsilon(z_0 + z_1) = \pi\kappa^2$.

⟦55⟧ $\Re\varepsilon(z_0 - \mu)$ is the distance of z_0 from the line P, that is, $d(J', P)$. Similarly $\Re\varepsilon(\mu - z_0') = d(J, P)$. For these distances to be at least $(4\kappa)^{-1}$ we require

$$\tfrac{1}{2}\kappa - \frac{1}{4\kappa} > \frac{1}{\sqrt{2}}(\mu - \mu_0) > \frac{1}{4\kappa} - \tfrac{1}{4}\kappa.$$

⟦56⟧ The left-hand side should be $\sum |h(p_{-k})|$, that is, $\sum \kappa u_{-k}$ (or alternatively the definition of u_k should be $u_k = |h(p_k)|$). Also, $\sqrt{2}\pi(K+1)/\kappa$ should be $-\sqrt{2}\pi(K+1)/\kappa$.

We suppose K is positive. To obtain the inequality, note that $|p_{-k}|^2 \geqslant \frac{1}{2}\mu^2$. Also, where $\eta = -\pi k/(\kappa\sqrt{2})$ and $\bar{\eta} = -\pi(K+1)/(\kappa\sqrt{2})$, note that $|2\mathrm{i}\sin(\pi p_{-k})| \geqslant \mathrm{e}^{-\eta} - \mathrm{e}^{\eta} \geqslant (1 - \mathrm{e}^{2\bar{\eta}})/\mathrm{e}^{\eta}$.

⟦57⟧ Let $\eta = -\pi(K+1)/(\kappa\sqrt{2})$. Then the inequality to be proved reduces to $\mu^{-\sigma}\mathrm{e}^{\eta}(1 - \mathrm{e}^{2\eta}) < |p_{-k-1}|^{-\sigma}/|2\mathrm{i}\sin(\pi p_{-k-1})|$.

The denominator is at most $\mathrm{e}^{\eta} + \mathrm{e}^{-\eta}$ and, since

$$\mathrm{e}^{\eta}(1 - \mathrm{e}^{2\eta})(\mathrm{e}^{\eta} + \mathrm{e}^{-\eta}) = 1 - \mathrm{e}^{4\eta} < 1$$

it suffices to prove that $|p_{-k-1}| < \mu$. This holds if and only if

$$K + 1 < \sqrt{2}\,\kappa\mu. \qquad (*)$$

Now we have $\Im p_{-k-1} < \mu\theta_0$, that is, $-\theta_0\mu\kappa\sqrt{2} < K + 1$. We also have $-\theta_0 < 1$ since $((1+\theta)^2 + \theta^2)(1 + 2\theta) - \alpha$ is negative when $\theta = -1$. So it is possible to choose the integer K to satisfy (*) provided $\sqrt{2}\,\kappa\mu(1 + \theta_0) > 1$.

Alternatively we could avoid (*) if we are prepared to assume $K + 1 \geqslant$

$\frac{1}{2}\kappa\mu\sqrt{2}$; for then we would have $|p_{-k}| \geqslant |p_{-k-1}|$ and the inequality would hold even with the term $2^{\sigma/2}$ deleted.

⟦58⟧ The analysis is similar but now we clearly have $|p_k| \geqslant |p_{K'+1}|$.

⟦59⟧ Let $E = \mu^{-1}p_{-k-1} - 1$. Then $E = -(K+1)(1+\mathrm{i})/\mu\kappa\sqrt{2}$. Now $-(K+1)/\kappa\sqrt{2} = \Im p_{-K-1} < \mu\theta$, where $((1+\theta)^2 + \theta^2)(1+2\theta) - \alpha = 0$. (We assume $\theta < 0$; if $\theta > 0$ the roles of K, K' should be interchanged.) Assume that K is the least such integer and define $K' = K$. When $\varrho = \kappa\tau^{-1/2}$ is small, $\mu \sim \tau^{-1/2}$ so $\alpha = \tau/\mu^2 \sim 1$ and hence $\theta \sim 0$. Thus $E \sim \theta(1+\mathrm{i})$, so E is small.

For later note that $p_{-K-1} \sim \mu(1+\theta(1+\mathrm{i}))$. Let $E' = \mu^{-1}p_{K'+1} - 1$. Then $E' = -E$ and $p_{K'+1} = \mu(1-E) \sim \mu(1-\theta(1+\mathrm{i}))$.

⟦60⟧ The approximate expression for u_k in the text should be multiplied by $(\mu/\sqrt{2})^{-\sigma}$.

By note ⟦59⟧, $p_{-K-1} \sim \mu(1+\theta(1+\mathrm{i})) = z$, say. Now

$$|h(p_{-K-1})| \sim |h(z)| = |\mathrm{e}^{\mathrm{i}\pi z^2} z^{-\mathrm{i}t}|\,|z^{-\sigma}|/|2\mathrm{i}\sin(\pi z)|.$$

Now $|z|^2 \geqslant \frac{1}{2}\mu^2$ so $|z^{-\sigma}| \leqslant (\mu/\sqrt{2})^{-\sigma}$. Next,

$$|\exp(\mathrm{i}\pi z^2)| = \exp(-2\pi\mu^2\theta(1+\theta)), \qquad |z^{-\mathrm{i}t}| = \exp(t \arg z)$$

and $\tan\arg z = \theta/(1+\theta) \sim \theta - \theta^2$ so we have

$$|\exp(\mathrm{i}\pi z^2)z^{-\mathrm{i}t}| \sim \exp(2\pi\mu^2\theta(-1-\theta+\alpha-\alpha\theta)) \sim \exp(-4\pi\mu^2\theta^2)$$

$$\sim \exp\left(-2\pi\frac{(K+1)^2}{\kappa^2}\right).$$

Now $|2\mathrm{i}\sin(\pi z)| \geqslant |\mathrm{e}^{\eta} - \mathrm{e}^{-\eta}|$, $\eta = \pi\mu\theta$ so we need to assume that $|\pi\mu\theta| \geqslant c > 0$ for some constant c or equivalently $(K+1)/\kappa \geqslant c' > 0$ in order to be able to majorize the expression $|2\mathrm{i}\sin(\pi z)|^{-1}$ by a constant. Similarly for $|h(p_{K'+1})|$ (by replacing θ by $-\theta$).

⟦61⟧ We have $\alpha = \tau/\mu^2$, $\mu \to \frac{1}{2}\kappa/\sqrt{2}$, $\alpha = 8/\varrho^2$, $\theta \to -\frac{1}{2}$. Now $-(K+1)/\kappa\sqrt{2} = \Im p_{-K-1} < \mu\theta$ and we may assume K is the smallest integer satisfying this inequality. Also $\mu\theta \sim -\frac{1}{2}\mu$ and

$$\Re p_{-K-1} = \mu - \frac{K+1}{\kappa\sqrt{2}} \geqslant \mu - \left(\frac{1}{\kappa\sqrt{2}} - \mu\theta\right) = \tfrac{1}{2}\mu - \frac{1}{\kappa\sqrt{2}} > 0, \quad \text{if } \kappa > 2.$$

Hence p_{-K-1} is in the fourth quadrant and therefore $|p_{-K-1}^{-\mathrm{i}t}| < 1$. The factor $2\mathrm{i}\sin(\pi z)$ may be neglected as in note ⟦60⟧ since we now have $|\pi\mu\theta| \sim \pi\kappa/4\sqrt{2} \geqslant c > 0$.

The argument just given does not work for $p_{K'+1}$ but to bound $|h(p_{K'+1})|$ we use that $\varphi(p_{K'+1}) \leqslant \varphi(p_{-K})$, where $\varphi(z) = \exp(\mathrm{i}\pi z^2)z^{-\mathrm{i}t}$;

this is because $\mu\theta \sim -\frac{1}{2}\mu < 0$ so $\Im p_{-K-1} < \mu\theta < \Im p_{-K} \leqslant \Im p_{K'+1}$. The sine factor may be neglected and $|p_{K'+1}| \leqslant (\mu/\sqrt{2})^{-\sigma}$.

⟦62⟧ The roots are $v_1 = -K/\kappa$ and $v_2 = K'/\kappa$ so $\frac{3}{2}\kappa = v_2 - v_1 = (K'+K)/\kappa \sim (K'+K+1)/\kappa = \tau/\kappa$, that is, $\tau \sim \frac{3}{2}\kappa^2$.

⟦63⟧ Take $K = K' = 1$. Now $\alpha = \tau/\mu^2 = 1$ so $\theta = 0$. K and K' obviously satisfy their defining conditions. Also $\varrho = \kappa\tau^{-1/2} \leqslant \frac{1}{2}$ so (5.20) is available. From (5.20), $|R| < a\tau^{-1/4} + b\tau^{1/2}\exp(-c\tau^{1/2})$, where a, b, c are positive constants so $|R| < a'\tau^{-1/4}$ for t large enough.

We have seen that the error $|R^*|$ is of the form $c\mu^{-\sigma} = c\tau^{-1/4}$, where c is a function of κ, K, K' so c is now a constant.

It remains to consider the error arising by replacing the factors $F_r = (1 - \exp E_r)^{-1}$, $E_r = 2\pi\kappa\varepsilon(r-\mu)$, by 0 or 1 in the series $\sum r^{-s}F_r$. Specifically we replace $F_1, \ldots, F_{m-1}$ by 1 and $F_{m+2}, F_{m+3}, \ldots$ by 0, where $m < \mu < m+1$. Thus instead of $\sum r^{-s}F_r$ we take

$$\sum_{r=1}^{m-1} r^{-s} + m^{-s}F_m + (m+1)^{-s}F_{m+1}.$$

We want to show that the absolute value of this error is at most $c\mu^{-\sigma} = c\tau^{-1/4}$ for some constant c. The error is

$$\sum_{r=1}^{m-1} r^{-s}(F_r - 1) + \sum_{r=m+2}^{\infty} r^{-s}F_r.$$

We show that the absolute value of this error is at most $c\mu^{-\sigma}$ where c depends only on κ. Majorizing $\sum_{m+2}^{\infty} |F_r|$ by a geometric series and using $|r^{-s}| \leqslant (m+2)^{-\sigma} \sim \mu^{-\sigma}$ we find $|\sum_{m+2}^{\infty} r^{-s}F_r| \leqslant c\mu^{-\sigma}$. Now let $h = [\frac{1}{2}\mu]$, $S_1 = \sum_1^b r^{-s}(F_r - 1)$ and $S_2 = \sum_{h+1}^{m-1} r^{-s}(F_r - 1)$. Write $T_r = \sqrt{2}\pi\kappa(\mu - r)$. Then for $1 \leqslant r \leqslant m-1$, $|F_r - 1| \leqslant (\exp T_r - 1)^{-1}$. We have $|S_1| \leqslant \sum_1^h (\exp T_r - 1)^{-1} \leqslant h/(\exp T_h - 1) < c\mu^{-\sigma}$, $|S_2| \leqslant (h+1)^{-\sigma}S_3$, where $S_3 = \sum_{h+1}^{m-1} (\exp T_r - 1)^{-1}$. Let $J = \exp(\sqrt{2}\pi\kappa)$ and $d = m - h - 1$. Then $S_3 < 2\sum \exp(-T_r) = 2(J^d - 1) \cdot (J-1)^{-1}\exp(-T_{h+1}) < 2(J-1)^{-1}\exp(\sqrt{2}\pi\kappa(m-\mu)) < 2/(J-1)$. Thus $|S_2| < c(h+1)^{-\sigma} < c(\frac{1}{2}\mu)^{-\sigma} = c'\mu^{-\sigma}$.

⟦64⟧ The conclusion is that, for this case,

$$\Im(s) = \sum_{k=-1}^{1} \varepsilon\kappa^{-1}h(p_k) - \sum_{1}^{m-1} r^{-s} - m^{-s}F_m - (m+1)^{-s}F_{m+1} + E,$$

$$\text{where } m = [\mu] = \left[\left(\frac{t}{2\pi}\right)^{1/2}\right] \text{ and } |E| \leqslant 0.0044\,\tau^{-1/4}.$$

If $t < 1000$, then $m - 1 \leqslant 11$.

Recall that on the critical line

$$\zeta(s)\Gamma(\tfrac{1}{2}s)\pi^{-s/2} = -2\Re(\Gamma(\tfrac{1}{2}s)\pi^{-s/2}\Im(s)).$$

The same conclusion holds if we take not $\mu=\tau^{1/2}$, but $\tau^{1/2}-\frac{1}{4}\leqslant \mu\leqslant\tau^{1/2}+\frac{1}{4}$. (Since $0<\mu_0/\tau^{1/2}-1<\frac{1}{3}\varrho^2$, we have $|\mu_0-\tau^{1/2}|<\frac{1}{3}\varrho^2\tau^{1/2}=\frac{1}{3}\kappa^2/\tau^{1/2}$. Thus if $|\mu-\tau^{1/2}|\leqslant\frac{1}{4}$, then $|\mu-\mu_0|\leqslant\frac{1}{4}+\frac{1}{3}\kappa^2/\tau^{1/2}<\frac{1}{2}$.) This would allow us to choose μ to be an integer or half an odd integer. The constant 0.0044 would have to be increased.

1948 *Rounding-off Errors in Matrix Processes*

Summary

A denotes an $n\times n$ matrix; (all matrices are real). In this paper, methods are discussed for calculating $\mathbf{A}^{-1}$ and solving $\mathbf{Ax}=\mathbf{b}$, where **A** is nonsingular and **b** is an $n\times 1$ matrix. **L** denotes a unit lower triangular matrix, that is, the entries above the diagonal are zero while each entry on the diagonal is 1. **U** denotes a unit upper triangular matrix and **D** a diagonal matrix. Turing first proves the theorem that if the principal minors of **A** are nonsingular — actually this condition involves no loss of generality — then there exist unique **L**, **D**, **U** such that $\mathbf{A}=\mathbf{LDU}$.

The method of proof is important: the required matrices **L**, **D**, **U** are determined step by step by considering in turn the expressions for

$$a_{11}, a_{12}, \ldots, a_{1n}, a_{21}, \ldots, a_{2n}, \ldots, a_{nn}.$$

A number of methods for calculating $\mathbf{A}^{-1}$ or solving $\mathbf{Ax}=\mathbf{b}$ are described but three are singled out:

The standard (Gauss) elimination process for $\mathbf{Ax}=\mathbf{b}$, reducing **A** to upper triangular form, that is, the form **DU**, by row operations. Each row operation corresponds to premultiplication by a matrix $\mathbf{J}_r$. We have $\mathbf{J}_n\cdots\mathbf{J}_1\mathbf{A}=\mathbf{DU}$ and if $\mathbf{J}_n\cdots\mathbf{J}_1\mathbf{b}=\mathbf{c}$, then $\mathbf{x}=(\mathbf{DU})^{-1}\mathbf{c}$. Moreover $\mathbf{J}_n\cdots\mathbf{J}_1$ has the form $\mathbf{L}^{-1}$, so $\mathbf{A}=\mathbf{LDU}$.

Jordan's method. This is like the previous method except that we reduce **A** to the unit matrix by row operations. Here $\mathbf{J}_n\cdots\mathbf{J}_1\mathbf{A}=\mathbf{I}$, so $\mathbf{J}_n\cdots\mathbf{J}_1=\mathbf{A}^{-1}$.

Unsymmetrical Choleski method. Find **L**, **D**, **U** as in the proof of the theorem; actually it is better to find **L** and **DU**. Then find $\mathbf{L}^{-1}$ and $(\mathbf{DU})^{-1}$. Then $\mathbf{A}^{-1}=(\mathbf{DU})^{-1}\mathbf{L}^{-1}$.

Write $M(\mathbf{A})=\max|a_{ij}|$, $N(\mathbf{A})=(\sum a_{ij}^2)^{1/2}$ and $B(\mathbf{A})=\max\|\mathbf{Ax}\|/\|\mathbf{x}\|$.

A statistical argument is given which suggests that $N(\mathbf{A})N(\mathbf{A}^{-1})/n$ may be taken as a measure of the ill-conditioning of a matrix. $nM(\mathbf{A})M(\mathbf{A}^{-1})$ is also suggested (presumably because $M(\mathbf{X})\leqslant N(\mathbf{X})\leqslant nM(\mathbf{X})$ for all $n\times n$ matrices **X**). Roughly speaking, if **A** is ill-conditioned, small percentage errors in the coefficients of $\mathbf{A}$ lead to large percentage errors in the solution of $\mathbf{Ax}=\mathbf{b}$.

'If the coefficients of a matrix are chosen at random from a normal population we shall get N-condition numbers of the order of $n^{1/2}$ and M-condition numbers about $\log n$ times greater.' However, 'The matrices which occur in practical problems are by no means random in this sense'.

If $\mathbf{A}$ is the unit lower triangular $n \times n$ matrix with all entries below the diagonal equal to -1, then $M(\mathbf{A}^{-1})=2^{n-2}$ and hence $nM(\mathbf{A})M(\mathbf{A}^{-1})=n2^{n-2}$.

If $\mathbf{B}$ is a reputed inverse of $\mathbf{A}$, let $\mathbf{E}=\mathbf{I}-\mathbf{AB}$. If $M(\mathbf{E})<1/n$, then

$$M(\mathbf{B}-\mathbf{A}^{-1})\leqslant\frac{nM(\mathbf{B})M(\mathbf{E})}{1-nM(\mathbf{E})}.$$

§11. Errors in Jordan's method

(a) Absolute error.

In carrying out Jordan's method we make errors, so that instead of $\mathbf{J}_n\cdots\mathbf{J}_1\mathbf{A}=\mathbf{I}$ we have $\mathbf{J}_n[\cdots\{\mathbf{J}_2(\mathbf{J}_1\mathbf{A}+\mathbf{S}_1)+\mathbf{S}_2\}\cdots]+\mathbf{S}_n=\mathbf{I}$ and instead of $\mathbf{J}_n\cdots\mathbf{J}_1=\mathbf{A}^{-1}$ we have

$$\mathbf{J}_n[\cdots\{\mathbf{J}_2(\mathbf{J}_1+\mathbf{S}_1')+\mathbf{S}_2'\}\cdots]+\mathbf{S}_n' = \mathbf{E}.$$

Thus $\mathbf{E}$ is the calculated value of $\mathbf{A}^{-1}$. Supposing $M(\mathbf{S}_r)\leqslant\varepsilon$, $M(\mathbf{S}_r')\leqslant\varepsilon'$, $r=1,\ldots,n$, we find

$$M(\mathbf{E}-\mathbf{A}^{-1})\leqslant n\varepsilon'+\frac{n(n-1)}{2}M(\mathbf{A}^{-1})\left\{\varepsilon+\varepsilon'+\frac{2n-1}{3}\varepsilon M(\mathbf{A}^{-1})\right\}.$$

Estimates for the error in B-measure and in N-measure are also given.

(b) Statistical error.

Let $F=(\mathbf{E}-\mathbf{A}^{-1})_{ij}$, where i and j are fixed. Each entry of $\mathbf{S}_r$ is assumed to be a random variable with mean zero and standard deviation η, and similarly for $\mathbf{S}_r'$. We find

$$\text{variance of } F\leqslant\eta^2P+\eta'^2Q,$$

where P, Q are polynomials in $M(\mathbf{A}^{-1})$ and n (depending on i, j); these polynomials are given explicitly. The leading term of the standard deviation of F is at most $\eta M(\mathbf{A}^{-1})^2\varphi$, where φ is $O(n^{3/2})$.

§12. Errors in the Gauss elimination process

Here there are errors

(i) in the reduction of $\mathbf{A}$ to the form $\mathbf{DU}$:

$$\mathbf{J}_n[\cdots\{\mathbf{J}_2(\mathbf{J}_1\mathbf{A}+\mathbf{S}_1)+\mathbf{S}_2\}\cdots]+\mathbf{S}_n = \widetilde{\mathbf{DU}};$$

(ii) performing the corresponding row operations on **b**:

$$\mathbf{J}_n[\cdots\{\mathbf{J}_2(\mathbf{J}_1\mathbf{b}+\mathbf{S}_1')+\mathbf{S}_2'\}\cdots]+\mathbf{S}_n'=\tilde{\mathbf{c}};$$

(iii) working out $(\mathbf{DU})^{-1}\mathbf{c}$ $(=\mathbf{x})$ or rather $(\widetilde{\mathbf{DU}})^{-1}\tilde{\mathbf{c}}$.

It is shown that the errors under (iii) are much smaller than errors under (i). We find that due to (i), (ii)

$$|\text{error in } x_m| \leqslant O(n^2)M(\mathbf{A}^{-1})\varepsilon' + O(n^4)M(\mathbf{A}^{-1})^2 M(\mathbf{b})\varepsilon,$$

where x_m is the mth coordinate of the vector $\mathbf{x}$.

§13. Errors in the unsymmetrical Choleski method

Let $\mathbf{W}=\mathbf{DU}$. Let the computed values of $\mathbf{L}$, $\mathbf{W}$ be $\mathbf{L}^*$, $\mathbf{W}^*$; let those of $\mathbf{L}^{*-1}$, $\mathbf{W}^{*-1}$ be $\mathbf{K}$, $\mathbf{V}$ and that of $\mathbf{VK}$ be $\mathbf{E}$ so that again $\mathbf{E}$ is the computed value of $\mathbf{A}^{-1}$. Then $\mathbf{L}^*\mathbf{W}^*=\mathbf{A}-\mathbf{S}$, $\mathbf{L}^*\mathbf{K}=\mathbf{I}-\mathbf{S}'$, $\mathbf{VW}^*=\mathbf{I}-\mathbf{S}''$ and $\mathbf{VK}=\mathbf{E}-\mathbf{S}'''$. Assume that $M(\mathbf{S})<\varepsilon$, $M(\mathbf{S}')<\varepsilon'$, $M(\mathbf{S}'')<\varepsilon'$ and $M(\mathbf{S}''')<\varepsilon''$. Then it is proved that

$$M(\mathbf{E}-\mathbf{A}^{-1})\leqslant n^2\varepsilon M(\mathbf{A}^{-1})^2+2n\varepsilon' M(\mathbf{A}^{-1})+\varepsilon''.$$

This suggests that this method is better than the Jordan or Gauss method.

The overall conclusion is that exponential build up of errors 'need not' occur in computing **A** or solving $\mathbf{Ax}=\mathbf{b}$ by any of the three methods considered.

A review by Bodewig (1949) is critical of one or two points in Turing's paper.

Notes

⟦1⟧ In this paper, all matrices are real. Turing is only interested in solving $\mathbf{Ax}=\mathbf{b}$ when it is given that **A** is nonsingular.

⟦2⟧ In the second equation, the minus sign should be deleted; this equation follows from $\mathbf{x}=\mathbf{A}^{-1}\mathbf{Ax}$, $x_i=\sum_{j,k}(\mathbf{A}^{-1})_{ij}a_{jk}x_k$.

⟦3⟧ If **A** is a nonsingular matrix, then, after a suitable permutation of its rows, the principal minors of **A** are nonsingular. Thus the condition is no real restriction.

⟦4⟧ More elegant proofs of the theorem exist but this is beside the point: the actual method of proof is what is important later.

The idea is to consider the expressions for

$$a_{11}, a_{12}, \ldots, a_{1n}, a_{21}, \ldots, a_{2n}, \ldots, a_{n1}, \ldots, a_{nn}.$$

Each successive expression determines one more 'unknown' (d_i, u_{ij} or l_{ij}).

⟦5⟧ This should be 'Suppose now that we have found the values of l_{ij}, u_{ij}, d_i whenever $i<i_0$'.
⟦6⟧ $\mathbf{J}_i$ is unit lower triangular and differs from $\mathbf{I}$ only in the ith column. Let $\mathbf{A}'=\mathbf{I}+\sum_{r=1}^{n-1}(\mathbf{I}-\mathbf{J}_r)$. Then the same row operations applied to $\mathbf{A}'$ that were applied to $\mathbf{A}$ reduce $\mathbf{A}'$ to $\mathbf{I}$. Hence $\mathbf{J}_{n-1}\cdots\mathbf{J}_1\mathbf{A}'=\mathbf{I}$ and $\mathbf{A}'=\mathbf{L}$.
⟦7⟧ There are $\sum_{m=2}^{n}(m-1)(m+2)$ multiplications of which $\sum_{m=2}^{n}(m-1)$ involve $\mathbf{b}$.
⟦8⟧ Call the equations (1), …, (5). We have $(\mathbf{J}_i-\mathbf{I})_{pq}=0$ for $q\neq i$, $(\mathbf{J}_i-\mathbf{I})_{pq}=0$ for $p\leqslant i$, $\mathbf{A}^r-\mathbf{A}^{r-1}=(\mathbf{J}_r-\mathbf{I})\mathbf{A}^{r-1}$. Hence $\mathbf{A}_{ij}^r-\mathbf{A}_{ij}^{r-1}=\sum_k(\mathbf{J}_r-\mathbf{I})_{ik}\mathbf{A}_{kj}^{r-1}=(\mathbf{J}_r-\mathbf{I})_{ir}\mathbf{A}_{rj}^{r-1}=0$ if $i\leqslant r$; and $=(\mathbf{J}_r)_{ir}\mathbf{A}_{rj}^{r-1}$ if $i>r$, proving (1).

Next, $\mathbf{A}_{ij}^r=0$ if $j\leqslant r$ and $i>j$, so $(\mathbf{J}_r\mathbf{A}^{r-1})_{ir}=\mathbf{A}_{ir}^r=0$ if $i>r$. Let $i>r$. Then $\sum_k(\mathbf{J}_r)_{ik}\mathbf{A}_{kr}^{r-1}=0$. Now $(\mathbf{J}_r)_{pq}=0$ unless $p=q$ or both $q=r$ and $p>r$, so $(\mathbf{J}_r)_{ii}\mathbf{A}_{ir}^{r-1}+(\mathbf{J}_r)_{ir}\mathbf{A}_{rr}^{r-1}=0$, proving (2) when $i>r$.

If $i>r$, then $\sum_{s=1}^{r}(\mathbf{J}_s)_{is}\mathbf{A}_{sj}^{s-1}=\sum_{s=1}^{r}\mathbf{A}_{ij}^s-\mathbf{A}_{ij}^{s-1}=\mathbf{A}_{ij}^r-\mathbf{A}_{ij}$, proving (3).

Let $i\leqslant q$. Then $\mathbf{A}_{ij}^q=\mathbf{A}_{ij}^{q-1}$, so $\mathbf{A}_{ij}^{n-1}=\cdots=\mathbf{A}_{ij}^{i-1}$. Now we have $\mathbf{A}_{ij}^{n-1}-\mathbf{A}_{ij}=\sum_{s=1}^{n-1}\mathbf{A}_{ij}^s-\mathbf{A}_{ij}^{s-1}=\sum(\mathbf{J}_s)_{is}\mathbf{A}_{sj}^{s-1}=\sum_{s=1}^{i-1}(\mathbf{J}_s)_{is}\mathbf{A}_{sj}^{n-1}$. Alternatively the same expression $\mathbf{A}_{ij}^{n-1}-\mathbf{A}_{ij}$ equals $\sum_{s=1}^{n-1}(\mathbf{J}_s-\mathbf{I})_{is}\mathbf{A}_{sj}^{n-1}$ since $\mathbf{A}_{sj}^{s-1}=\mathbf{A}_{sj}^{n-1}$. Thus (4) would appear to be incorrect and one of the alternative equations just given should be taken instead. Incidentally, note that $\mathbf{A}^n$ does not exist.

Let $i\geqslant r$. We have $\mathbf{A}_{ij}^{r-1}-\mathbf{A}_{ij}=\sum_{s=1}^{r-1}\mathbf{A}_{ij}^s-\mathbf{A}_{ij}^{s-1}=\sum(\mathbf{J}_s)_{is}\mathbf{A}_{sj}^{s-1}$, so

$$\mathbf{A}_{ir}^{r-1}=\mathbf{A}_{ir}+\sum_{s=1}^{r-1}(\mathbf{J}_s)_{is}\mathbf{A}_{sr}^{s-1}. \tag{6}$$

Now (5) follows from (6) and (2); moreover, both occurrences of $\mathbf{A}_{sr}^{s-1}$ in (5) could be replaced by $\mathbf{A}_{sr}^{n-1}$ (and this was probably intended).
⟦9⟧ We have $\mathbf{A}_{ij}^r=\sum(\mathbf{J}_r)_{ik}\mathbf{A}_{kj}^{r-1}$. Now $\sum_{k\neq r}(\mathbf{J}_r)_{ik}\mathbf{A}_{kj}^{r-1}=\sum_{k\neq r}\delta_{ik}\mathbf{A}_{kj}^{r-1}=0$ if $i=r$; and $=\mathbf{A}_{ij}^{r-1}$ if $i\neq r$. Thus we have

$$\mathbf{A}_{ij}^r=\begin{cases}\mathbf{A}_{ij}^{r-1}+(\mathbf{J}_r)_{ir}\mathbf{A}_{rj}^{r-1}, & i\neq r,\\ (\mathbf{J}_r)_{ir}\mathbf{A}_{rj}^{r-1}, & i=r,\end{cases}$$

and also $\mathbf{A}_{rj}^r=(\mathbf{J}_r)_{rr}\mathbf{A}_{rj}^{r-1}$.

These results remain true if '$\mathbf{A}$' is replaced by '$\mathbf{X}$' everywhere.

Let $i\neq r$. Taking $j=r$ we get $\mathbf{A}_{ir}^r=\mathbf{A}_{ir}^{r-1}+(\mathbf{J}_r)_{ir}\mathbf{A}_{rr}^{r-1}$, but since $i\neq r$ we have $\mathbf{A}_{ir}^r=0$, so $(\mathbf{J}_r)_{ir}=-\mathbf{A}_{ir}^{r-1}/\mathbf{A}_{rr}^{r-1}$, $i\neq r$.
⟦10⟧ This equation is subject to the condition $i\neq r$.
⟦11⟧ We show that $\mathbf{M}$ exists. Since $\mathbf{A}$ is nonsingular, $\mathbf{A}^*\mathbf{A}$ is a positive definite matrix, so $\mathbf{A}^*\mathbf{A}=\mathbf{LDU}$ for some $\mathbf{L},\mathbf{D},\mathbf{U}$. Hence $\mathbf{LDU}=\mathbf{U}^*\mathbf{D}^*\mathbf{L}^*$ so that $\mathbf{L}=\mathbf{U}^*$ by uniqueness. Thus $\mathbf{D}=\mathbf{N}^*\mathbf{A}^*\mathbf{A}\mathbf{N}$, where $\mathbf{N}=\mathbf{U}^{-1}$. $\mathbf{D}$ is positive definite so its entries are nonnegative and we may write $\mathbf{D}=\mathbf{D}'^2$.

Then $\mathbf{M}=\mathbf{N}\mathbf{D}'^{-1}$ satisfies the requirements.

⟦12⟧ Let $\mathbf{B}=\mathbf{A}^*\mathbf{A}$. From the (1,1)-st entry we get $m_{11}^2=b_{11}^{-1}$, so we know m_{11} and hence the first column of $\mathbf{M}$. Assume that we know the first r columns of $\mathbf{M}$. Then we know the first r rows of $\mathbf{M}^*$, hence the first r rows of $\mathbf{M}^*\mathbf{B}$. From the $(i,r+1)$-st entry we get $\sum(\mathbf{M}^*\mathbf{B})_{ik}\mathbf{M}_{k,r+1}=0$, $i=1,\ldots,r$, so $\mathbf{M}_{i,r+1}$, $i=1,\ldots,r+1$, are known apart from a multiplicative factor. The $(r+1,r+1)$-st entry gives an equation from which the factor can be determined.

⟦13⟧ $N(\mathbf{XY})^2 = \sum_{ij}(\sum_k x_{ik}y_{kj})^2 \leqslant \sum_{ij}\sum_k x_{ik}^2\sum_l y_{lj}^2 = \sum_{ik}x_{ik}^2\sum_{jl}y_{lj}^2 = N(\mathbf{X})^2N(\mathbf{Y})^2$, proving (7.6).

Equations (7.7), (7.8) are trivial. In the definition of $B(\mathbf{A})$, taking $\mathbf{x}$ to be the qth column of the unit matrix we obtain $(\sum_i a_{iq}^2)^{1/2}\leqslant B(\mathbf{A})$. Hence $|a_{pq}|\leqslant B(\mathbf{A})$ and (7.9) follows. Squaring and summing over q we obtain $N(\mathbf{A})^2\leqslant nB(\mathbf{A})^2$, proving (7.12). Next,

$$(\mathbf{Ax},\mathbf{Ax}) = \sum_i\Big(\sum_k a_{ik}x_k\Big)^2 \leqslant \sum_i\Big(\sum_k a_{ik}^2\Big)\Big(\sum_k x_k^2\Big),$$

hence $B(\mathbf{A})^2\leqslant\sum_{ik}a_{ik}^2=N(\mathbf{A})^2$, proving (7.11).

⟦14⟧ By (7.7), $B(\mathbf{A})\leqslant nM(\mathbf{A})$. This is not the same as (7.10), but (7.10) can be seen to be false from the example

$$\mathbf{A}=\begin{pmatrix}1 & 1\\ 1 & 1\end{pmatrix}.$$

Taking $\mathbf{x}=(1\ \ 1)^{\mathrm{T}}$ in the definition of $B(\mathbf{A})$, we obtain $2\leqslant B(\mathbf{A})$. If (7.10) were correct we would have $2\leqslant\sqrt{2}$.

⟦15⟧ Let $\mathbf{B}=\mathbf{A}^{-1}+\mathbf{A}^{-1}\mathbf{S}\mathbf{A}^{-1}$. Since, to the first order in $\mathbf{S}$, $(\mathbf{A}-\mathbf{S})\mathbf{B}=\mathbf{I}=\mathbf{B}(\mathbf{A}-\mathbf{S})$ we have that $(\mathbf{A}-\mathbf{S})^{-1}$ is approximately $\mathbf{B}$.

Multiplying by $\mathbf{b}$ we see that $\mathbf{x}$ is approximately $\mathbf{x}_0+\mathbf{A}^{-1}\mathbf{S}\mathbf{x}_0$.

⟦16⟧ R.M.S. means 'root mean square' or standard deviation.

One may also argue nonstatistically: $|\mathbf{x}-\mathbf{x}_0|/|\mathbf{x}_0| = |\mathbf{A}^{-1}\mathbf{S}\mathbf{x}_0|/|\mathbf{x}_0|\leqslant \mathbf{B}(\mathbf{A}^{-1}\mathbf{S})$, which is at most $B(\mathbf{A})B(\mathbf{A}^{-1})B(\mathbf{S})/B(\mathbf{A})$ and which is less than or equal to $N(\mathbf{A}^{-1}\mathbf{S})\leqslant N(\mathbf{A})N(\mathbf{A}^{-1})N(\mathbf{S})/N(\mathbf{A})$, which suggests taking $B(\mathbf{A})B(\mathbf{A}^{-1})$ or $N(\mathbf{A})N(\mathbf{A}^{-1})$ as measures.

⟦17⟧ In connexion with later results (Sections 11, 12, 13) it is perhaps worth noting that $M(\mathbf{A}^{-1})=M((\mathbf{B}-\mathbf{A}^{-1})-\mathbf{B})\leqslant nM(\mathbf{B})M(\mathbf{E})/(1-nM(\mathbf{E}))+M(\mathbf{B})=M(\mathbf{B})/(1-nM(\mathbf{E}))$ and similarly $M(\mathbf{A}^{-1})\leqslant M(\mathbf{B}_2)/(1-n^2\{M(\mathbf{E})^2\})$.

⟦18⟧ This expression equals $\mathbf{A}_{ij}^{(r)}$ if $i\neq r$; similarly for the expression below involving $\mathbf{X}$'s.

⟦19⟧ Since $\mathbf{X}^{(n)}\mathbf{A}=\mathbf{A}^{(n)}=\mathbf{I}$, we have $\Xi=(\mathbf{A}^{-1}\mathbf{X}^{(n)-1})\mathbf{A}^{-1}(\mathbf{X}^{(n)-1}\Xi)$, giving (11.3).

⟦20⟧ Consider the property 'equals **I** in all columns after the rth'. The product of two matrices with this property has this property. Each of $\mathbf{J}_1, \ldots, \mathbf{J}_r$ has the property, hence so does $\mathbf{X}_r$.
⟦21⟧ $\mathbf{X}_r(\mathbf{A}\mathbf{I}_r + \mathbf{I} - \mathbf{I}_r) = \mathbf{I}_r + \mathbf{X}_r(\mathbf{I} - \mathbf{I}_r) = \mathbf{I}_r + \mathbf{I} - \mathbf{I}_r = \mathbf{I}$.
⟦22⟧ The minus sign in the right-hand side should be deleted and in (11.7), $-\mathbf{S}'_r$ should be $+\mathbf{S}'_r$.
⟦23⟧ Let **C** denote $\mathbf{A}^{-1}$ with its first r columns changed to zero. Then $M(\mathbf{A}^{-1}(\mathbf{I} - \mathbf{I}_r)) = M\mathbf{C} \leqslant M(\mathbf{A}^{-1})$.
⟦24⟧ $|(\mathbf{I} - \mathbf{I}_r)\mathbf{X}| = (x_{r+1}^2 + \cdots + x_n^2)^{1/2} \leqslant |\mathbf{X}|$, so $B(\mathbf{I} - \mathbf{I}_r) \leqslant 1$. Similarly, $B(\mathbf{I}_r) \leqslant 1$. Now $B(\mathbf{S}_r) \leqslant nM(\mathbf{S}_r) \leqslant n\varepsilon$ (see note ⟦14⟧) and $B(\mathbf{S}'_r) \leqslant n\varepsilon'$. Hence in (11.9), $n^{3/2}$ should be n^2.
⟦25⟧ $(1-r)^{1/2}$ should be $(n-r)^{1/2}$.
⟦26⟧ $\sum_{r=1}^{n} (n-r)^{1/2} \leqslant \sum n^{1/2} \leqslant \frac{2}{3}(n+1)^{3/2}$ and $\frac{2}{3}n(n+1)^{3/2} \leqslant \frac{2}{3}(n+1)^{5/2}$.
⟦27⟧ Write

$$\mathbf{A}^{-1} = \begin{pmatrix} \mathbf{X} & \mathbf{Y} \\ \mathbf{Z} & \mathbf{T} \end{pmatrix},$$

where **X** is $r \times r$ and write all the other matrices involved in (11.7) in similar block form. Then we have

$$\mathbf{S}_r = \begin{pmatrix} \mathbf{0} & \mathbf{B} \\ \mathbf{0} & \mathbf{D} \end{pmatrix}, \qquad \mathbf{S}'_r = \begin{pmatrix} \mathbf{A} & \mathbf{0} \\ \mathbf{C} & \mathbf{0} \end{pmatrix}$$

and for example $\mathbf{I}_r\mathbf{S}_r\mathbf{A}^{-1}$ equals

$$\begin{pmatrix} \mathbf{BZ} & \mathbf{BT} \\ \mathbf{0} & \mathbf{0} \end{pmatrix}.$$

Since B has $n-r$ columns and **Z**, **T** have $n-r$ rows,

$$M(\mathbf{I}_r\mathbf{S}_r\mathbf{A}^{-1}) \leqslant (n-r)\varepsilon M(\mathbf{A}^{-1}).$$

Proceeding in this way we obtain (11.11).
⟦28⟧ We have $(\mathbf{S}_r)_{kl} = 0$ if $l \leqslant r$. Now $\sum$ (coeff. of $(\mathbf{S}_r)_{kl})^2$ over r, k, l such that $l > r$ is $\sum_1 + \sum_2$, where $\sum_1 = \sum (\mathbf{A}^{-1})_{lj}^2$ over r, l such that $i \leqslant r < l$; $\sum_2 = \sum (\mathbf{A}^{-1})_{ik}^2 (\mathbf{A}^{-1})_{lj}^2$ over r, k, l such that $r < k$ and $r < l$. Hence $\sum_1 = \sum (l-i)(\mathbf{A}^{-1})_{lj}^2$ over $l > i$ and $\sum_2 = \sum \min(k-1, l-1)(\mathbf{A}^{-1})_{ik}^2 (\mathbf{A}^{-1})_{lj}^2$ over k, l.

This accounts for the first two summands, that is, the coefficient of η^2 in Turing's equation (except that instead of K we have $K-1$).

The coefficient of η'^2 is $\sum$(coeff. of $(\mathbf{S}'_r)_{kj})^2$ over r, k with $r \geqslant j = \sum [(\mathbf{I}_r)_{ik} + (\mathbf{A}^{-1})_{ik}(\mathbf{I} - \mathbf{I}_r)_{kk}]^2$ over r, k with $r \geqslant j$ $= (n+1) - \max(i,j) + \sum_{k>j} (k-j)(\mathbf{A}^{-1})_{ik}^2$, which is somewhat different from the expression in the text. However, since $\sum \min(k-1, l-1) = \mathrm{O}(n^3)$, $\sum_{l>i} (l-i) = \mathrm{O}(n^2)$

and $\sum_{k>j}(k-j)=O(n^2)$ we still get that the leading term in the R.M.S. error in $(\mathbf{A}^{-1})_{ij}$ is $\eta\{M(\mathbf{A}^{-1})\}^2\mathbf{E}$, where $\mathbf{E}=O(n^{3/2})$.

⟦29⟧ This is because the coefficient of η^2 is $O(n^3)$ but the coefficient of η'^2 is only $O(n^2)$; thus the error $\mathbf{S}_r$ dominates the error $\mathbf{S}'_r$.

⟦30⟧ This comes from the inequality (11.8), with ε' replaced by $\delta' M(\mathbf{A}^{-1})$ and ε by $\delta M(\mathbf{A})$.

⟦31⟧ To see this, recall the proof of (11.8).

⟦32⟧ This should be $\mathbf{s}'_1, \mathbf{s}'_2, \ldots, \mathbf{s}'_n$.

⟦33⟧ $\mathbf{J}_{n-1}\cdots\mathbf{J}_2\mathbf{J}_1\mathbf{A}$ is upper triangular $(=\mathbf{DU})$. Define $\mathbf{J}_n=\mathbf{I}$. Then $\mathbf{J}_n\cdots\mathbf{J}_1\mathbf{A}=\mathbf{DU}$ and $\mathbf{X}_n=\mathbf{J}_n\cdots\mathbf{J}_1=\mathbf{L}^{-1}$. Let $\mathbf{J}_n\cdots\mathbf{J}_1\mathbf{b}=\mathbf{c}$. Then $(\mathbf{DU})^{-1}\mathbf{c}=\mathbf{A}^{-1}\mathbf{b}=\mathbf{x}$. Put $\mathbf{J}_n(\cdots(\mathbf{J}_2(\mathbf{J}_1\mathbf{A}+\mathbf{S}_1)+\mathbf{S}_2)\cdots)+\mathbf{S}_n=\mathbf{E}$ and $\mathbf{J}_n(\cdots(\mathbf{J}_2(\mathbf{J}_1\mathbf{b}+\mathbf{s}'_1)+\mathbf{s}'_2)\cdots)+\mathbf{s}'_n=\mathbf{F}$. Then the error is $\mathbf{E}^{-1}\mathbf{F}-\mathbf{A}^{-1}\mathbf{b}$. Now $\mathbf{X}_n(\mathbf{A}+\sum\mathbf{X}_r^{-1}\mathbf{S}_r)=\mathbf{E}$ and similarly for $\mathbf{F}$, so the error is $(\mathbf{A}+\sum\mathbf{X}_r^{-1}\mathbf{S}_r)^{-1}(\mathbf{b}+\sum\mathbf{X}_r^{-1}\mathbf{s}'_r)-\mathbf{A}^{-1}\mathbf{b}$. If $\mathbf{G}$ is small, $(\mathbf{A}+\mathbf{G})^{-1}$ is approximately $\mathbf{A}^{-1}-\mathbf{A}^{-1}\mathbf{G}\mathbf{A}^{-1}$ so the error is approximately

$$\mathbf{A}^{-1}\sum\mathbf{X}_r^{-1}(\mathbf{s}'_r-\mathbf{S}_r\mathbf{A}^{-1}\mathbf{b}), \qquad (*)$$

which equals $\mathbf{U}^{-1}\mathbf{D}^{-1}\mathbf{X}_n\sum\mathbf{X}_r^{-1}(\mathbf{s}'_r-\mathbf{S}_r\mathbf{U}^{-1}\mathbf{D}^{-1}\mathbf{X}_n\mathbf{b})$.

⟦34⟧ The equation is proved by a method similar to that in note ⟦6⟧. Since $M(\mathbf{J}_i)=1$, $i=1,\ldots,n$, we have $M(\mathbf{X}_r^{-1})=1$.

⟦35⟧ This follows by (*) in note ⟦33⟧. Parentheses are missing in the next line.

⟦36⟧ The right-hand side of this inequality should be divided by n.

⟦37⟧ This should be $\mathbf{d}_r^{-1}[(\mathbf{L}^{-1}\mathbf{b})_r-\sum_{i>r}(\mathbf{DU})_{ri}x_i]$.

⟦38⟧ Let the calculated value of x_r be x'_r. Then $x'_r=\mathbf{d}_r^{-1}[(\mathbf{L}^{-1}\mathbf{b})_r-\sum_{i>r}(\mathbf{DU})_{ri}x'_i]+t_r=\mathbf{d}_r^{-1}[(\mathbf{L}^{-1}\mathbf{b}+\mathbf{Dt})_r-\sum_{i>r}(\mathbf{DU})_{ri}x'_i]$. That is, $\mathbf{DUx}'=\mathbf{L}^{-1}\mathbf{b}+\mathbf{Dt}$. So $\mathbf{DU}(\mathbf{x}'-\mathbf{U}^{-1}\mathbf{t})=\mathbf{L}^{-1}\mathbf{b}$ and the error in $\mathbf{x}$ is $\mathbf{U}^{-1}\mathbf{t}$.

⟦39⟧ $\mathbf{L}=\mathbf{X}_n^{-1}$ so $M(\mathbf{L})=1$. Now $(\mathbf{A}^{-1}\mathbf{LDt})_m=\sum_{k,p}(\mathbf{A}^{-1})_{mk}\mathbf{L}_{kp}\mathbf{d}_p\mathbf{t}_p$, which is in modulus at most $\sum_{k,p}M(\mathbf{A}^{-1})\varepsilon\leqslant n^2\varepsilon M(\mathbf{A}^{-1})$, if $|\mathbf{d}_p\mathbf{t}_p|\leqslant\varepsilon$.

⟦40⟧ The (n,n) entry is $d_n\sum_j(\mathbf{A}^{-1})_{nj}\sum_r(\mathbf{X}_r^{-1}\mathbf{S}_r)_{jn}$. Now $(\mathbf{X}_r^{-1}\mathbf{S}_r)_{jn}=\sum_k(\mathbf{X}^{-1})_{jk}(\mathbf{S}_r)_{kn}\leqslant n\varepsilon$, so the (n,n) entry does not exceed $n^3 d_n M(\mathbf{A}^{-1})\varepsilon$ (not n^2) in absolute value.

⟦41⟧ $\mathbf{L}_{ij}=\sum(\mathbf{J}_1^{-1})_{ir_1}(\mathbf{J}_2^{-1})_{r_1r_2}\cdots(\mathbf{J}_n^{-1})_{r_{n-1}j}$. Each $\mathbf{J}_s^{-1}$ has at most two nonzero entries in each row. Hence there are at most two choices for r_1. For each of these there are at most two choices for r_2, and so on. Thus the number of nonzero summands is at most 2^{n-1}. Since $M(\mathbf{J}_s^{-1})=1$ we have $|\mathbf{L}_{ij}|\leqslant 2^{n-1}$, that is, $M(\mathbf{L})\leqslant 2^{n-1}$.

⟦42⟧ The argument given can be recast as follows. Let the computed values of $\mathbf{L}$, $\mathbf{W}$ be $\mathbf{L}^*$, $\mathbf{W}^*$; let those of $\mathbf{L}^{*-1}$, $\mathbf{W}^{*-1}$ be $\mathbf{K}$, $\mathbf{V}$ and that of $\mathbf{VK}$ be $\mathbf{J}$. Then $\mathbf{L}^*\mathbf{W}^*=\mathbf{A}-\mathbf{S}$, $\mathbf{L}^*\mathbf{K}=\mathbf{I}-\mathbf{S}'$, $\mathbf{VW}^*=\mathbf{I}-\mathbf{S}''$ and $\mathbf{VK}=\mathbf{J}-\mathbf{S}'''$. The error in the computed value of $\mathbf{A}^{-1}$ is

$$\mathbf{J}-\mathbf{A}^{-1}=(\mathbf{I}-\mathbf{S}'')(\mathbf{A}-\mathbf{S})^{-1}(\mathbf{I}-\mathbf{S}')+\mathbf{S}'''-\mathbf{A}^{-1}$$
$$=\mathbf{A}^{-1}\mathbf{S}\mathbf{A}^{-1}-\mathbf{S}''\mathbf{A}^{-1}-\mathbf{A}^{-1}\mathbf{S}'+\mathbf{S}'''$$

to the first order. Assume that $M(\mathbf{S})<\varepsilon$, $M(\mathbf{S}')<\varepsilon'$, $M(\mathbf{S}'')<\varepsilon'$ and $M(\mathbf{S}''')<\varepsilon''$. Then the error is at most

$$n^2\varepsilon M(\mathbf{A}^{-1})^2+2n\varepsilon' M(\mathbf{A}^{-1})+\varepsilon''.$$

1950 *The Word Problem in Semi-groups with Cancellation*

Summary

A *semi-group with cancellation* is a set with an associative product satisfying the cancellation laws

$$ab=ac \text{ implies } b=c, \qquad ba=ca \text{ implies } b=c.$$

If $\mathfrak{S}$ is a set of symbols, an $\mathfrak{S}$-*word* (or briefly a *word*) is a finite sequence of symbols from $\mathfrak{S}$. The product of two words is obtained by writing one after the other.

A *presentation* consists of a set $\mathfrak{S}$ of symbols and a set D of ordered pairs of $\mathfrak{S}$-words. We say that two words U, V are *equivalent* if there is a sequence of pairs of words $P_1, \ldots, P_n$, where P_n is (U, V) and, for each i, P_i either (i) has the form (W, W) or belongs to D, or (ii) arises from some previous P_j by one of the operations

$$(A,B)\to(B,A), \qquad (A,B)\rightleftarrows(As,Bs), \qquad (A,B)\rightleftarrows(sA,sB),$$

where $s\in\mathfrak{S}$, or (iii) is say (A,C) where two previous pairs are (A,B), (B,C) for some B.

It is easily verified that the set of all words modulo this equivalence relation is a semi-group with cancellation which we denote by $\mathrm{sc}(\mathfrak{S},D)$. The product is such that $[A][B]=[AB]$ for all words A,B.

We say that *the word problem is solvable for* $(\mathfrak{S},D)$ if there is an algorithm which determines for any two words whether or not they are equivalent. (Clearly, two words are equivalent if and only if the corresponding classes are equal in $\mathrm{sc}(\mathfrak{S},D)$.)

Theorem 2. *There exists a presentation* $(\mathfrak{S},D)$, *where* $\mathfrak{S}$ *and* D *are finite sets, for which the word problem is not solvable.*

The proof is by reduction to the following known result:
There is a Turing machine $\mathfrak{B}$ *such that there is no algorithm which determines for any initial complete configuration whether or not B will eventually halt with a blank tape.*

A finite presentation $(\mathfrak{S}, D)$ and mappings

$$\varphi_1, \varphi_2 : (\text{complete configuration } C \text{ of } \mathfrak{B}) \mapsto \mathfrak{S}\text{-word}$$

are constructed, for which it is proved that the next theorem holds.

Theorem 1. *$\varphi_1(C), \varphi_2(C)$ are equivalent words if and only if $\mathfrak{B}$ with initial configuration C eventually halts with a blank tape.*

Theorem 2 follows immediately from Theorem 1.

Notes

⟦1⟧ An explicit description of the machine $\mathfrak{B}$ is given in *An Analysis of Turing's "The Word Problem in Semi-groups with Cancellation"* by BOONE (1958). This article is reprinted in the present volume.

Those reading Turing's paper for the first time are strongly advised to do so under the guidance of Boone's article.

⟦2⟧ In each of these relations there should be no commas anywhere in between σ_m and τ_m. Each comma should be replaced by a prime (dash) attached to the previous letter.

⟦3⟧ 'h_h' should be 'n_h'.

⟦4⟧ For 'φ_1' read 'φ_2'.

⟦5⟧ 'υ_i' should be 'r_i' here and two lines later.

⟦6⟧ 'r_h' should be 'n_h' here and three lines later.

⟦7⟧ An alternative argument for the induction step is as follows.

We have $\psi_1(H_{r-1}) = C_{r-1} = AUB$. Now U has one of the forms rs or sl; we have omitted subscripts. Hence for some X, Y, H_{r-1} is $X\upsilon TkY$ or $XjSuY$, respectively. Here T is a word in the τ's and S is a word in the σ's. Also, in both cases we have $\psi_1(X) = A$, $\psi_1(Y) = B$.

Instead of the notation (C, D) for a relation, let us write $C = D$. By the commutation relations,

$$\begin{aligned} H_{r-1} &= X\sigma(T)\upsilon kY \quad \text{or} \quad Xju\tau(S)Y, \quad \text{respectively} \\ &= X\sigma(T)\varphi_1(U)Y \quad \text{or} \quad X\varphi_1(U)\tau(S)Y, \quad \text{respectively.} \end{aligned}$$

Let $H_r = X\sigma(T)\sigma_m\varphi_1(V)\tau_m Y$ or $X\sigma_m\varphi_1(V)\tau_m\tau(S)Y$, respectively. Then in both cases $H_{r-1} = H_r$ and

$$\psi_1(H_r) = \psi_1(X)V\psi_1(Y) = AVB = C_r.$$

Also, since H_{r-1} is normal, so is H_r.

⟦8⟧ From this point onwards, no nontrivial editorial comment is given,

because the article by BOONE (1958) reprinted in this volume, provides all that is necessary; (see note ⟦1⟧).
⟦9⟧ For remarks on this notation and on Lemma 5, see Boone's article.
⟦10⟧ For '(A_1B)' read '(A, B)'.
⟦11⟧, ⟦12⟧, ⟦13⟧ See Boone's article.
⟦14⟧ The last occurrence of 'G_6' should be 'G_4'.
⟦15⟧ For '$G_1 G_5 G_6$' read 'G, G_5, G_6'.
⟦16⟧ '$G_1 F$' should be '$G_2 F$'.
⟦17⟧, ⟦18⟧, ⟦19⟧, ⟦20⟧, ⟦21⟧ See Boone's article.

1953 *Some Calculations of the Riemann Zeta-function*

Summary

In June 1950 the Manchester University computer was used to investigate the zeros of the Riemann zeta-function $\zeta(s)$, $s=\sigma+it$, for

$$2\pi \cdot 63^2 \leqslant t \leqslant 2\pi \cdot 64^2$$

in the hope of finding (nontrivial) zeros off the critical line $\sigma=\frac{1}{2}$. The tentative conclusion was that all the zeros of $\zeta(s)$ in this region are simple zeros on the critical line.

The range $1414 \leqslant t \leqslant 1608$ was also investigated in an attempt to extend Titchmarsh's investigation. He had investigated $0<t\leqslant 1468$ (TITCHMARSH 1936). Due to a computing error, the conclusion here was only that there are no zeros off the critical line for $0<t<1540$.

PART I. GENERAL

Turing first states a version of a result given in Chapter XV of TITCHMARSH (1951), which arises from the approximate functional equation for the zeta-function. This is

Theorem 1. *Let* $\tau \geqslant 64$, $m=[\tau^{1/2}]$, $\tau^{1/2}=m+\xi$,

$$\kappa(\tau) = \frac{1}{4\pi i}\log\frac{\Gamma(\frac{1}{4}+\pi i\tau)}{\Gamma(\frac{1}{4}-\pi i\tau)} - \tfrac{1}{2}\tau\log\tau,$$

$$Z(\tau) = \zeta(\tfrac{1}{2}+2\pi i\tau)\exp(2\pi i\kappa(\tau)),$$

$$\kappa_1(\tau) = \tfrac{1}{2}(\tau\log\tau-\tau-\tfrac{1}{8}),$$

$$h(\xi) = \frac{\cos 2\pi(\xi^2-\xi-\frac{1}{16})}{\cos 2\pi\xi}.$$

Then $Z(\tau)$ is real and

$$Z(\tau)=2\sum_{n=1}^{m} n^{-1/2}\cos 2\pi\{\tau\log n-\kappa(\tau)\}+(-1)^{m+1}\tau^{-1/4}h(\xi)+\Theta(1.09\,\tau^{-3/4}),$$

where $\Theta(\alpha)$ indicates a number whose absolute value is at most α.
Also, $\kappa(\tau)=\kappa_1(\tau)+\Theta(0.006\tau^{-1})$.

Next, to facilitate the computation of $Z(\tau)$, the author wishes to replace $h(\xi)$ by a quadratic expression and relax the definition of m, ξ to: $\tau^{1/2}=m+\xi$, $|\xi-\frac{1}{2}|<0.53$, $m\in\mathbb{Z}$.

Theorem 2. *If $|\xi-\frac{1}{2}|<\frac{1}{2}$, then*

$$h(\xi)=0.373+2.16(\xi-\tfrac{1}{2})^2+\Theta(0.0153).$$

If $|\xi-\frac{1}{2}|<0.53$, then $h(\xi)=0.373+2.16(\xi-\frac{1}{2})^2+\Theta(0.0243)$.

Theorem 3. *Theorem 1 is valid with the error term replaced by $\Theta(1.15\,m^{-3/2})$ if m, ξ are defined instead by $\tau^{1/2}=m+\xi$, $|\xi-\frac{1}{2}|<0.53$, $m\in\mathbb{Z}$.*

Now let $N(\tau)$ be the number of zeros of $\zeta(\sigma+it)$ for $0<t<\tau$, $0\leqslant\sigma\leqslant 1$. Let $S(t)$ be $1/\pi\arg\zeta(\frac{1}{2}+it)$, suitably defined. Then $N(\tau)=2\kappa(\tau/2\pi)+1+S(\tau)$.

Let $S_1(t)=\int_0^t S(u)\,du$.

Theorem 4. *Let $t_2>t_1>168\pi$. Then*

$$S_1(t_2)-S_1(t_1)=\Theta\left(2.3+0.128\log\frac{t_2}{2\pi}\right).$$

The final theorem implies that, in certain cases, $N(2\pi\tau_0)$ can be found exactly.

Theorem 5. *Let $64<\tau_{-R_1}<\cdots<\tau_0<\cdots<\tau_{R_2}$ and let $\kappa_1(\tau_r)=c_r$, $\delta_r=c_r-c_0-\frac{1}{2}r$. Assume that $\delta_{R_2}=\delta_{-R_1}=0$ and $Z(\tau_r)Z(\tau_{r+1})<0$ for $1-R_1\leqslant r\leqslant R_2-2$. Then*

$$-\frac{1}{2}+\frac{2}{R_1}\sum_{1-R_1}^{-1}\delta_r-\frac{0.006}{\tau_{-R_1}}-\frac{2}{R_1}\{0.184\log\tau_0+0.0103(\log\tau_0)^2\}$$

$$\leqslant N(2\pi\tau_0)-2\kappa(\tau_0)-1$$

$$\leqslant\frac{1}{2}+\frac{2}{R_2}\sum_{1}^{R_2-1}\delta_r+\frac{0.006}{\tau_0}+\frac{2}{R_2}\{0.184\log\tau_{R_2}+0.0103(\log\tau_{R_2})^2\}.$$

As an example of the use of Theorem 5 Turing proves that there is a number τ in the interval $|\tau - 551| < 0.05$ for which $N(2\pi\tau_0) = 1103$.

With the aid of Theorem 5 we can find various pairs (t_0, t_1) for which the number of zeros of $\zeta(s)$ in the region $t_0 \leqslant t \leqslant t_1$, namely $N(t_0) - N(t_1)$, is known exactly. To attempt to prove that all zeros of $\zeta(s)$ in one such region are simple zeros on the line $\sigma = \frac{1}{2}$ we could proceed as follows. Calculate the sign of $Z(t/2\pi)$ for several points t of the interval (t_0, t_1), taking more points if necessary in the hope that eventually the number of changes of sign obtained is equal to the known number of zeros. If we succeed the desired result is proved.

Part II. The Computations

The author gives a brief description of the Manchester computer and states some of the practical computing strategy employed for calculating $Z(\tau)$. The values of τ taken, say $\tau_0, \tau_1, \tau_2, \ldots,$ were obtained automatically by a simple formula expressing τ_{n+1} in terms of τ_n, $\kappa(\tau_n)$. The definition ensures that $2\kappa(\tau_0)$ is near an integer p while $2\kappa(\tau_n)$ is near the integer $p+n$. More values of τ were then considered near those points for which $Z(\tau)\cos 2\pi\kappa(\tau) < 0$.

The computing method naturally requires that τ should not be too large. The range of τ chosen was $63^2 \leqslant \tau \leqslant 64^2$; for such τ, $m = 63$ and the error in computing $Z(\tau)$ is then less than 0.02.

Remark

In his paper, Lehman (1970) gives an exposition, following Turing's method, of a slightly modified version of Theorem 4:

If $t_2 > t_1 > 168\pi$, then

$$|S_1(t_2) - S_1(t_1)| \leqslant 1.91 + 0.114 \log\left(\frac{t_2}{2\pi}\right).$$

The main points of interest are as follows.

(i) A justification is provided of Turing's Lemma 3, or rather of a slightly modified form of it.

(ii) The author detects and puts right an error in the interchange of limits in the proof of Turing's Lemma 9.

Notes

⟦1⟧ Instead of '$-\frac{1}{4}$' read '$-\frac{1}{2}$'.

⟦2⟧ Delete the minus sign in the exponent.

⟦3⟧ Replace '$-\frac{1}{2}$' by '$-\frac{1}{8}$'.
⟦4⟧ Apart from the error terms, the equations for $Z(\tau)$ and $\kappa(\tau)$ and the fact that $Z(\tau)$ is real, follow from 15.3 of TITCHMARSH (1951); Turing's $Z(\tau)$ is the $Z(2\pi\tau)$ of Titchmarsh and we write $\tau = t/2\pi$. Moreover, $\kappa(\tau)$ is real.
⟦5⟧ In this paper we are almost entirely concerned with regarding $Z(\tau)$ as a (real valued) function of a real variable τ. However, at one point the case of complex τ is considered; there is an obvious one-one correspondence between the zeros of $Z(\tau)$ in the region $\mathscr{R}\tau > 0$, $|\mathscr{I}\tau| \leqslant \frac{1}{4}\pi$ and the zeros of ζ in the region $0 \leqslant \sigma \leqslant 1$, $t > 0$.
⟦6⟧ One may justify this remark as follows. In Turing **(6)**, we have for s on the critical line

$$\zeta(s)\Gamma(\tfrac{1}{2}s)\pi^{-s/2} = -2\mathscr{R}(\Gamma(\tfrac{1}{2}s)\pi^{-s/2}I(s)),$$

where $I(s) = -\sum_{n=1}^{m} 1/n^s + \cdots$ and m is approximately $\tau^{1/2}$. Let $B(s) = \Gamma(\frac{1}{2}s)\pi^{-s/2}$. Then $\exp 2\pi i\kappa(\tau) = lB(s)$ where l is real; this can be proved either directly or as follows: $\zeta(s)B(s)$ is real and so is $\zeta(s)\exp 2\pi i\kappa(\tau)$; hence their quotient is real. Thus Turing **(6)** implies that $Z(\tau) = \zeta(s)lB(s) = -2\mathscr{R}(\exp 2\pi i\kappa(\tau)(-\sum 1/n^s + \cdots)) = 2\sum_{n=1}^{m} n^{-1/2}\cos 2\pi(\tau\log n - \kappa(\tau)) + \cdots$.

I have not investigated Turing's claim that the formula of his paper **(6)** is more accurate [in what sense?].
⟦7⟧ Let $J(\xi) = 0.373 + 2.16(\xi - \frac{1}{2})^2$. Then $f = h - J$. Now h and J are symmetrical about $\xi = \frac{1}{2}$, hence so is f. $f(\xi)$ is small for $\xi = \frac{1}{30}n$ when $15 \leqslant n \leqslant 32$ and hence when $-2 \leqslant n \leqslant 32$. Let $|\xi - \frac{1}{2}| < 0.53$. Then n exists with

$$-2 \leqslant n-1 < n \leqslant 30\xi < n+1 < n+2 \leqslant 32.$$

Take $\xi_i = \frac{1}{30}(n+i-2)$, $i = 1,2,3,4$. To prove $|P(\xi)|$ is small, proceed as follows. $P(\xi_i) = f(\xi_i) = \eta_i$, say. Let $d = \frac{1}{30}$. Then

$$P(\xi) = \eta_1 + \frac{u_1(\xi-\xi_1)}{d} + \frac{u_2(\xi-\xi_1)(\xi-\xi_2)}{2d^2} + \frac{u_3(\xi-\xi_1)(\xi-\xi_2)(\xi-\xi_3)}{6d^3}.$$

Since the η_i are small, so are the first, second and third differences u_1, u_2, u_3 and we have $|P(\xi)| < |\eta_1| + 2|u_1| + |u_2| + |\frac{1}{3}u_3|$.
⟦8⟧ $(-)^m$ should be $(-)^{m+1}$.
⟦9⟧ The new error is

$$-2(m+1)^{-1/2}\cos 2\pi[\tfrac{1}{2}m^2 + m - \tfrac{1}{2} - E - e] + (-1)^{m+1}\tau^{-1/4}2\cos 2\pi E,$$

where $E = \xi^2 - 2\xi - \frac{1}{16}$ and $|e| \leqslant (\xi-1)^3/(3m+3) + 0.006(m+1)^{-2}$. One then has to consider the cases m even, m odd.
⟦10⟧ $N(\tau)$ is odd or even according as $Z(\tau/2\pi)$ is positive or negative; for $\arg Z(\tau) = \arg\zeta(\frac{1}{2} + 2\pi i\tau) + 2\pi\kappa(\tau) = \pi(S(2\pi\tau) + 2\kappa(\tau)) = \pi(N(2\pi\tau) - 1)$. So

if for a given τ the sign of $Z(\tau/2\pi)$ is known, $N(\tau)$ and hence $S(\tau)$ are known modulo 2.

⟦11⟧ The total 'detour' may be taken to consist of line segments parallel to the x-axis and parts of small circles with centres at the zeros of $\zeta(s)$. The real part of the integral of $\log \zeta(s)$ over a line segment y = constant is independent of the branch of log, while the real part of the integral over one of the small circles tends to zero with the radius.

⟦12⟧ We want to use Theorem 1, so $0 \leqslant \xi < 1$. If $\xi \neq 0$, then $h(\xi) < 0.373 + \frac{1}{4}(2.16) + 0.0153$ while $h(0) = \cos \frac{1}{8}\pi$. Hence $h(\xi) < 0.95$,

$$|\zeta(\tfrac{1}{2} + 2\pi i\tau)| = |Z(\tau)| < 2 \sum r^{-1/2} + 0.95\tau^{-1/4} + 1.09\tau^{-3/4}.$$

⟦13⟧ Replace the first 'τ' by 'ζ' and '1.2' by '1.09'.

⟦14⟧ (a) From the series expansions we have, if $\sigma > 1$. $|\zeta(s)| \leqslant \zeta(\sigma)$, $|\log \zeta(s)| \leqslant \log \zeta(\sigma)$ but

$$\left|\frac{\zeta'(s)}{\zeta(s)}\right| \leqslant -\frac{\zeta'(\sigma)}{\zeta(\sigma)}.$$

(b) The integral from $1.25 + it$ to ∞ is intended to be $1.25 + it$ to $\infty + it$.

⟦15⟧ The exponent should be '$-\frac{3}{8} + \frac{1}{4}s$'.

⟦16⟧ By Lemma 2, if $t \geqslant 128\pi$, then $|\zeta(\frac{1}{2} + it)| < 4(t/2\pi)^{1/4} < 4t^{1/4}$. Hence $|f(\frac{1}{2} + it)| < 4$.

⟦17⟧ We use that $\tan^{-1} x \leqslant x$ for $x \geqslant 0$.

⟦18⟧ For $\sigma \geqslant \frac{1}{2}$, $t \geqslant 1$ we have $|\zeta(s)| < 9t^{1/2} = 9(128\pi)^{1/2}$.

⟦19⟧ An appropriate version of the Phragmén–Lindelöf theorem is that if $|\varphi(s)| \leqslant M$ on the half lines $\sigma = \pm c$, $t \geqslant 0$ and on the line segment $-c \leqslant \sigma \leqslant c$ and if $|\varphi(s)| < A \exp(\exp t\beta)$, where $0 < \beta < \pi/2c$ and $A > 0$, then $|\varphi(s)| \leqslant M$ throughout the infinite strip bounded by the half lines and the line segment.

In our case $c = \frac{3}{8}$; take $\beta = 1$.

⟦20⟧ We use here that $t > 168\pi$.

⟦21⟧ This follows from $|\zeta(s)| \leqslant 4(\exp 0.1)(t^2 + (\sigma - \frac{1}{2})^2)^a$, where $a = \frac{1}{2}(\frac{3}{8} - \frac{1}{4}\sigma)$.

⟦22⟧ We need $a \neq 0$.

⟦23⟧ As an alternative to the first part of the proof (reduction to the real case) one has the following:

$$\mathscr{R} \int_{a-1}^{a} \log \frac{z}{z+1} \, dz = \int_0^1 \log \left|\frac{a-\lambda}{a-\lambda+1}\right| d\lambda = F(\alpha, \beta), \quad \text{say},$$

where $a = \alpha + i\beta$. One now proves that $F(\alpha, \beta) \geqslant F(\alpha, 0)$. This is easy if $\alpha \geqslant \frac{1}{2}$, for then $(\alpha + 1 - \lambda)^2 \geqslant (\alpha - \lambda)^2$. When $0 < \alpha < \frac{1}{2}$ put $\mu = \lambda - \frac{1}{2} - \alpha$, calling the new integrand E. Split the interval of integration into $[-\frac{1}{2} - \alpha, 0]$ and $[0, \frac{1}{2} - \alpha]$. For the second interval put $\nu = -\mu$. We now get $\int_{-\frac{1}{2}-\alpha}^{\alpha-\frac{1}{2}} E \, d\mu$.

Here $(\frac{1}{2}-\mu)^2 \geqslant (\mu+\frac{1}{2})^2$ so the result follows.

⟦24⟧ At the larger solution the function $\psi(a)+k/a$, a real, has a *maximum*. As $a \to 1+$ or $a \to 1-$, the function tends to $k-2\log 2>0$. As $a \to +\infty$ the function tends to 0. Thus it is only necessary to show that the function is positive at the smaller solution.

⟦25⟧ The sign before the integral should be changed.

⟦26⟧ Let $\mu>1$, $x>0$, $z=x-\mathrm{i}\mu x$. Then $\mathscr{R}z>0$ but the integrand has a pole at $u=\mu x+\mathrm{i}x$, which lies in between the old and new lines. However, if we change the hypothesis of Lemma 8 to $\mathscr{R}z>0$, $\mathscr{I}z>0$, the lemma is true and it is adequate for later applications.

⟦27⟧ The term $-\frac{5}{2}/(t^2+\frac{25}{4})$ should be replaced by an upper bound for $\sigma/(t^2+4\sigma^2)$, for example $\frac{9}{4}/(t^2+\frac{81}{4})$. Further, the numerator 2 of the third term should be replaced by 1.

⟦28⟧ Replace '$-\frac{1}{2}\log\frac{1}{2}t$' by '$-\frac{1}{2}\log\frac{1}{2}t+c$', where c is a suitable small numerical constant. (See note ⟦27⟧.)

⟦29⟧ The integral equals $\int_{w-\varrho-1}^{w-\varrho}\log(z/z+1)\,\mathrm{d}z$, where $w=\frac{3}{2}+\mathrm{i}t$. Since $0<\mathscr{R}\varrho<1$, Lemma 7 is available. Since $\mathscr{R}(k/\varrho)>0$ we have, noting that the resulting series converge,

$$E \underset{\text{def}}{=} \mathscr{R}\left(\sum \int_{w-\varrho-1}^{w-\varrho} \log\frac{z}{z+1}\,\mathrm{d}z\right) \geqslant -1.49\,\mathscr{R}\sum\left(\frac{1}{w-\varrho}+\frac{1}{\varrho}\right)$$

$$= e_2 - 1.49\,\mathscr{R}\left[\left(\frac{\zeta'}{\zeta}\right)(w) - \tfrac{1}{2}\log\pi + \tfrac{1}{2}\left(\frac{\Gamma'}{\Gamma}\right)(\tfrac{1}{2}w+1)\right],$$

where $e_2=-1.49\,\mathscr{R}(-b_1+1/(w-1))$. e_2 is 'small' since b_1 is approximately -0.023 by INGHAM (1931), p. 58, and $t>50$. Hence by Lemma 8, Lemma 3 and the inequality $\frac{1}{4}\log(\frac{1}{4}t^2+\frac{49}{16})<\frac{1}{2}\log t/2\pi$ we obtain

$$E \geqslant e_2 - 1.49\left[2.62 + \tfrac{1}{2}\log\frac{t}{2\pi} + e_3\right],$$

where $e_3=(\pi^2\,|\frac{1}{4}t^2-\frac{49}{16}|)^{-1}$.

⟦30⟧ Using these lemmas I obtained $s_1(t_2)-s_1(t_1)=\Theta(2.36+0.116\log t_2/2\pi)$.

⟦31⟧ The right-hand side should be the first expression plus $1/2\pi$ times the integral.

⟦32⟧ The following change should be made in the hypothesis: define c_r to be $\kappa_1(\tau_r)$. (The statement $\delta_{R_2}=\delta_{-R_1}=0$ is an assumption, not a definition.) Replace $N(2\pi\tau_0)-2c_0-1$ by $N(2\pi\tau_0)-2\kappa(\tau_0)-1$.

⟦33⟧ $2/R_1$ should be $2/R_2$ (in this line only).

⟦34⟧ R_2-1 should be R_2-2.

If $Z(\tau_r)Z(\tau_{r+1})<0$, then $Z(\tau)=0$, that is, $\zeta(\frac{1}{2}+2\pi\mathrm{i}\tau)=0$ for some τ such that $\tau_r<\tau<\tau_{r+1}$; hence $N(2\pi\tau_{t+1})\geqslant N(2\pi\tau_r)+1$.

In the interval $(\tau_{R_2-1}, \tau_{R_2})$ we only have $N(2\pi\tau) \geqslant N(2\pi\tau_{R_2-1}) \geqslant N(2\pi\tau_0) + (R_2 - 1)$.

In the next part of the proof, R means R_2.

⟦35⟧ $(c_R - c_{R-1})$ should be subtracted from the right-hand side. Hence in the next line $\sum \delta_r$ should be replaced by

$$(\sum \delta_r) + \delta_R - \delta_{R-1} + \tfrac{1}{2} \quad \left(= \tfrac{1}{2} + \sum_{r=1}^{R-2} \delta_r\right)$$

and the same replacement should be made in the conclusion of Theorem 5.

⟦36⟧ 0.006 should be 0.012 and, in the next line, 0.003 should be 0.006.

If $D(\tau) = \kappa(\tau) - \kappa_1(\tau)$, then $|D(\tau)| \leqslant 0.006\tau^{-1}$. Also, $c_R - c_0 = \tfrac{1}{2}R$, since $\delta_R = 0$.

⟦37⟧ In particular since $c_8 - c_7 = c_0 + 4 - c_7 \simeq 551 + 4 - 554.5 > 0$, we have $\tau_8 > \tau_7$.

⟦38⟧ We only need $|N(2\pi\tau_0) - 2c_0 - 1| < 1.9$ for what follows.

⟦39⟧ Recall that $N(2\pi\tau)$ is odd or even according as $Z(\tau)$ is positive or negative. Also, $\cos 2\pi\kappa(\tau_0) \simeq \cos(2\pi(551)) = 1$.

⟦40⟧ $\delta = 2\kappa(\tau) - n$, $n \in \mathbb{Z}$, not $n - 2\kappa(\tau)$.

In the next line, replace $\kappa(\tau)$ by $2\kappa(\tau)$, so that $|\delta| < 0.125$.

Actually the conclusion that $2\kappa(\tau')$ differs from an integer by less than 0.125 is only true if we ignore a very small quantity. To make the result exact, replace 0.1 by 0.09 (say) or (in both the hypothesis and the conclusion) replace 0.125 by 0.126 (say).

⟦41⟧ (i) E is defined later.

(ii) Is this criterion correctly stated? The computed value of $Z(\tau)$ may be much less than $-E$ or much greater than E and H may be positive or negative (not respectively). Of these four cases the criterion holds for two but fails for the other two. I suspect that it should be $|Z(\tau)H(\kappa)| > 0.31E$.

(iii) Note that $H(\kappa)\cos 2\pi\kappa < 0$ for all κ.

⟦42⟧ Using 0.09 instead of 0.1, as suggested in note ⟦40⟧, ensures that if the initial value of $2\kappa(\tau)$ differs from an integer by less than 0.125, then all subsequent values will differ from an integer by less than 0.12.

Now if $2\kappa = n + e$, where $|e| < 0.12$ and $n \in \mathbb{Z}$, then $\kappa - \tfrac{1}{4}$ = integer + λ, where $|\lambda| < \tfrac{1}{4} + \tfrac{1}{2}(0.12) = 0.31$.

Also note that $|H(\kappa)| = |\lambda| > \tfrac{1}{4} - \tfrac{1}{2}(0.12) = 0.19$.

Turing's Work on the Zeta-function

[Thanks are due to Dr D.R. Heath-Brown for the following assessment of Turing's papers on the zeta-function.]

In the first place, it says a lot for Turing that he should produce this work, of a highly specialist and difficult kind, in an area that was something of a side-line for him. The 1943 paper is distinctly unpleasant reading, even compared with other papers on the zeta-function, but that is partly because of the numerical nature of the work. Titchmarsh's paper, to which he refers, is similar.

As to the paper's significance, the 1943 paper is now forgotten: it is superseded by Titchmarsh's method. The latter has at least a more elegant result, if the proof is more disagreeable. One has formulae for $\zeta(\frac{1}{2}+it) =$ sum + error, in which the error is $O(t^{-c})$, c constant. Typically $|\zeta(\frac{1}{2}+it)|$ is no smaller than $O(\exp(-\text{constant}(\log t)^{1/2}))$. [In fact $\log|\zeta(\frac{1}{2}+it)|/(\frac{1}{2}\log\log t)^{1/2}$ has a standard $N(0,1)$ distribution in a suitable sense.] So an error, $O(t^{-1/4})$ even, should usually be good enough to detect sign changes of $Z(t)$. The problem is that for 'small' t ($\leqslant 1000$, say) the constants in the 0-terms are not very good. However now, with the calculations having passed the 1.5×10^9-th zero, such problems are irrelevant. Turing's method could be interpreted loosely as follows: approximating $Z(\tau)$ by $2\sum_1^{[\sqrt{\tau}]}$ is unsound, as the sum is discontinuous: use instead $2\sum_{n=1}^{\infty} W(n/\tau^{1/2})\ldots$, where $W(x)$ is a smooth weight, such as e^{-x}. In fact one can do this with various weights—much nicer ones than Turing produces, and get very good approximations. Such methods are not used for numerical calculations, because (i) the error in Titchmarsh's formula is small enough for most practical purposes, and (ii) $W(n/\tau^{1/2})$ is an awkward extra quantity to compute.

The 1953 paper is distinctly more interesting. Number theory has always been a popular area for computer 'number crunching'. This paper is the first such instance for the zeta-function. Note that even here the method for computing $Z(\tau)$ is Titchmarsh's. Undoubtedly the best bit of the paper is in Theorems 4 and 5; showing that the zeros found on the critical line are indeed *all* the zeros to that height. This is much easier and more elegant than the technique used by Titchmarsh (which was rather hit or miss—the new method is fail-safe). It is the method adopted in all recent computations.

Roger HEATH-BROWN

1954 *Solvable and Unsolvable Problems*

Notes

⟦1⟧ See CROWELL and FOX (1977).
⟦2⟧ See note ⟦18⟧.
⟦3⟧ Another name for 'substitution puzzle' is 'semi-Thue system'.

⟦4⟧ This statement needs to be qualified a little. Proving a theorem in group theory, in the sense that this is ordinarily understood by a mathematician, would not be an example of a puzzle. If, however, one is restricted to proofs within first-order logic, this would be an example of a puzzle.
⟦5⟧ References are: CHURCH (1936 A and B); KLEENE (1935/1936); POST (1936); TURING (1937 A).
⟦6⟧ This statement is still valid today.
⟦7⟧ Few would disagree with this but there are interesting questions. Given the existence of an algorithm for some decision problem, we may ask, for example, does there exist an algorithm which can be carried out in polynomial time? Complexity theory is now a very active branch of mathematics.
⟦8⟧ GÖDEL (1931).
⟦9⟧ See note ⟦5⟧. (The Entscheidungsproblem is concerned with the existence of an algorithm for deciding for any sentence of a first-order language whether or not it is logically valid. The result that no such algorithm exists is known as Church's theorem.)
⟦10⟧ This is not a deep result and it is fairly easy to prove.
⟦11⟧ For example, the halting problem for Turing machines.
⟦12⟧ This is a consequence of Church's theorem. For if A is a first-order sentence then the axiom 'Not A' leads to a contradiction if and only if A is logically valid.
⟦13⟧ This was proved by Turing. His proof is included in this volume.
⟦14⟧ The result that there exists a finitely presented semi-group with unresolvable word problem was proved by MARKOV (1947) and POST (1947).

The corresponding result, that there exists a finitely presented group with unsolvable word problem was proved by NOVIKOV (1955).

This significant result is often jointly credited to Boone, who wrote a series of papers in 1954–1957; for references see BOONE (1959).

The method sketched in the earlier announcement by Novikov in 1952 was inadequate (NOVIKOV (1952)).

Biographical Remark

When Novikov's 1952 announcement appeared, Newman translated it and gave a copy to Turing. After studying it for a few days, Turing declared that it could not be right.

(With the hindsight of reading Novikov's full and correct 1955 proof, I studied the 1952 version and concluded that the method was inadequate (BRITTON (1958)).)

⟦15⟧ The precise result is that if $n \geqslant 6$, there exists a set of 102 matrices, $n \times n$ over the integers, such that there is no algorithm for deciding of an

arbitrary matrix with integral entries whether or not it is a product with each factor one of the given matrices (Markov (1951)).
⟦16⟧ A specialization of this is the following. As mentioned in note ⟦14⟧, there is a finitely presented group whose word problem is unsolvable. This raises the question: is there an algorithm which will decide of a given finite presentation whether or not its word problem is unsolvable? That no such algorithm exists was shown by Adyan (1957).
⟦17⟧ Markov has shown that there is no algorithm for deciding of two 4-manifolds whether or not they are homeomorphic. The proof is by reduction to the unsolvability of the isomorphism problem for groups, that is, to the result that there is no algorithm for deciding of any two finite presentations of groups whether or not the groups are isomorphic (Adyan (1957); Markov (1958); Rabin (1958)).
⟦18⟧ (a) Hemion has shown that the decision problem for knots is solvable and that the homeomorphism problem for a large class of 3-manifolds is solvable (Hemion (1979)).

(b) Another decision problem which Turing might have included is Hilbert's 10th problem: is there an algorithm which decides for any polynomial $p(x_1,\dots,x_n)$ over the integers whether or not there exist integers $a_1,\dots,a_n$ such that $p(a_1,\dots,a_n)=0$?

This was shown to be unsolvable by Matijasevič (1970); his proof was based on the work of Davis, Putnam and Robinson (1961).

A Note on Normal Numbers

Notes

⟦1⟧ In fact, Champernowne (1933) gave an example of a normal number.
⟦2⟧ More accurately, γ is $u_1,\dots,u_r$ where $0\leqslant u_i<t$, $i=1,\dots,r$, and $u_1,\dots,u_r$ are all different.
⟦3⟧ The second summation should be from 0 to R. The right-hand side should be $(R-r+1)t^{-r}$; however this formula is not needed in the sequel.
⟦4⟧ Fix R, T and r and simplify the notation to $N(\gamma,n), S(\alpha,\gamma)$. Let J be the set of all R-figure integers in scale t and for $a\in(0,1)$ let $\varphi(a)\in J$ be the first R figures after the decimal point in the expansion of a in scale t.

If $\xi\in J$, let $S'(\xi,\gamma)$ be the number of occurrences of γ in ξ. We have $N(\gamma,n)$ = number of elements of J in which γ occurs exactly n times. Now

$$\text{number of pairs } (\xi,\gamma) \text{ such that } S'(\xi,\gamma)=x$$
$$=\sum_{\gamma} (\text{number of pairs } (\xi,\gamma) \text{ such that } S'(\xi,\gamma)=x)$$

$$= \sum_{\gamma} \text{(number of } \xi \text{ such that } \gamma \text{ occurs } x \text{ times in } \xi)$$

$$= \sum_{\gamma} N(\gamma, x).$$

Let $\Delta < 0.3$. The number of pairs such that $|S'(\xi, \gamma) - Rt^{-r}| > R/t^r\Delta$ equals $\sum N(\gamma, x)$ over γ and x where x satisfies $|x - Rt^{-r}| > R/t^r\Delta$. This sum is less than $2t^R \exp(-k^2t^r/4R)$. The set of all a such that $\varphi(a)$ is a given ξ is an interval of length t^{-R}. Also, $S(a, \gamma) = S'(\varphi(a), \gamma)$. Hence the measure of the set of all a in (0, 1) such that $|S(a, \gamma) - Rt^{-r}| > R/t^r\Delta$ is less than $2 \exp(-k^2t^r/4R)$, that is, $2 \exp(-Rt^{-r}/4\Delta^2)$. The required inequality follows.

⟦5⟧ Replace T^{L+1} by $2T^{L+1}$.

⟦6⟧ A_{k+n} is a finite sum (i.e., union) of intervals since each $B(\Delta, \gamma, t, R)$ is; this is essentially because in the notation of note ⟦4⟧, $\varphi^{-1}(\xi)$ is an interval for each ξ in J.

Moreover the end-points of these intervals are rational.

⟦7⟧ The proof of this theorem that is given is certainly inadequate. Indeed I suspect that the theorem is false.

In the theorem let $\theta(p) = 0$ for all p or, more generally, let θ be recursive. Then in the notation of the proof we can recursively enumerate $m_1, m_2, \ldots$, so the intervals $I_n = (m_n/2^n, (m_n + 1)/2^n)$ are recursively enumerable. Assuming that the intersection of these intervals is a single point x and noting that $m(I_n) = 2^{-n}$ we see that x is not Martin-Löf random (see CHAITIN (1987)).

⟦8⟧ There seems to be a serious gap in the argument giving the construction of these intervals. It is trivial to find such intervals nonconstructively but we have to proceed constructively.

Note that if m and n are given we can calculate $a_{n,m}$ and $b_{n,m}$ (see note ⟦9⟧). Suppressing n, we have that (a_m) is a decreasing sequence with limit $a \geqslant 0$ (unknown) and similarly $b_m \to b$ (unknown). Let $c = a + b$; then $c > 0$. Assume that $0 < c' \leqslant c \leqslant c''$, where c' and c'' are known and $c'' < 2c'$. Choose x and y such that $0 < x < c'$, $0 < y < c'$, $c'' < x + y$. Then some $a_m < x$ or some $b_m < x$ or for some m both $a_m < y$ and $b_m < y$; otherwise, for all m, $a_m + b_m \geqslant x + y$ so $c \geqslant x + y > c''$.

In the first case $a < x$ so $b > c - x \geqslant c' - x > 0$; similarly in the third case $a < y$, $b < y$ so $a > c' - y > 0$ and $b > c' - y > 0$. (This analysis shows that perhaps '$>1/k(k+n+1)$' should be '$<1/k(k+n+1)$'.)

In the first case we choose the right half-interval, in the second case the left one and in the third case we make a choice determined by θ.

If something of the form above is what the author had in mind for obtaining I_{n+1} from I_n, I see no way of defining x, y, c', c'' as functions of n which would allow the construction of all the intervals I_n.

⟦9⟧ The symbol 'm' on the left clearly denotes the measure of the intersection of the two sets. The right-hand side depends on n and another variable, unfortunately called m, so evidently $c(k, n)$ should be $c(k, m)$.

⟦10⟧ What is to exclude the case when, from some point onwards, always the left half-interval is chosen? In this case the intersection of the intervals is empty. (However, if the intersection is empty, then the point in common to all the corresponding closed intervals is rational.)

Assuming the intersection is nonempty, why is the number normal? (Consider the analogous situation: $I_n = (\frac{1}{2} - \frac{1}{3}/n, \frac{1}{2} + \frac{1}{3}/n)$, $E = (0, \frac{1}{2}) \cup (\frac{1}{2}, 1)$. Then $I_n \cap E$ has positive measure, $\bigcap I_n = \{\frac{1}{2}\}$ but $\frac{1}{2}$ is not in E.)

⟦11⟧ Replace 2^{-n-1} by 2^{-n} (and replace $1 - 2/k$ by $1 - 3/k$)?

⟦12⟧ There are three reference numbers in the text but there are no corresponding references.

For further information on normal numbers, see KUIPERS and NIEDERREITER (1974).

The Word Problem in Compact Groups

Notes

⟦1⟧ Let L be the first-order language with equality having one binary relation symbol $<$ and the binary function symbols $-$ and $\times$. A sentence A of L is said to be *true for* $\mathbb{R}$ if it is true for the real numbers when $=, <, -, \times$ have their usual meanings. Tarski's theorem may be expressed by saying that there is an algorithm which given any sentence A determines whether or not A is true for $\mathbb{R}$.

Note for later that we can express $y = 0$ in L as $y - y = y$. Also,

$$yy = y \quad \text{and not } y - y = y$$

expresses $y = 1$, $z = x - ((x - x) - y)$ expresses $z = x + y$, and the formula $zy = x$ and not $y - y = y$ expresses $z = xy^{-1}$.

⟦2⟧ For procedure (a) to make sense, the group G must be given in terms of generators and defining relations, the set of generators must be recursively enumerable and the set of defining relations must also be recursively enumerable. In particular G must be countable; this rules out, for example, the unitary group $U(n)$.

⟦3⟧ Before 'matrices' insert 'complex nonsingular'.

'order r' means of course that the matrices are $r \times r$.

⟦4⟧ This means that, if the statement in quotation marks is denoted by S_r, then there is a sentence A_r in the first-order language in note ⟦1⟧ such that

S_r is true if and only if A_r is true for $\mathbb{R}$. (To see this, begin by expressing the (p,q)-th entry of E_k in the form $x_{pqk}+\mathrm{i}y_{pqk}$, where x_{pqk}, y_{pqk} are variables.)

⟦5⟧ Observe that both M and N have to be finite for A_r to be a sentence of first-order logic. Thus the main theorem is really that the word problem is solvable in finitely presented compact groups.

⟦6⟧ Tarski (1951) is a 'second edition' of this RAND report.

For more details of Tarski's work, see Doner and Hodges (1988).

On Permutation Groups

Summary

We consider permutations of T objects. Let S and A denote the symmetric and alternating group, respectively. R will denote a fixed T-cycle, $R=(a_1,a_2,\dots,a_T)$.

For any permutation U, $J(U)$ or J denotes the subgroup of S generated by R and U; $H(U)$ or H denotes the normal subgroup of J consisting of all expresions of the form

$$R^{t_0}UR^{t_1}\dots UR^{t_p},\quad \sum t_i=0.$$

We are interested in H rather than J. In studying J and H the permutation U is called the *upright*.

A subgroup of S is *unexceptional* if it is A or S.

Theorem I. *Let $U\neq 1$, $T\neq 4$. Then H is unexceptional if and only if J is unexceptional.*

We say that U is *unexceptional* if H is unexceptional. The problem considered is that of finding all exceptional H.

Theorem II. *Let $T>4$. If m is prime to T and J contains a permutation of the form $(\alpha,R^m\alpha)$ or $(\alpha,R^m\alpha,\beta)$ or $(\alpha,R^m\alpha)(\beta,\gamma)$, where in the third case the permutation does not commute with $R^{T/2}$ if T is even, then J is unexceptional: it contains all 3-cycles and, in the first case, also all 2-cycles.*

Note. If J contains (α,R^m,α) where m is not prime to T, then H is an intransitive group (and hence is exceptional). The same applies if J contains a 3-cycle $(\alpha,R^m\alpha,R^{m+p}\alpha)$, where some prime number divides all of m,p,T. In the third case above, if the permutation does commute with $R^{T/2}$, then every member of H does and hence H is exceptional.

In practice we express $U, UR, UR^2, \ldots$ in cycles and hope to find an appropriate power of one of these which satisfies the hypothesis of the theorem. If we are successful, then U is unexceptional and can be rejected.

Note that

if $U' = R^m UR^n$ for some m, n then $J(U') = J(U)$;

if V commutes with the group (R), then $H(VUV^{-1}) \cong H(U)$;

if $U' = U^m$ and $U = U'^n$, then $H(U') = H(U)$.

We say that U *has a beetle* [sic] if some $R^m UR^n$ fixes two or more objects.

Define the vector $(f(1), \ldots, f(T))$ by

$$R^{f(i)} Ua_i = Ua_{i+1}, \quad i = 1, \ldots, T,$$

where a_{T+1} means a_1.

If the numbers

$$f(1) + \cdots + f(n) - n, \quad n = 0, 1, \ldots, T-1,$$

are all different, then T is odd. If, as is usually the case, these numbers are not all different, then U has a beetle.

The vector for $R^m UR^n$ turns out to be just a cyclic permutation of that for U. A suitable cyclic permutation of the vector for U is called the *invariant* of U.

If U has a beetle, we may assume without loss of generality that U fixes two or more objects, one of which is a_1. It is shown that the second object fixed may be taken to be $R^t a_1$ $(= a_{t+1})$, where t divides T and $1 \leqslant t < T$.

The author now embarks on a lengthy, detailed and laborious search for exceptional groups, covering the cases $T = 1, 2, \ldots, 8$.

In particular when $T = 8$, it is found that there are 2144 exceptional uprights out of the $8! = 40\,320$ possible uprights. Among the exceptional groups there is a group of order 336 whose intersection with A is the simple group of order 168.

The last few pages of the paper are on a different but related topic.

Let H be a finite group of order h with generators $U_1, \ldots, U_k$, where these elements are all different. Let E be the set of all real-valued functions on H. For $\theta \in E$, let $\|\theta\| = (h^{-1} \sum \theta(a)^2)^{1/2}$.

$f \in E$ is defined as follows

$$f(a) = \begin{cases} k^{-1}, & \text{if } a \text{ is one of } U_1, \ldots, U_k, \\ 0, & \text{otherwise.} \end{cases}$$

g denotes an element of E such that $g(a) \geqslant 0$ for all a and $g(a) > 0$ for some a. R_f is defined by

$$(R_f g)(a) = \sum_h f(ab^{-1}) g(b).$$

Lemma. *$\|R_f g\| \leqslant \|g\|$ and equality holds only if $g(a^{-1}bx)/g(x)$ is independent of x for any a, b for which $f(a) \neq 0$ and $f(b) \neq 0$.*

If a subsequence of $g, R_f g, R_f^2 g, \ldots$ tends to a limit k, then k is called a *limiting distribution.*

H_1 denotes the normal subgroup of H consisting of all expressions $U_{r_1}^{m_1} \ldots U_{r_p}^{m_p}$, where $\sum m_i = 0$.

Theorem III. *Any limiting distribution k is constant on each coset of H_1. H/H_1 is cyclic. If g is f, then k is zero on each coset except one.*

Note. For any $n \geqslant 1$, $R_f^{n-1} f$ is zero on each coset of H_1 except one, namely $H_1 U_1^n$.

Corollary. *If H is the symmetric group S or the alternating group A on at least five symbols, then H_1 is A if all of $U_1, \ldots, U_k$ have the same parity, and S otherwise.*

Notes

⟦1⟧ Replace "$m' = k$" by "$m' = km$". If the prime p divides km and T, then it divides k, in contradiction to $2k \equiv 1$ (T).
⟦2⟧ See Hall (1959), §1.9, Example 4.
⟦3⟧ H is intransitive and therefore exceptional under the hypothesis that U is a 2-cycle of the described form. (Any element of H is a product of elements of the form $R^t U R^{-t}$ and each such element preserves the stated intransitivity sets.)

Is this the intended hypothesis? The use of the word 'generator' two lines later would suggest so.
⟦4⟧ Hence H is exceptional.
⟦5⟧ We do not need the fact that the distance NF is 8; it is clear that U^{22} does not commute with R^{13}.
⟦6⟧ If $U = R$, $m = 1$, $n = 0$ and T is even, then $U' = RU = R^2$ so $H(U) = (R)$ but $H(U') = R^2$; hence $H(U), H(U')$ are not isomorphic. However we always have $J(U) \cong J(U')$.
⟦7⟧ Note that if U is a power of R, then $f(n) = 1$ for all n.

A power of R has a beetle since $R^m U R^n = 1$ for some m, n.
⟦8⟧ Let $(VUV^{-1})R^{n+1}Z = R^{g(n)}(VUV^{-1})R^n Z$. Then

$$VUR^{-n-1}V^{-1}Z = R^{g(n)}VUR^{-n}V^{-1}Z,$$

$$R^{g(n)}UR^{-n-1}V^{-1}Z = UR^{-n}V^{-1}Z.$$

We have $V^{-1}Z = R^kZ$ for some k. Hence

$$R^{g(n)}UR^{-n-1+k}Z = UR^{-n+k}Z$$
$$= R^{f(-n+k-1)}UR^{-n+k-1}Z,$$

so $g(n) = f(-n+k-1)$, where $f(n)$ is defined for all n by the convention that $f(i) = f(j)$ if $i \equiv j$ (T) and similarly for $g(n)$.

⟦9⟧ Let the objects be $A_1, \dots, A_T$. Then Z is A_T; write A for A_1. We find that $R^{-n}UR^nA = R^SUA$, where $S = f(1) + \cdots + f(n) - n$, $0 \leqslant n \leqslant T-1$. If two of the numbers in the sequence are equal modulo T, then $R^{-i}UR^iA = R^{-j}UR^jA$, say, where $i - j = p > 0$. Hence $UR^pR^jA = R^pUR^jA$. Let $UA_{j+1} = A_d$. Then

$$UA_{p+j+1} = R^pUA_{j+1} = A_{p+d}.$$

Then $U' = R^{1+j-d}U$ fixes A_{j+1} and A_{j+p+1}, so U has a beetle.

⟦10⟧ The phrase 'other than where U is a power of R' is puzzling; see note ⟦7⟧.

⟦11⟧ Let $0 < t < T$ and let the highest common factor of T and t be d. We only need the result when s equals d. We first show there exists k such that

$$0 < k < T, \quad kd \equiv t \ (T) \quad \text{and} \quad (k, T) = 1.$$

If k exists, then $kd = t + hT$, say, so $k = t' + hT'$ where $t = t'd$, $T = T'd$. Each of

$$t', t' + T', \dots, t' + (d-1)T'$$

satisfies the first two of the required properties of k. Assume that none of them satisfies the third property. Then for each $h = 0, 1, \dots, d-1$ there is a prime p_h dividing $t' + hT'$ and $T'd$. It cannot divide T' otherwise it would also divide t'; hence it divides d. If $p_i = p_j = p$ say, where $i \neq j$, then p would divide $t' + iT'$ and $t' + jT'$ and hence would divide t' and T'. Thus $p_0, p_1, \dots, p_{d-1}$ are all different and divide d. This gives the contradiction that $d \geqslant 2^d$.

Since R^k has the same order as R, there is a permutation V such that $VA = A$ and $V^{-1}RV = R^k$. Hence $V^{-1}R^dV = R^{kd} = R^t$, so $VR^tV^{-1} = R^d$. Also $V^{-1}(R)V = (R)$ so $V(R)V^{-1} = (R)$.

Finally the permutation VUV^{-1} fixes A and R^dA since

$$VUV^{-1}R^d\text{A} = VUR^tV^{-1}\text{A} = VUR^t\text{A} = VR^t\text{A} = R^dV\text{A} = R^d\text{A}.$$

⟦12⟧ (a) There are five permutations with invariant 22222 and similarly for 33333 and 44444. The permutations with invariant 11111 form the cyclic group (R). The 'numatizer' of (R) in this case would appear to consist of these 20 permutations.

(b) The sentence 'These together prove the numatizer of (R)' seems nonsensical.

⟦13⟧ $R_f^{n-1}g$?

⟦14⟧ Replace '$g(ab^{-1}x)$' by '$g(a^{-1}bx)$'.

⟦15⟧ This sum equals $(\sum f(a))^2 \|g\|^2 = h^2\bar{f}^2 \|g\|^2$. So we really need $\bar{f}=h^{-1}$, i.e., $\sum f(a)=1$, rather than $\bar{f}=1$.

⟦16⟧ First note that

$$(R_f^m g)(a_1) = \sum f(a_1 a_2^{-1}) \cdots f(a_m a_{m+1}^{-1}) g(a_{m+1}).$$

Now let $F=R_f^{m-1}f$. We find that $R_F h = R_f^m h$ for any h. From $\|R_f^m k\| = \|k\|$ follows $\|R_F k\| = \|k\|$, so $F(a)\neq 0$, $F(b)\neq 0$ implies that $k(a^{-1}bx)/k(x)$ is independent of x (see note ⟦14⟧). Now $F(a_1)=(R_f^{m-1}f)(a_1)$ and if this is nonzero, then there exist generators $U_{r_1}, \ldots, U_{r_m}$ whose product is a_1 (since $f(b)\neq 0$ if and only if b is a generator), so $k(yx)/k(x)$ is independent of x when y has the form

$$(U_{r_1} \cdots U_{r_m})^{-1} U_{s_1} \cdots U_{s_m}.$$

⟦17⟧ The domain of definition L is generated by elements of the form $(U_{r_1} \cdots U_{r_m})^{-1} U_{s_1} \cdots U_{s_m}$, so L is contained in H_1. Also L is a normal subgroup of H, because if $h \in H$, $y \in L$, then $h^{-1}yh \in L$ since h may be written as a word in the generators in which no negative exponents occur. It is now easily proved by induction on length that if a word has exponentsum zero, then it belongs to L. Hence $L=H_1$.

⟦18⟧ In this line and the next, replace g by k.

⟦19⟧ This is clear, but we need to show that the last statement of Theorem III follows.

Let $R_f^{n_1}f, R_f^{n_2}f, \ldots \to k$, that is, $\|R_f^{n_r}f - k\| \to 0$ as $r \to \infty$. It follows that for each a in H, $(R_f^{n_r}f)(a) \to k(a)$. But $\sum (R_f^{n_r}f)(a)=1$ so $\sum k(a)=1$. For each r we have that $(R_f^{n_r}f)(a)\neq 0$ implies a belongs to a coset $C(r)$ of H_1. Passing to a suitable subsequence we may assume that $C(r)=C$, independent of r. If a is not in C, then $(R_f^{n_r}f)(a)=0$ for all r so $k(a)=0$. But k is constant on C so k is nonzero on C and zero off C.

⟦20⟧ $(R_f^{n-1}f)(a) = \sum f(U_{i_1}) f(U_{i_2}) \cdots f(U_{i_n})$ over $(i_1, \ldots, i_n)$ such that $a = U_{i_1} U_{i_2} \cdots U_{i_n}$. This equals k^{-n} times the number of such n-tuples. In the present case, $k=2$. When $n=1,2$ we obtain rows 1, 2 of the table. However when $n=3$ we get

$$0 \;\; \tfrac{1}{8} \;\; \tfrac{1}{8} \;\; 0 \;\; 0 \;\; \tfrac{3}{8} \;\; \tfrac{3}{8} \;\; 0$$

so that row 3 of the table would seem to be in error.

⟦21⟧ If all generators have the same parity, then, since a permutation and its inverse have the same parity, any element of H_1 is a product of an even number of permutations of the same parity, so it is even and $H_1 = A$. If however U_1 and U_2 have opposite parity, then $U_1^{-1} U_2 \in H_1$ and is odd; hence H_1 is the symmetry group.

The Difference $\psi(x) - x$

Notes

⟦1⟧ 'a' and 'b' should be replaced by '$\log a$' and '$\log b$', respectively.
⟦2⟧ Specifically, $r(t) = e^{-t/2}(-\frac{1}{2}\log(1 - e^{-2t}) - (\zeta'/\zeta)(0))$.
⟦3⟧ Let $b \geqslant a = \eta T$. Then

$$\int_a^b \cos\frac{\gamma v}{T}\, w(v)\, dv = w(a)\int_a^\xi \cos\frac{\gamma v}{T}\, dv$$

for some ξ between a and b, since $w(v)$ is positive and decreasing. The right-hand side equals $w(a)T\gamma^{-1}(\sin\gamma\xi/T - \sin\gamma a/T)$ which is at most $(2T/\gamma)\, w(\eta T)$.
⟦4⟧ Here we make use of the inequality

$$|a\cos\gamma w + b\sin\gamma w| \leqslant (a^2 + b^2)^{1/2}.$$

⟦5⟧ In this line we have $w + \frac{1}{137}$ and in the next line $w + \frac{1}{125}$. In fact these should be equal. It does not seem to matter whether both are $w + \frac{1}{137}$ or both are $w + \frac{1}{125}$.
⟦6⟧ After differentiation, we have to prove

$$\frac{\alpha^{-3}}{2\pi}\int_0^\infty \sin v\, f(v)\, dv = \frac{1}{2\sqrt{2\pi}}\exp(-\tfrac{1}{2}\alpha^2),$$

where $f(v) = v\exp(-\frac{1}{2}v^2/\alpha^2)$. Let $F(v) = \int_0^v f(v)\, dv = -\alpha^2\exp(-\frac{1}{2}v^2/\alpha^2)$. Then $\int_0^\infty \sin v\, f(v)\, dv = -\int_0^\infty \cos v\, F(v)\, dv$, which we may evaluate by making use of the identity

$$2\int_0^\infty \cos v\eta\, w(v)\, dv = \exp(-\tfrac{1}{2}\eta^2).$$

which is equivalent to an identity stated earlier in the text.
⟦7⟧ $T/2\pi$ should be added to the right-hand side.
⟦8⟧ Let $c = 2\pi/137$ and let $\gamma < 125\pi$. Then

$$\frac{\gamma(w + \frac{1}{125})}{2\pi} = N + \theta,$$

where N is an integer and $|\theta| < \frac{1}{137}$. Hence

$$-\sin\gamma w = \sin\left(\frac{\gamma}{125} - 2\pi\theta\right).$$

Now use that, since $|2\pi\theta| < c$, $\cos 2\pi\theta > \cos c$ and $|\sin 2\pi\theta| < c$.

⟦9⟧ This comes immediately from the definition of S by using the lower bound just obtained for $-\sin yw$ when $y < 125\pi$ and using that $-\sin yw \geqslant -1$ when $y \geqslant 125\pi$.

⟦10⟧ Replace 25/137 by $2\pi/137$.

⟦11⟧ The interval of integration should be changed to be from 0 to 125π.

⟦12⟧ The minus sign on the left should be replaced by a plus sign.

⟦13⟧ The three occurrences of '135' should be changed to '137'.

Indeed the line should probably read:

$$= \cos\frac{2\pi}{137} I_1 - I_2 - \frac{2\pi}{137} C + \frac{2\pi}{137} I_3 + \cos\frac{2\pi}{137} E_1 + E_2 + \frac{2\pi}{137} E_3.$$

⟦14⟧ The minus sign to the left of v^{-1} should be deleted; that is, the operation should be multiplication rather than subtraction.

⟦15⟧ $S(0) = -\frac{7}{8}$.

⟦16⟧ The equality sign should be replaced by a plus sign.

Has a term $S(125\pi)/125\pi$ been omitted?

[Thanks are due to Dr A.M. Cohen for the following comments.]

In view of the references it is very probable that this work comes chronologically before the paper (V) on $\pi(x) - \mathrm{li}\, x$ and probably even before Turing (1953) had extended Titchmarsh's computation of the zeta-function by using a computer.

The basic ideas in the manuscript are probably sound. However, a substantial amount of work would be required to tidy up the results. For instance, the inequality (i) is incorrect since the left-hand side has approximate value 0.011908 which is about 40% larger than Turing's value of 0.00851.

On a Theorem of Littlewood

Summary

The article by Cohen and Mayhew (1968) which appears in this volume is essentially a considerably corrected and amplified version of the major part of this unpublished script, namely of §1 to §5, that is, the part up to and including the proof of Theorem 2.

For a summary of this part of the script, see the first page of Cohen and Mayhew's article.

In §6, Skewes and Turing indicate briefly how an improved estimate should be possible by the use of a digital computer, assuming first that the computations give us say the first 300 nontrivial zeros of the zeta-function to seven places of decimals (all these zeros being on the critical line).

Then a statistical argument is given to show that there is an even chance that this approach would yield an x, $2<x<\exp 220000$, for which $\pi(x)>\mathrm{li}\,x$.

Finally in §7 the possibility is considered that a nontrivial zero is found off the critical line.

Theorem 3. *Suppose that* $\beta_1+\mathrm{i}\gamma_1$, $\beta_1>\frac{1}{2}$, $\gamma_1>0$, *is a zero of the zeta-function and for every other nontrivial zero* $\beta+\mathrm{i}\gamma$ *either* $\beta=\frac{1}{2}$ *or* $|\gamma-\gamma_1|>14$. *Then for some* x, $2<x<(16\gamma_1)^a$, $a=1.12/(\beta_1-\frac{1}{2})$, *we have* $\pi(x)>\mathrm{li}\,x$.

Notes

⟦1⟧ I have not been able to locate the orginal script but I had access to a photocopy of it.

The article is in typescript with all mathematical symbols in manuscript. In spite of the joint authorship, the handwriting and phrasing indicate that it was perhaps written by Turing alone.

There are handwritten comments by another person, probably A.E. Ingham. Such comments are here enclosed in square brackets [ ]. Where necessary an earlier $ sign indicates the place in the text where the remark applies. If the remark refers to a symbol or expression in the text, the relevant $ sign appears immediately before the symbol or expression.

It may be advisable to read this article with continual reference to COHEN and MAYHEW (1968) which paper is reproduced in this volume.

The lemmas of Skewes and Turing and Cohen and Mayhew are related as follows (Theorems 1, 2 of Skewes and Turing correspond to Theorems 1, 2 of Cohen and Mayhew, respectively):

S & T	1	2	3	4	5	6	7	8	9	10	–	12	13
C & M	1	4	3	5	(a)	2	6	7	(b)	8	–	–	9

(a): page 695; (b): page 708.

⟦2⟧ In the script, $\mu^{-1}t_0^{-3/2}$ is preceded by a crossed-out expression and followed by a left bracket and then a crossed-out expression.

⟦3⟧ Note that $\exp 220000$ is much greater than the value 1.65×10^{1165} obtained by R. Sherman Lehman in 1966; see the end of Cohen and Mayhew's article. However we may say that Turing guessed rightly that there is an x, $2<x<\exp 220000$ for which $\pi(x)>\mathrm{li}\,x$.

BIBLIOGRAPHY

[This bibliography does not contain the references given earlier in the volume in the two papers by I.J. Good.]

ADYAN, S.I.

1957 Unsolvability of some algorithmic problems in the theory of groups
Trudy Moskov. Mat. Obshch. **6**, 231–298

BACKLUND, R.J.

1918 *Acta Math.* **41**

BAER, R.

1934 Erweiterung von Gruppen und ihren Isomorphismen
Math. Z. **38**, 357–416

BARGMANN, V.
see VON NEUMANN, J.

BÉNOIT, Commandant

1924 Note sur une méthode, etc. (Procédé du Commandant Cholesky)
Bull. Géodésique (Toulouse) **2**, 5–77

BERNAYS, P.
see HILBERT, D.

BODEWIG, E.

1949 (Review)
Math. Rev. **10**, 405

BOONE, W.W.

1952 (Review)
J. Symbolic Logic **17**, 74–76

1958 An analysis of Turing's "The word problem in semi-groups with cancellation"
Ann. of Math. **67** (1), 195–202
(reprinted in this volume)

1959 The word problem
Ann. of Math. **70**, 207-265

BRITTON, J.L.
1958 (Reviews)
J. Symbolic Logic **23**, 50-54

CHAITIN, G.J.
1987 *Algorithmic Information Theory*
(Cambridge Univ. Press, Cambridge)

CHAMPERNOWNE, D.
1933 *J. London Math. Soc.* **8**

CHURCH, A.
1936 A A note on the Entscheidungsproblem
J. Symbolic Logic **1**, 40-41
1936 B An unsolvable problem of elementary number theory
Amer. J. Math. **58**, 345-363

COHEN, A.M. and M.J.E. MAYHEW
1968 On the difference $\pi(x) - \mathrm{li}\, x$
Proc. London Math. Soc. (3) **18**, 691-713
(reprinted in this volume)

CROWELL, R.H. and R.H. FOX
1977 *Introduction to Knot Theory*
Graduate Texts in Math. (Springer, New York)

DAVIS, M., H. PUTNAM and J. ROBINSON
1961 The decision problem for exponential Diophantine equations
Ann. of Math. **74**, 425-436

DONER, J. and W. HODGES
1988 Alfred Tarski and decidable theories
J. Symbolic Logic **53** (1), 20-35

EDWARDS, H.M.
1974 *Riemann's Zeta-function*
(Academic Press, New York)

EMDE
see JAHNKE

FOX, R.H.
see CROWELL, R.H.

FOX, L., H.D. HUSKEY and J.H. WILKINSON
1924 Notes on the solution of algebraic linear simultaneous equations (Procédé du Commandant Cholesky)
Bull. Géodésique (Toulouse) **2**, 149–173

GÖDEL, K.
1931 Über formal unentscheidbare Sätze der Principia Mathematica und verwandter Systeme I
Monatsh. Math. Phys. **38**, 173–198

GOLDSTINE, H.H.
see VON NEUMANN, J.

HALL, M.
1959 *The Theory of Groups*
(Macmillan, New York)

HASELGROVE, C.B. and J.C.P. MILLER
1960 Tables of the Riemann zeta-function
Roy. Soc. Math. Tables **6**

HEMION, G.
1979 On the classification of homeomorphisms of 2-manifolds and the classification of 3-manifolds
Acta Math. **142** (1–2), 123–155

HILBERT, D. and P. BERNAYS
1940 *Grundlagen der Mathematik*
(Berlin) (two volumes)

HODGES, A.
1983 *Alan Turing – The Enigma of Intelligence*
(Burnett/Hutchinson, London); also (Unwin, London, 1985)

HODGES, W.
see DONER, J.

HOTELLING, H.
1943 Some new methods in matrix calculation
Ann. Math. Statist. **14**, 34

HUSKEY, H.D.
see FOX, L.

INGHAM, A.E.
1932 *The Distribution of Prime Numbers*
Cambridge Math. Tracts, No. 30; also (Hafner, New York, 1971)
1936 A note on the distribution of primes
Acta Arith. **1**, 202–211

IVIC, A.
1985 *The Riemann Zeta Function*
(Wiley, New York)

JAHNKE and EMDE
1948 *Tafeln höherer Funktionen*
(Teubner, Leipzig) 324

KARRASS, A.
see MAGNUS, W.

KLEENE, S.C.
1935/1936 General recursive functions of natural numbers
Math. Ann. **112**, 727–742
1936 λ-definability and recursiveness
Duke Math. J. **2**, 340–353
1952 *Introduction to Metamathematics*
(North-Holland, Amsterdam)

KUIPERS, L. and H. NIEDERREITER
1974 *Uniform Distribution of Sequences*
(Wiley Interscience, New York)

LEHMAN, R.S.
1966 On the difference $\pi(x) - \operatorname{li} x$
Acta Arith. **11**, 397–410
1970 On the distribution of zeros of the Riemann zeta-function
Proc. London Math. Soc. (3) **20** (2), 303–320

LEHMER, D.H.
1956 A On the roots of the Riemann zeta function
Acta Math. **95**, 291–298
1956 B Extended computation of the Riemann zeta function
Mathematika **3**, 102–108

LINDEBERG
1922 *Math. Z.* **15**.

LITTLEWOOD, J.E.
1914 Sur la distribution des nombres premiers
Comptes Rendus **158**, 1869–1872
1924 On the zeros of the Riemann zeta-function
Proc. Cambridge Phil. Soc. **22**, 295–318

LOOMIS, L.H.
1953 *An Introduction to Abstract Harmonic Analysis*
(Van Nostrand, New York)

MAGNUS, W., A. KARRASS and D. SOLITAR
1966 *Combinatorial Group Theory*
(Wiley, New York)

MARKOV, A.
1947 *Dokl. Akad. Nauk SSSR* **55**, 587–590
1951 On an unsolvable problem concerning matrices
Dokl. Akad. Nauk **78**, 1089–1092
1956 *Math. Rev.* **17**, 706
1958 The insolubility of the problem of homeomorphy
Dokl. Akad. Nauk **121**, 218–220

MATIJASEVIČ, Y.
1970 Enumerable sets are Diophantine
Dokl. Akad. Nauk **191**, 279–282

MAYHEW, M.J.E.
see COHEN, A.M.

MILLER, J.C.P.
see HASELGROVE, C.B.

MONTGOMERY, D.
see VON NEUMANN, J.

MORRIS, J.
1946 An escalator method for the solution of linear simultaneous equations
Philos. Mag. (7) **37**, 106

NIEDERREITER, H.
see KUIPERS, L.

NOVIKOV, P.S.
1952 On algorithmic undecidability of the problem of identity
Dokl. Akad. Nauk SSSR **85**, 709–712
1955 On the algorithmic unsolvability of the word problem in group theory
Trudy Mat. Inst. Steklov. **44** (in Russian)

PATTERSON, S.J.
1988 *Introduction to the Theory of the Riemann Zeta-function*
(Cambridge Univ. Press, Cambridge)

PONTRYAGIN, L.S.
1939 *Topological Groups*
(Princeton Univ. Press, Princeton, NJ)

POST, E.L.
1936 Finite combinatory processes – formulation 1
J. Symbolic Logic **1**, 103–105
1947 Recursive unsolvability of a problem of Thue
J. Symbolic Logic **12**, 1–11

PUTNAM, H.
see DAVIS, M.

RABIN, M.O.
1958 Recursive unsolvability of group theoretic problems
Ann. of Math. (2) **67**, 172–194

REIDEMEISTER, K.
1926 Knoten und Gruppen
Hamb. Abhandl. **5**, 8–23

ROBINSON, J.
see DAVIS, M.

SCHREIER, O.
1926 Über die Erweiterung von Gruppen
Monatsh. Math. Phys. **34**, 165–180

SIEGEL, C.L.
1931 Über Riemanns Nachlass zur analytischen Zahlentheorie
Quell. Gesch. Math. **B2**, 45–80

SKEWES, S.
1933 On the difference $\pi(x) - \operatorname{li} x$ (I)
J. London Math. Soc. **8**, 277–283
1955 On the difference $\pi(x) - \operatorname{li} x$ (II)
Proc. London Math. Soc. (3) **5**, 48–70

SOLITAR, D.
see MAGNUS, W.

SPEISER, A.
1927 *Theorie der Gruppen von endlicher Ordnung*
(Berlin, 2nd ed.)

TARSKI, A.
1948 A decision method of elementary algebra and geometry
RAND project report R-109
1951 *A Decision Method for Elementary Algebra and Geometry*
(Univ. of California Press, Berkeley, CA)

TITCHMARSH, E.C.
1930 *The Zeta-function of Riemann*
Cambridge Math. Tracts, No. 26
1935 The zeros of the Riemann zeta-function
Proc. Roy. Soc. (A) **151**, 234–255
1936 The zeros of the zeta-function
Proc. Roy. Soc. (A) **157**, 261–263

1937 *Introduction to the Theory of Fourier Integrals*
(Oxford)

1951 *The Theory of the Riemann Zeta-function*
(Clarendon Press, Oxford) (2nd ed., rev. by D.R. HEATH-BROWN 1987)

TURING, A.M.

1937 A On computable numbers, with an application to the Entscheidungsproblem
Proc. London Math. Soc. (2) **42**, 230–265

1937 B (Correction to the above)
Proc. London Math. Soc. (2) **43**, 544–546

1937 C Computability and λ-definability
J. Symbolic Logic **2**, 153–163

VON NEUMANN, J.

1933 Die Einführung analytischer Parameter in topologischen Gruppen
Ann. of Math. **34**, 170–190

1934 A Almost periodic functions in a group
Trans. Amer. Math. Soc. **36**, 445–492

1934 B Zum Haarschen Mass in topologischen Gruppen
Compositio Math. **1**, 106–114

VON NEUMANN, J., V. BARGMANN and D. MONTGOMERY

1946 *Solution of Linear Systems of High Order*
lithographed (Princeton)

VON NEUMANN, J. and H.H. GOLDSTINE

1947 Numerical inverting of matrices of high order
Bull. Amer. Math. Soc. **53**, 1021–1099

WATSON, G.N.
see WHITTAKER, E.T.

WHITTAKER, E.T. and G.N. WATSON

1946 *Modern Analysis*
(Cambridge Univ. Press, Cambridge, 4th ed.)

WILKINSON, J.H.
see FOX, L.

INDEX

Printed and bound by CPI Group (UK) Ltd, Croydon, CR0 4YY
03/10/2024
01040330-0016